In Pólya's Footsteps

Miscellaneous Problems and Essays

The inspiration for many of these problems came from the Olympiad Corner of *Crux Mathematicorum*, now *Crux Mathematicorum with Mathematical Mayhem,* published by the Canadian Mathematical Society with support from the University of Calgary, Memorial University of Newfoundland, and the University of Ottawa. Full attribution can be found with each problem.

Library of Congress Catalog Card Number 97-70506

Complete Set ISBN 0-88385-300-0
Vol. 19 ISBN 0-88385-326-4

Printed in the United States of America

Current printing (last digit):
10 9 8 7 6 5 4 3 2 1

The Dolciani Mathematical Expositions

NUMBER NINETEEN

In Pólya's Footsteps

Miscellaneous Problems and Essays

Ross Honsberger
University of Waterloo

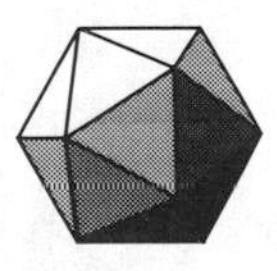

Published and Distributed by
THE MATHEMATICAL ASSOCIATION OF AMERICA

THE
DOLCIANI MATHEMATICAL EXPOSITIONS

Published by
THE MATHEMATICAL ASSOCIATION OF AMERICA

The DOLCIANI MATHEMATICAL EXPOSITIONS series of the Mathematical Association of America was established through a generous gift to the Association from Mary P. Dolciani, Professor of Mathematics at Hunter College of the City University of New York. In making the gift, Professor Dolciani, herself an exceptionally talented and successful expositor of mathematics, had the purpose of furthering the ideal of excellence in mathematical exposition.

The Association, for its part, was delighted to accept the gracious gesture initiating the revolving fund for this series from one who has served the Association with distinction, both as a member of the Committee on Publications and as a member of the Board of Governors. It was with genuine pleasure that the Board chose to name the series in her honor.

The books in the series are selected for their lucid expository style and stimulating mathematical content. Typically, they contain an ample supply of exercises, many with accompanying solutions. They are intended to be sufficiently elementary for the undergraduate and even the mathematically inclined high-school student to understand and enjoy, but also to be interesting and sometimes challenging to the more advanced mathematician.

---

1. *Mathematical Gems,* Ross Honsberger
2. *Mathematical Gems II,* Ross Honsberger
3. *Mathematical Morsels,* Ross Honsberger
4. *Mathematical Plums,* Ross Honsberger (ed.)
5. *Great Moments in Mathematics (Before 1650),* Howard Eves
6. *Maxima and Minima without Calculus,* Ivan Niven
7. *Great Moments in Mathematics (After 1650),* Howard Eves
8. *Map Coloring, Polyhedra, and the Four-Color Problem,* David Barnette
9. *Mathematical Gems III,* Ross Honsberger
10. *More Mathematical Morsels,* Ross Honsberger
11. *Old and New Unsolved Problems in Plane Geometry and Number Theory,* Victor Klee and Stan Wagon
12. *Problems for Mathematicians, Young and Old,* Paul R. Halmos
13. *Excursions in Calculus: An Interplay of the Continuous and the Discrete,* Robert M. Young
14. *The Wohascum County Problem Book,* George T. Gilbert, Mark Krusemeyer, and Loren C. Larson
15. *Lion Hunting and Other Mathematical Pursuits: A Collection of Mathematics, Verse, and Stories by Ralph P. Boas, Jr.,* edited by Gerald L. Alexanderson and Dale H. Mugler
16. *Linear Algebra Problem Book,* Paul R. Halmos
17. *From Erdős to Kiev: Problems of Olympiad Caliber,* Ross Honsberger
18. *Which Way Did the Bicycle Go? . . . and Other Intriguing Mathematical Mysteries,* Joseph D. E. Konhauser, Dan Velleman, and Stan Wagon
19. *In Pólya's Footsteps: Miscellaneous Problems and Essays,* Ross Honsberger

Preface

Just as a recording of a Mozart concerto makes no pretense of teaching one to compose music, these mathematical performances are not motivated by a desire to teach mathematics; they are offered *solely for your enjoyment.* There is no denying that a certain degree of concentration is required for the appreciation of their beautiful ideas, but it is hoped that a leisurely pace and generous explanations will make them a pleasure to read. The technical demands are very modest; a high school graduate should be well equipped to handle many of the topics and a university undergraduate in mathematics ought to be perfectly comfortable throughout.

I hope you will find something exciting in each of these topics—a surprising result, an intriguing approach, a stroke of ingenuity—and that you will approach them as entertainment. You are certainly not required to attempt these problems before going through the solutions, but if you are able to give them a little thought first, I'm sure you will find them all the more exciting.

It is a pleasure to acknowledge the great debt this volume owes the undergraduate problems journal *Crux Mathematicorum,* now *Crux Mathematicorum with Mathematical Mayhem,* published by the Canadian Mathematical Society with support from the University of Calgary, Memorial University of Newfoundland, and the University of Ottawa. For interesting elementary problems, this publication is in a class by itself. I came across the great majority of the problems discussed in the present volume in the Olympiad Corner columns of *Crux Mathematicorum.* Sometimes a solution given in the present collection has been based on a solution published in this column, but unless otherwise acknowledged, the solutions here are based on my own work (with one exception) and in all cases the responsibility for shortcomings in presentation is mine alone. The exception is the essay The Infinite Checkerboard, which was written probably twenty years ago, and with my deepest apologies, I cannot recall where I encountered the topic. The sections may be read in any order. No attempt has been made to bring together all the problems on a specific subject; in fact, this has been avoided in the hope of encouraging a feeling of spontaneity throughout the work.

I would like to extend my warmest thanks to chairman Bruce Palka of the Dolciani Subcommittee and to the members Christine Ayoub and Irl Bivens for their generous reception of this manuscript and their perceptive reviews. Finally, it has again been my great good fortune to have had Beverly Ruedi and Elaine Pedreira see this work through publication; their unfailing geniality and excellence in all phases of this trying process are deeply appreciated.

CONTENTS

Praise God from Whom
all blessings flow.

Four Engaging Problems

1. (From the 1990 U.S.A. Olympiad; Crux Mathematicorum, 1990, 162)

A certain state wishes to issue license plates consisting of 6 digits from 0 through 9. If each plate is to differ from every other plate in *at least two places,* what is the maximum number of different license plates the state can issue?

Our solution is based on the notion of a check-digit. We begin by considering a license plate to consist of two parts. With 6 digits in all, you might expect each part to have 3 digits. However, the appropriate division in this case is a lopsided 5-digit first part and a single-digit second part. Since we have 10 choices for each digit, there are 10^5 different first parts, and if more than 10^5 plates were issued, some two of them would have to have identical first parts, leaving them to differ only in their sixth digits, not the necessary two places. Thus not more than 10^5 plates are possible. However, 10^5 plates *can* be issued under the following ingenious scheme.

Let each of the 10^5 different 5-digit numbers be extended into a license plate by attaching as its sixth digit *the last digit of the sum of its digits;* thus 1 2 3 4 5 would extend into 1 2 3 4 5 5 since $1 + 2 + 3 + 4 + 5 = 15$. It is easy to see that each plate would then differ from every other plate in at least two places as follows.

If the 5-digit initial sections of two plates themselves differ in two or more places, the required condition is satisfied. Consider, then, two plates whose first parts differ in only one place (being different 5-digit numbers, the first parts must differ in at least one place). In this case, the sum of the 5 digits in these first sections cannot possibly end in the same digit, for their sums are made up of equal subtotals obtained from the four common digits supplemented by different fifth digits: the only way to get final sums with the same last digit would be by adding the same digit to the equal subtotals or by adding 0 to one and 10 to the other, neither of which is the case. Thus the check-digit at the end provides a second place of difference between the plates. For example, 7 6 3 **8** 1 and 7 6 3 **3** 1 have sums $17 + 8 = 25$ and $17 + 3 = 20$, giving plates 7 6 3 **8** 1 **5** and 7 6 3 3 1 **0**, which differ again in their last digits.

2. (From the 1988 Ibero-American Olympiad, proposed by Cuba; *Crux Mathematicorum,* 1989, 163)

> Consider three concentric circles, center P, with radii 3, 5, and 7. Show that, of all the triangles ABC which have one vertex on each circle, a triangle of *maximum perimeter* must have the point P as its *incenter.*

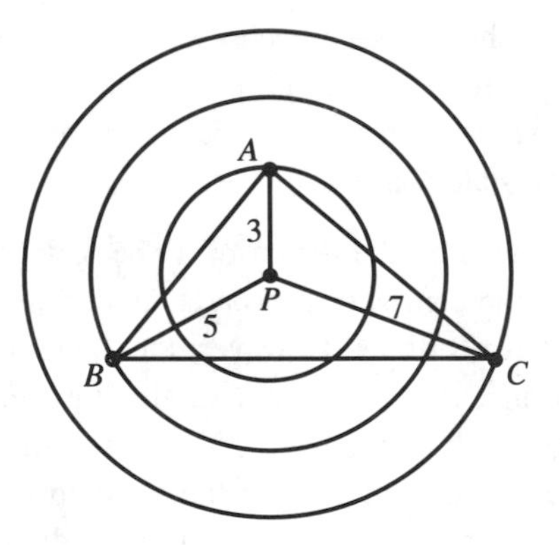

FIGURE 1

Let E be the ellipse having foci B and C and passing through A. Then, for all points Q on E, the sum of the focal radii $QB + QC$ is equal to $AB + AC$. Now, A is free to range on the circle $P(3)$ (i.e., center P, radius 3), and if any part of this circle were to lie outside E, then A could be shifted to any point A' on the exterior arc to yield a triangle $A'BC$ having greater perimeter than $\triangle ABC$ (recall that E can be drawn by pinning a loop of string at B and C and keeping the loop taut with a pencil at a moving point Q; in order for such an ellipse to reach a point A' outside E, a longer string would have to be used, implying a greater sum of focal radii).

When $\triangle ABC$ has maximum perimeter, then, it must be that E contains the entire circle $P(3)$. Now, this would not be the case unless E and the circle had a common tangent at their point of intersection A (otherwise the circle

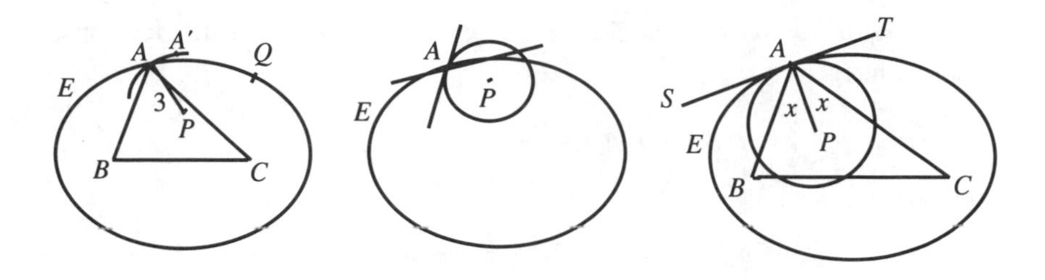

FIGURE 2

would protrude beyond E). This common tangent SAT is perpendicular to the radius AP at the point of contact, making AP a common normal. Since the reflector property of the ellipse declares that the angles x between the focal radii and the normal are equal, AP must be the bisector of angle A in $\triangle ABC$. Similarly, BP and CP bisect angles B and C and the conclusion follows.

3. (From the 1989 Bulgarian Olympiad; *Crux Mathematicorum,* 1990, 134; an alternative solution is given in 1992, 9)

If p and q are prime numbers such that

$$\sqrt{p^2+7pq+q^2}+\sqrt{p^2+14pq+q^2}$$

is an *integer,* prove that p and q must be equal.

If $\sqrt{p^2+7pq+q^2}+\sqrt{p^2+14pq+q^2}=k$, an integer, then

$$\sqrt{p^2+14pq+q^2}=k-\sqrt{p^2+7pq+q^2},$$

and

$$p^2+14pq+q^2=k^2-2k\sqrt{p^2+7pq+q^2}+p^2+7pq+q^2.$$

At this point it is difficult to resist cancelling terms and simplifying, but there is no need to do that for the real value of this equation lies in the revelation that, in solving for $\sqrt{p^2+7pq+q^2}$, we would obtain a *rational* number. But if the square root of an integer is rational, then it must actually be an integer:

> if $\sqrt{m}=\frac{a}{b}$ in lowest terms, then $mb^2=a^2$; thus $b^2|a^2$ and $b|a$, and since a and b are relatively prime, this implies $b=1$, making $\sqrt{m}=a$.

Thus $p^2+7pq+q^2$, and similarly $p^2+14pq+q^2$, must be perfect squares.

Now, $p^2 + 7pq + q^2$ is clearly greater than $(p + q)^2$. Thus, for some positive integer r,

$$p^2 + 7pq + q^2 = (p + q + r)^2,$$

that is,

$$p^2 + 7pq + q^2 = p^2 + q^2 + r^2 + 2pq + 2pr + 2qr,$$

and

$$5pq = r(r + 2p + 2q).$$

Accordingly, r divides $5pq$, and since 5, p, and q are all prime numbers, the only possible values of r are 1, 5, p, q, $5p$, $5q$, pq, and $5pq$. But r can't be very big. For example, since $5pq/r = r + 2p + 2q$, $r = 5p$ gives $q = 5p + 2p + 2q$, which is obviously false; similarly, each of $r = 5q$, pq, and $5pq$ is impossible.

Also, $r = 1$ yields

$$5pq = 1 + 2p + 2q,$$

that is,

$$pq + 2pq + 2pq = 1 + 2p + 2q,$$

which cannot be true since each term on the left is greater than the corresponding term on the right.

It remains only to check $r = p$, q, and 5.

(i) $r = p$: in this case,

$$5q = p + 2p + 2q,$$

giving

$$3q = 3p \quad \text{and} \quad p = q.$$

(ii) $r = q$: similarly, this gives

$$5p = q + 2p + 2q$$

and $p = q$.

(iii) $r = 5$: here we have

$$pq = 5 + 2p + 2q,$$

$$pq - 2p - 2q + 4 = 9,$$

$$(p - 2)(q - 2) = 9.$$

Without loss of generality, let $p \geq q$; then either $p - 2 = q - 2 = 3$ and the obvious $p = q$, or $p - 2 = 9$ and $q - 2 = 1$, making $p = 11$ and $q = 3$; recalling that $p^2 + 14pq + q^2$ must also be a perfect square, this would require $121 + 14(33) + 9 = 130 + 462 = 592$ to be a perfect square, which it isn't. Hence there is no escaping $p = q$.

4. (From the 1989 International Olympiad; *Crux Mathematicorum,* 1989, 196)

Prove that, for each positive integer n, there exist n consecutive positive integers none of which is an integral power of a prime.

I am sure that everyone participating in an international olympiad would be well aware that, for $n \geq 1$, a string of n consecutive *composite* positive integers is given by

$$S = \{(n+1)! + 2, (n+1)! + 3, \ldots, (n+1)! + (n+1)\}$$

(the first is divisible by 2 and is bigger than 2, the second is divisible by 3 and is bigger than 3, and so forth).

For $n = 2$, this gives $\{3! + 2, 3! + 3\} = \{8, 9\}$, both of which are composite, but unfortunately each is also a power of a prime. Thus the set S is not strong enough to satisfy the present requirement.

However, with a slight adjustment, S can be made to qualify. Replacing $(n+1)!$ with $\left[(n+1)!\right]^2$ gives the satisfactory set

$$S' = \left\{ \left[(n+1)!\right]^2 + 2, \left[(n+1)!\right]^2 + 3, \ldots, \left[(n+1)!\right]^2 + (n+1) \right\}.$$

Each integer in S' is of the form $\left[(n+1)!\right]^2 + k$, where $k \geq 2$ and k is a divisor of $(n+1)!$. Thus k^2 divides $\left[(n+1)!\right]^2$ and for some positive integer t we have

$$\left[(n+1)!\right]^2 + k = k^2 t + k = k(kt + 1),$$

which displays the integer as the product of two *relatively prime* factors k and $kt + 1$. Since each of these factors is greater than 1, they introduce at least two different prime divisors into $\left[(n+1)!\right]^2 + k$, making it impossible for the number to be a power of a single prime.

A Problem from the 1991 Asian Pacific Olympiad†

Suppose 997 distinct points are chosen in the plane. If each pair of these points is joined by a segment and its midpoint is colored red, prove that at least 1991 different points must be colored. Can you find a special set of 997 points which yields exactly 1991 red points?

FIGURE 3

Clearly any set of 3 points gives rise to 3 segments and 3 midpoints. Thus let us take as an induction hypothesis the proposition that, for $k \geq 1$, any set of $2k+1$ points in the plane gives rise to at least $4k-1$ red midpoints.

Consider, then, a set S of $2(k+1)+1 = 2k+3$ points in the plane, $k \geq 1$. Since S is finite, the segments which join a pair of points in S determine only a finite number of directions. Suppose L is a straight line which is not parallel to any segment determined by S. Consequently, if we begin on one side of S and let L sweep across the plane, it will pass over the points of S *one at a time.* Since $k \geq 1$, S has at least 5 points, and suppose that the first

† (*Crux Mathematicorum,* 1991, 130)

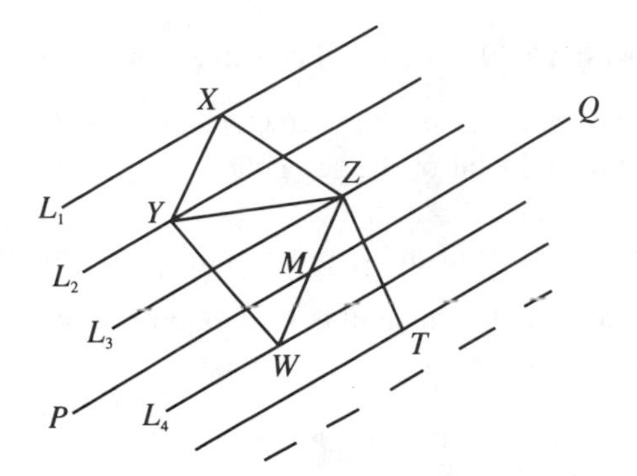

FIGURE 4

four points of S that L passes over are X, Y, Z, and W, in that order (figure 4).

The set $S - X - Y$ contains $2k + 1$ points, where $k \geq 1$, and therefore by the induction hypothesis gives rise to at least $4k - 1$ red points. One of these is the midpoint M of ZW (figure 4), which lies on the midline PQ of the strip between L_3 and L_4, the positions of L as it passes over Z and W respectively. With all the points of $S - X - Y$, except Z and W, on the side of L_4 opposite Z (e.g., T), none of its red points can be closer to L_3 than the midline PQ. Therefore the midpoints of XY, XZ, YZ, and YW are all on the wrong side of PQ to be red points of $S - X - Y$ and hence they constitute four new red points when the entire set S is considered (it could happen that Z itself is the midpoint of YW, but even so it is a new red point since Z is not a red point of $S - X - Y$). Hence S generates at least

$$(4k - 1) + 4 = 4(k + 1) - 1$$

red points and it follows by induction that $2k + 1$ points, $k \geq 1$, always generates at least $4k - 1$ red midpoints. For $k = 498$, then, it follows that a set of 997 points provides at least $4 \cdot 498 - 1 = 1991$ red points.

The minimum case is given by any set of $2k + 1$ equally-spaced points along a straight line. Consider the 997 integer points $0, \pm 1, \pm 2, \ldots, \pm 498$ on the x-axis.

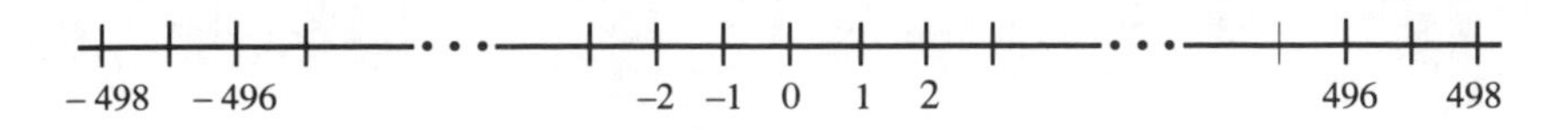

The integers from -497 to $+497$ themselves constitue 995 red midpoints:

-497 is the midpoint of the segment joining -498 to -496,
-496 is the midpoint of the segment joining -497 to -495,

....

497 is the midpoint of the segment joining 496 to 498.

In addition, there are the 996 midpoints of the unit intervals between consecutive integers, comprising the set

$$H = \{-497.5, -496.5, \ldots, 497.5\}.$$

It is easy to see that this collection of $995 + 996 = 1991$ points comprises all the red midpoints of this set of points: the midpoint of the segment joining integers s and t is itself a red integer when s and t have the *same* parity and is one of the points in H when s and t have *opposite* parity.

Four Problems from the First Round of the 1988 Spanish Olympiad†

1. The integers $1, 2, 3, \ldots, n^2$ are arranged to form the $n \times n$ matrix A as shown:

$$A = \begin{pmatrix} 1 & 2 & \ldots & n \\ n+1 & n+2 & \ldots & 2n \\ 2n+1 & 2n+2 & \ldots & 3n \\ & \ldots & \ldots & \\ (n-1)n+1 & \ldots & \ldots & n^2 \end{pmatrix}$$

Now a sum S is constructed as follows.

The first term x_1 in S is chosen at random from the entries in A, after which the rest of x_1's row and column are deleted. The second term x_2 is then chosen at random from the remaining entries in A, after which what is left of x_2's row and column are deleted. Similar random selections and deletions are carried out until A is exhausted.

Prove that the sum S always builds up to the same total no matter what entries might be taken as $x_1, x_2, \ldots$.

Clearly S contains one term from each row of A, for a total of n terms. But these terms also occur one in each column. Thus S is the sum of n elements of A no two of which are in the same row or in the same column, that is, the sum of the n terms in a *transversal* of A, and thus our problem reduces to showing that the sum of the entries in a transversal is always the same.

---

† (*Crux Mathematicorum,* 1990, 225)

If a_i denotes the entry that S receives from row i, then

$$S = a_1 + a_2 + \cdots + a_n.$$

Now consider the $n \times n$ matrix B, each of whose rows is simply $1, 2, 3, \ldots n$, or equivalently, whose ith column consists of n i's:

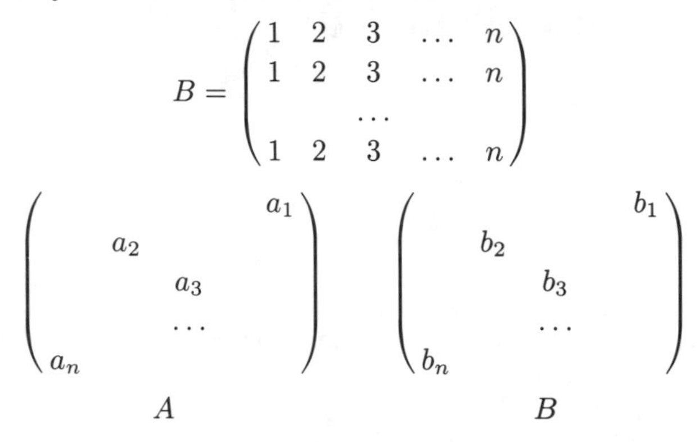

Let $b_1, b_2, \ldots, b_n$ be the elements of B that occur in the positions corresponding to those occupied by the chosen elements $a_1, a_2, \ldots, a_n$ in A. Now, A and B have the same first row, and each term in the second row of A is n more than the corresponding term in B; similarly, each term in the third row of A is $2n$ more than the corresponding term in B, and generally, each term in the ith row of A is $(i-1)n$ more than the corresponding term in B. Hence

$$a_1 = b_1,\ a_2 = b_2 + n,\ a_3 = b_3 + 2n, \ldots, a_n = b_n + (n-1)n.$$

However, the sum $b_1 + b_2 + \cdots + b_n$, containing an entry from each *column* of B, must in some order be the sum of the integers $1, 2, \ldots, n$; hence

$$b_1 + b_2 + \cdots + b_n = 1 + 2 + \cdots + n = \frac{n(n+1)}{2}.$$

Therefore

$$\begin{aligned}
S &= a_1 + a_2 + a_3 + \cdots + a_n \\
&= b_1 + (b_2 + n) + (b_3 + 2n) + \cdots + \big[b_n + (n-1)n\big] \\
&= (b_1 + b_2 + \cdots + b_n) + n\big[1 + 2 + \cdots + (n-1)\big] \\
&= \frac{n(n+1)}{2} + n\frac{(n-1)n}{2} \\
&= \frac{n}{2}(n + 1 + n^2 - n) \\
&= \frac{n(n^2+1)}{2}, \text{ a constant.}
\end{aligned}$$

(Essentially the same solution is given in 1992, 38.)

2. What is the maximum value of the function

$$f(x) = |x||x-1||x-2||x-3||x-4||x-5||x-6||x-7|,$$

for x in the closed interval $[3,4]$?

Since $|x-y| = y-x$ for $y \geq x$, we have for $x \in [3,4]$ that

$$\begin{aligned} f(x) &= |x||x-1||x-2||x-3||x-4||x-5||x-6||x-7| \\ &= x(x-1)(x-2)(x-3)(4-x)(5-x)(6-x)(7-x). \end{aligned}$$

Consequently, substituting $x = \frac{7}{2} + y$, where $-\frac{1}{2} \leq y \leq \frac{1}{2}$ in order to place x in $[3,4]$, we get

$$\begin{aligned} f(x) &= \left(\tfrac{7}{2}+y\right)\left(\tfrac{5}{2}+y\right)\left(\tfrac{3}{2}+y\right)\left(\tfrac{1}{2}+y\right)\left(\tfrac{1}{2}-y\right)\left(\tfrac{3}{2}-y\right)\left(\tfrac{5}{2}-y\right)\left(\tfrac{7}{2}-y\right) \\ &= \left[\left(\tfrac{7}{2}\right)^2 - y^2\right]\left[\left(\tfrac{5}{2}\right)^2 - y^2\right]\left[\left(\tfrac{3}{2}\right)^2 - y^2\right]\left[\left(\tfrac{1}{2}\right)^2 - y^2\right] \\ &\leq \left(\tfrac{7}{2}\cdot\tfrac{5}{2}\cdot\tfrac{3}{2}\cdot\tfrac{1}{2}\right)^2 \\ &= \left(\tfrac{105}{16}\right)^2, \end{aligned}$$

with equality if and only if $y = 0$. Hence the maximum value is $(\frac{105}{16})^2$ and it occurs only when $x = \frac{7}{2}$, which we might have expected from the symmetry connecting the domain of x and the factors in $f(x)$. (An alternative solution is given in 1992, 39.)

3. Let n be a positive integer > 2 and let $a, b, \ldots, t$ denote the positive integers $< n$ which are relatively prime to n. For example, for $n = 8$, the set $\{a, b, \ldots, t\} = \{1, 3, 5, 7\}$.

Observe that

$$1 + 3 + 5 + 7 = 16 \quad (= 2 \cdot 8),$$

$$1^3 + 3^3 + 5^3 + 7^3 = 1 + 27 + 125 + 343 = 496 \quad (= 62 \cdot 8),$$

and

$$1^5 + 3^5 + 5^5 + 7^5 = 1 + 243 + 3125 + 16897 = 20176 \quad (= 2522 \cdot 8)$$

are all multiples of 8.

Prove that, for every positive integer $n > 2$, the sum

$$S = a^m + b^m + \ldots + t^m$$

is a multiple of n for *all* odd exponents m.

It is difficult to represent the integers $\{a, b, \ldots, t\}$ which are relatively prime to n in more descriptive terms than these uninformed symbols. However, we can observe that when k belongs to the set so does $n - k$: in the event the greatest common divisor $(n, n-k) = d > 1$, then d would divide both n and k and give the contradiction $(n, k) \geq d > 1$. Therefore the terms in the sum S go together in pairs $k^m + (n-k)^m$. Now, since m is odd, the final term in the expansion of $(n-k)^m$ is negative, giving

$$k^m + (n-k)^m = k^m + n^m - mn^{m-1}k + \cdots + mnk^{m-1} - k^m,$$

in which the k^m's cancel, leaving an expression with a common factor n. Holding for each pair in the sum, it follows that S, too, is a multiple of n. (Essentially the same solution is given in 1992, 39.)

4. Sometimes a positive even number has a prime divisor that is greater than its square root; for example, $\sqrt{22} < 5$, and the prime 11 divides 22. Prove, for all positive even integers n which have a prime divisor $> \sqrt{n}$, that neither of the odd numbers $n - 1$ or $n^3 - 1$ can be the product of two *consecutive* odd numbers.

(a) This is a chance to use the rarely decisive fact that the product of two consecutive odd numbers (or even numbers) is always one less than a perfect square:

$$(m-1)(m+1) = m^2 - 1.$$

Hence, for $n - 1$ equal to the product of two consecutive odd numbers, we would have

$$n - 1 = (m-1)(m+1) = m^2 - 1,$$

giving $n = m^2$, implying that the prime divisors of n are given by the prime divisors of m. But $m = \sqrt{n}$, and therefore we have the contradiction that no prime divisor of n can exceed $\sqrt{n}$.

(b) Finally, suppose $n^3 - 1$ is the product of two consecutive odd numbers:

$$n^3 - 1 = (m-1)(m+1) = m^2 - 1.$$

Then $n^3 = m^2$. In this case, the common value k of n^3 and m^2 must be a perfect sixth power:

in the prime decomposition of k, the exponent of a prime must be a multiple of 3 since k is a cube, and also a multiple of 2 since k is a square; being therefore an *even* multiple of 3, it must be a multiple of 6.

Hence

$$n^3 = q^6,$$

for some integer q, and

$$n = q^2,$$

which, as we saw in part (a), yields the contradiction of not allowing n to have a prime divisor that exceeds $\sqrt{n}$.

(An alternative solution is given in 1992, 37.)

Problem K797 from Kvant†

It is well known that a perfect square cannot end in 2, 3, 7, or 8. But what about the digits immediately preceding the final one? Can these be prescribed arbitrarily? That is to say, given any finite sequence of digits $abc \ldots j$, is there a perfect square k^2 with these digits immediately preceding its last digit x:

$$k^2 = \ldots abc \ldots jx?$$

With the enormous capabilities and versatility of the natural numbers, one isn't inclined to dismiss something like this out of hand and it wouldn't be surprising to discover that there is even an infinite sequence of ever-increasing squares with any prescribed penultimate string of digits.

An affirmative answer is supported by the ease of finding squares with a prescribed tens digit:

$$1\underline{0}0,\ \underline{1}6,\ \underline{2}5,\ \underline{3}6,\ \underline{4}9,\ 2\underline{5}6,\ \underline{6}4,\ 5\underline{7}6,\ \underline{8}1,\ 1\underline{9}6, \ldots.$$

Even 2-digit prescriptions are readily determined:

$$\underline{10}0,\ 2\underline{11}6\ (= 46^2),\ \underline{12}1,\ 3\underline{13}6\ (= 56^2),\ \underline{14}4,\ 1\underline{15}6\ (= 34^2),\ \underline{16}9, \ldots.$$

But this is just Arithmetic's idea of a joke, for our efforts to turn up a square for the penultimate sequence 101 are bound to fail because it can be proved that no such square exists.

Proceeding indirectly, suppose

$$k^2 = \ldots 101x,$$

† (*Crux Mathematicorum,* 1991, 34; solution due to Andy Liu, University of Calgary.)

that is, for some nonnegative integer y,

$$k^2 = 10^4 y + 1010 + x.$$

Then (mod 4) we would have

$$k^2 \equiv 2 + x,$$

where x is 0, 1, 4, 5, 6, or 9 (i.e., not 2, 3, 7, or 8), making

$$k^2 \equiv 2, 3, 2, 3, 0, \text{ or } 3.$$

But squares are congruent to 0 or 1 (mod 4): $n \equiv 0, 1, 2, 3$ gives $n^2 \equiv 0, 1, 0, 1$. Hence the only feasible case is

$$k^2 \equiv 0 \pmod 4,$$

given exclusively by $x = 6$, and we have

$$k^2 = 10^4 y + 1016.$$

Also, this shows k^2 is even, making k even, and hence for some positive integer h, we have $k = 2h$. Therefore

$$k^2 = 4h^2 = 10^4 y + 1016,$$

implying

$$h^2 = 2500y + 254 \equiv 2 \pmod 4,$$

an impossibility. The conclusion follows.

An Unused Problem from the 1990 International Olympiad†

Proposed by Mexico

Determine for which positive integers k the set

$$S = \{1990, 1991, 1992, \ldots, 1990 + k\}$$

can be partitioned into disjoint subsets A and B such that the sum of the members of A is the same as the sum of the members of B.

I expect this lovely problem was given serious consideration for a place on this olympiad.

In order for the sums of the members of A and B to be the same, the sum of all the numbers in S would have to be an even number, and this means that S would have to contain an even number of odd integers. Therefore the string of integers in S must end either at one of the appropriate odd numbers itself or at the even integer just past such an odd number, like

$$\{1990, 1991, 1992, \mathbf{1993}\} \quad \text{or} \quad \{1990, 1991, 1992, \mathbf{1993}, 1994\}.$$

As a result, the number of integers in S is either $4n$ or $4n + 1$, for some positive integer n, and since the number of integers in S is $k + 1$, either

$$k = 4n - 1 \quad \text{or} \quad k = 4n.$$

Now, it is easy to see that if any value of k is acceptable, so is $k + 4$: increasing k to $k + 4$ simply extends S to include the next four consecutive integers, $\{a, a+1, a+2, a+3\}$, which can be assigned to the existing subsets A and B so as to preserve equal sums by putting the first and last in one subset and the middle two in the other, thus increasing each sum by $2a + 3$.

---

† (*Crux Mathematicorum,* 1991, 196; an alternative solution is given in 1993, 9)

As an initial example of this device, when $k = 3$, $S = \{1990, 1991, 1992, 1993\}$, and we may take $A = \{1990, 1993\}$ and $B = \{1991, 1992\}$; accordingly, $k = 3, 7, 11, \ldots$, that is, $k = 4n - 1$ is acceptable for all $n \geq 1$.

Suppose, then, that $k = 4n$, thus providing S with $4n + 1$ integers. Since $4n + 1$ is odd, the subsets A and B can't contain the same number of integers, and one of them must contain at least $2n + 1$ of the integers. Now, in the event that the $2n + 1$ smallest integers in S were to "outweigh" the $2n$ largest ones, no partition of S would be acceptable, for a subset containing any $2n + 1$ or more of the integers would weigh more than half the total and hence outweigh the complementary subset. This would be the case if

$$\begin{aligned} 1990 + 1991 + \cdots + (1990 + 2n) &> (1990 + 2n + 1) + (1990 + 2n + 2) \\ &\quad + \cdots + (1990 + 4n) \qquad (1) \\ 1990 + (1 + 2 + \cdots + 2n) &> (2n + 1) + (2n + 2) + \cdots + (4n) \\ 1990 + n(2n + 1) &> n(6n + 1) \\ 1990 &> 4n^2 \\ 44.6\ldots &> 2n \\ n &\leq 22. \end{aligned}$$

Thus $k = 4n$ must fail for $n \leq 22$.

For $n = 23$, however, we have $4n^2 > 1990$, and there is a chance that subsets A and B of equal weight might be attainable. Taking our cue from the inequality (1) above, let's put all the integers on the left side into A and all the integers on the right into B; for $n = 23$, this yields

$$A = \{1990, 1991, \ldots, 2036\}, \quad B = \{2037, 2038, \ldots, 2082\}.$$

From the above calculations on (1), we know that B outweighs A by

$$4n^2 - 1990 = 4 \cdot 23^2 - 1990 = 126.$$

Therefore if we switch the 1990 in A with the 2053 in B, thus increasing A by 63 and decreasing B by the same amount, a balance would be attained:

$$\begin{aligned} A &= \{1991, 1992, \ldots, 2036, 2053\}, \\ B &= \{1990, 2037, \ldots, 2052, 2054, \ldots, 2082\}. \end{aligned}$$

Thus $n = 23$ is acceptable, and consequently so are all values $k = 4n$ for $n \geq 23$.

Altogether, then, the acceptable values of k are

(i) $$4n - 1, \quad \text{for } n \geq 1,$$

and

(ii) $$4n, \quad \text{for } n \geq 23.$$

A Problem from the 1990 Nordic Olympiad†

This problem is concerned with the construction of positive integers in decimal notation. New integers may be obtained from any of those in hand only by applications of the following three operations:
(1) putting a zero at the end (i.e., multiplying by 10),
(2) putting a 4 at the end (multiplying by 10 and adding 4), and
(3) dividing it in half if it is even.

Prove that, given the integer 4 to start things off, *all* positive integers can be generated in this way.

It is easy to produce the first ten positive integers (let $x \to y$ indicate that an appropriate operation applied to x yields y; it will always be evident from the context which operation is intended):

$$4 \to 2 \to \mathbf{1};$$

$$4 \to \mathbf{2};$$

$$4 \to 2 \to 24 \to 12 \to 6 \to 3;$$

$$\mathbf{4};$$

$$4 \to 2 \to 1 \to 10 \to \mathbf{5};$$

$$4 \to 2 \to 24 \to 12 \to \mathbf{6};$$

$$4 \to 2 \to 1 \to 14 \to \mathbf{7};$$

$$4 \to 2 \to 24 \to 12 \to 6 \to 64 \to 32 \to 16 \to \mathbf{8};$$

$$4 \to 2 \to 1 \to 14 \to 144 \to 72 \to 36 \to 18 \to \mathbf{9};$$

$$4 \to 2 \to 1 \to \mathbf{10}.$$

† (*Crux Mathematicorum*, 1990, 259)

Therefore, in the event that it is *not* possible to generate all positive integers, there must exist a *smallest unattainable* integer n which is bigger than 10. Suppose this is the case and that $n = 10k + r$; thus $k \geq 1$ and $0 \leq r \leq 9$.

Now, all positive integers less than n are attainable, in particular the integers k, $2k$, $2k + 1$, $4k$, and $4k + 2$. Clearly, however, if $n = 10k + \mathbf{0}$, then the contradiction would follow that n could be produced by multiplying the attainable number k by 10. Hence r can't be 0. Similarly, it is easy to see that r cannot be any of the digits 1, 2, 4, 5, 6, or 7:

$$\begin{aligned}
4k &\to 10(4k) + 4 = 40k + 4 \to 20k + 2 \to 10k + \mathbf{1}, \\
2k &\to 10(2k) + 4 = 20k + 4 \to 10k + \mathbf{2}, \\
k &\to 10k + \mathbf{4}, \\
2k + 1 &\to 10(2k + 1) = 20k + 10 \to 10k + \mathbf{5}, \\
4k + 2 &\to 10(4k + 2) + 4 = 40k + 24 \to 20k + 12 \to 10k + \mathbf{6}, \\
2k + 1 &\to 10(2k + 1) + 4 = 20k + 14 \to 10k + \mathbf{7}.
\end{aligned}$$

If $n = 10k + \mathbf{3}$, then $8k + 2$ would be less than n, and hence attainable, and we would have the contradiction

$$8k+2 \to 10(8k+2)+4 = 80k+24 \to 40k+12 \to 20k+6 \to 10k+\mathbf{3} = n.$$

Thus $r \neq \mathbf{3}$. Also, if $r = \mathbf{8}$, then $8k + 6 < 10k + 8 = n$, making $8k + 6$ attainable and leading to the contradiction

$$8k+6 \to 10(8k+6)+4 = 80k+64 \to 40k+32 \to 20k+16 \to 10k+\mathbf{8} = n.$$

Therefore r can't be any digit < 9, and the proof of the nonexistence of n is completed by showing that r can't be $\mathbf{9}$ either.

Suppose, then, that $n = 10k + \mathbf{9}$. Note first that if $16k + 14$ is attainable, so is $10k + 9$:

$$\begin{aligned}
16k + 14 &\to 10(16k + 14) + 4 = 160k + 144 \to 80k + 72 \\
&\to 40k + 36 \to 20k + 18 \to 10k + \mathbf{9}.
\end{aligned}$$

The trouble is that $16k + 14$ is greater than n and we don't have any guarantee that it is attainable. To see that $16k + 14$ must in fact be attainable, consider half of it. Clearly $8k + 7 < 10k + 9 = n$, and is therefore attainable. But $8k + 7$ is *odd,* and hence the final step in producing it must be taking half of $16k + 14$ (attaching either a 0 or a 4 to a number can't produce an odd integer). Therefore $16k + 14$ must be attainable, and from it the number $n = 10k + 9$, a final contradiction.

Three Problems from the 1991 AIME†

The highlight of this paper is its last question, which is truly a gem. Unfortunately, I have written about this problem elsewhere and so it can't be included here. However, I hope you will enjoy three other engaging problems from this examination, two of which relate to the year of the contest, the first one in a most unusual way.

> **1.** Let $[x]$ denote the integer part of x, that is, the greatest integer $\leq x$. If r is a real number such that
>
> $$S = \left[r + \frac{19}{100}\right] + \left[r + \frac{20}{100}\right] + \cdots + \left[r + \frac{91}{100}\right] = 546,$$
>
> what is the value of $[100r]$?

Suppose $n \leq r < n + 1$, so that $[r] = n$. Since all 73 of the numbers $r + \frac{t}{100}$, $t = 19, 20, \ldots, 91$, lie between r and $r + 1$, their integer parts are either n or $n+1$ (figure 5). If k of them are equal to n and $73 - k$ are $(n+1)$,

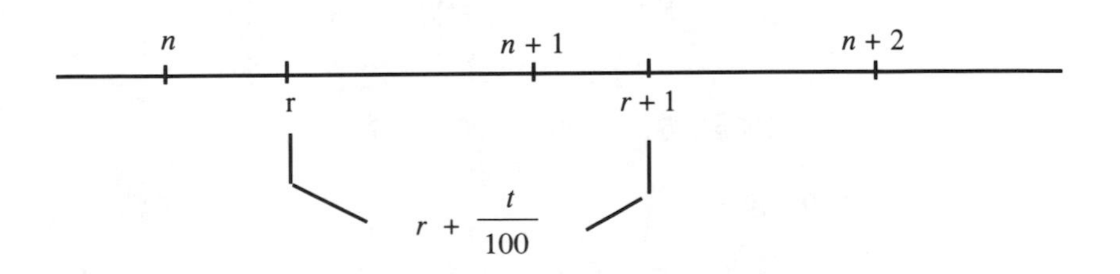

FIGURE 5

† (*Crux Mathematicorum,* 1991, 99)

then

$$S = kn + (73 - k)(n+1) = 546$$
$$73n + 73 - k = 546$$
$$73n + 73 = 546 + k,$$

revealing that $546+k$ is divisible by 73. Since $0 \leq k \leq 73$, the only possibility is $k = 38$:

$$546 + k = 511 + (35 + k)$$
$$= 7 \cdot 73 + (35 + k),$$

implying 73 divides $(35 + k)$, and hence $k = 38$. Therefore

$$546 + k = 584$$
$$= 8 \cdot 73,$$

and

$$73n + 73 = 8 \cdot 73,$$

making $n = 7$.

Thus the integer parts jump from $n = 7$ to $n + 1 = 8$ as we pass from the 38th to the 39th term in S, that is, from

$$\left[r + \frac{56}{100}\right] \quad \text{to} \quad \left[r + \frac{57}{100}\right]$$

Hence

$$r + \frac{56}{100} < 8 \leq r + \frac{57}{100},$$

and

$$100r + 56 < 800 \leq 100r + 57.$$

Turning this inside out gives

$$800 - 57 \leq 100r < 800 - 56,$$
$$743 \leq 100r < 744,$$

and we have

$$[100r] = 743.$$

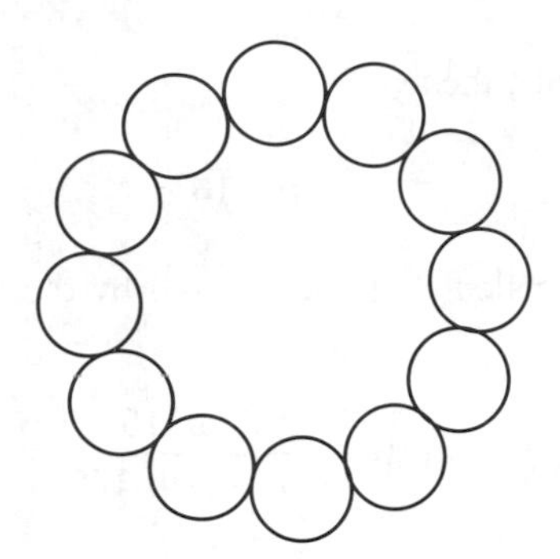

FIGURE 6

2. A ring of twelve identical disks is arranged on top of a unit circle C so as to cover C completely, each disk being tangent to its two neighbors as in figure 6. If the sum of the areas of the twelve disks is expressed in the form $\pi(a - b\sqrt{c})$, where a, b, and c are positive integers and $\sqrt{c}$ is irrational, what is the value of $a+b+c$?

Since the common tangent to adjacent disks at their point of contact X goes through the center O of C and the segment YZ joining the centers of the disks is perpendicular to the common tangent, the circle C does *not* go through the centers of the disks (figure 7).

Because there are twelve disks in the ring, YZ subtends an angle of $30°$ at O and hence the radius $r = XZ$ of a disk subtends an angle of $15°$ at O.

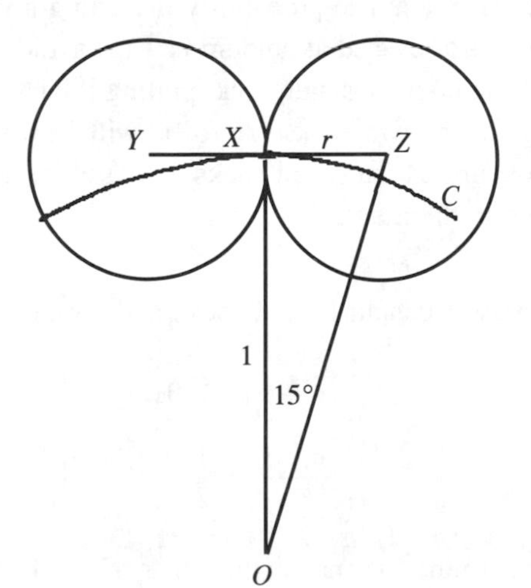

FIGURE 7

Since C has unit radius, then

$$r = \tan 15^\circ,$$

which we can readily calculate to be $2 - \sqrt{3}$ by the following standard procedure: from

$$\tan 30^\circ = \frac{2\tan 15^\circ}{1 - \tan^2 15^\circ},$$

we have

$$\frac{1}{\sqrt{3}} = \frac{2r}{1 - r^2} \quad \text{and} \quad r^2 + 2\sqrt{3}r - 1 = 0,$$

giving

$$r = \frac{-2\sqrt{3} \pm \sqrt{12+4}}{2} = -\sqrt{3} \pm 2,$$

and since r is positive, $r = 2 - \sqrt{3}$.

Hence the total area of the disks is

$$12(\pi r^2) = 12\pi(2 - \sqrt{3})^2 = 12\pi(7 - 4\sqrt{3}) = \pi(84 - 48\sqrt{3}),$$

and we have $a + b + c = 84 + 48 + 3 = 135$.

3. A drawer contains a mixture of red socks and blue socks, at most 1991 socks in all. The probability of getting a matching pair when two socks are selected at random is $\frac{1}{2}$ (to avoid the complications arising from taking a single sock, putting it back, and possibly drawing it again, the two socks are to be withdrawn together). If the drawer contains as many red socks as possible, how many red socks are there in the drawer?

Suppose the drawer contains r red socks and b blue ones. Then

$$r + b \leq 1991 \qquad (1),$$

$$\frac{\binom{r}{2} + \binom{b}{2}}{\binom{r+b}{2}} = \frac{1}{2} \qquad (2),$$

and we want to determine the maximum value of r which is permitted by these relations.

From (2) we have

$$\begin{aligned}
\frac{r(r-1)}{2}+\frac{b(b-1)}{2} &= \frac{1}{2}\cdot\frac{(r+b)(r+b-1)}{2}\\
2r(r-1)+2b(b-1) &= (r+b)^2-(r+b)\\
2r^2-2r+2b^2-2b &= r^2+2rb+b^2-r-b\\
r^2-2rb+b^2 &= r+b\\
(r-b)^2 &= r+b,
\end{aligned}$$

revealing that the total number of socks $r+b$ must be a perfect square.

Conversely, any perfect square k^2 yields a pair of positive integers (r, b) which satisfies condition (2): if

$$r+b=(r-b)^2=k^2,$$

then, considering the case $r>b$ in order to maximize r, we have $r-b=k$ (rather than $b-r=k$), and the easy solution gives

$$r=\frac{k^2+k}{2}, \quad b=\frac{k^2-k}{2}.$$

Thus k and r increase together. By condition (1), however,

$$k^2=r+b\leq 1991,$$

making

$$k^2\leq 1936=44^2,$$

the greatest square not exceeding 1991, and therefore the maximum value of

$$r=\frac{44^2+44}{2}=\frac{1980}{2}=990.$$

An Elementary Inequality

Here is a challenging little problem that has at least two beautiful solutions.

If a, b, and c are positive real numbers that add up to 1, prove that

$$P = \left(1 + \frac{1}{a}\right)\left(1 + \frac{1}{b}\right)\left(1 + \frac{1}{c}\right) \geq 64.$$

For example, when $a = \frac{1}{2}$, $b = \frac{1}{3}$, and $c = \frac{1}{6}$, we have

$$P = (1+2)(1+3)(1+6) = 3 \cdot 4 \cdot 7 = 84.$$

Solution 1. Simply multiplying out, we get

$$P = 1 + \left(\frac{1}{a} + \frac{1}{b} + \frac{1}{c}\right) + \left(\frac{1}{ab} + \frac{1}{bc} + \frac{1}{ca}\right) + \frac{1}{abc}.$$

From the arithmetic mean-geometric mean inequality, we have

$$\frac{\frac{1}{a} + \frac{1}{b} + \frac{1}{c}}{3} \geq \left(\frac{1}{abc}\right)^{1/3},$$

and setting $(\frac{1}{abc})^{1/3} = q$ gives $\frac{1}{a} + \frac{1}{b} + \frac{1}{c} \geq 3q$. In the same way we have

$$\frac{1}{ab} + \frac{1}{bc} + \frac{1}{ca} \geq 3\left(\frac{1}{a^2b^2c^2}\right)^{1/3} = 3q^2.$$

Since $\frac{1}{abc} = q^3$, then

$$P \geq 1 + 3q + 3q^2 + q^3,$$

and we get the possibly surprising result that $P \geq (1+q)^3$.

Applying the A.M.-G.M. inequality to a, b, and c themselves, gives

$$\frac{a+b+c}{3} \geq (abc)^{1/3} = \frac{1}{q},$$

from which

$$q \geq \frac{3}{a+b+c} = 3,$$

since $a+b+c=1$. Hence

$$P \geq (1+q)^3 \geq (1+3)^3 = 64,$$

with equality if and only if $a=b=c=\frac{1}{3}$.

A nice feature of this approach is that the generalization to n variables is at hand with virtually no extra work: if n positive real numbers $a, b, \ldots, k$ add up to 1, then

$$P = \left(1+\frac{1}{a}\right)\left(1+\frac{1}{b}\right)\cdots\left(1+\frac{1}{k}\right) \geq (1+n)^n;$$

we need only observe that the coefficients $\binom{n}{r}$ in the expansion of $(1+q)^n$ parallel the numbers of terms in the groups of fractions in the expression for P: for example,

$$\left(\frac{1}{ab}+\frac{1}{bc}+\cdots+\frac{1}{ka}\right) \quad \text{containing } \binom{n}{2} \text{ terms.}$$

Solution 2. (Due to my colleague Paul Schellenberg)

Clearly

$$P = \left(\frac{1+a}{a}\right)\left(\frac{1+b}{b}\right)\left(\frac{1+c}{c}\right).$$

Now, one would certainly expect that the value of the information "$a+b+c=1$" would reside in substituting 1 for the more involved expression $a+b+c$. However, Paul very craftily proceeds to complicate things by doing just the opposite, to obtain

$$P = \frac{(a+b+c+a)(a+b+c+b)(a+b+c+c)}{abc}.$$

When multiplied out, the numerator N in this fraction contains $4\cdot 4\cdot 4 = 64$ terms, each containing three factors from $\{a, b, c\}$. If the product of these 64 terms is M, the A.M.-G.M. inequality gives

$$\frac{N}{64} \geq M^{1/64}.$$

But the symmetry of the factors in N shows that no variable is favored and that when multiplied out each of a, b, c will occur the same total number of times throughout the expression. Since each of the 64 terms of N contains 3 factors, for a total of 192, each of a, b, and c must occur 64 times, and we have that the product

$$M = a^{64}b^{64}c^{64}.$$

Hence

$$\frac{N}{64} \geq M^{1/64} = (a^{64}b^{64}c^{64})^{1/64} = abc,$$

from which the desired

$$P = \frac{N}{abc} \geq 64$$

follows immediately.

Clearly this solution also generalizes effortlessly to the case of n variables.

Six Geometry Problems

1. A Problem from Sweden. (an unused international olympiad problem)

In figure 8, E trisects diameter AB of a circle X with center O, and Y is the circle on diameter AE. C is an arbitrary point on X and AC crosses Y at P. OP and BC meet at D. Prove that C is the midpoint of BD.

Since BE is one-third of diameter AB, it is two-thirds of radius OB. Now, because AB and AE are diameters, both BC and EP are perpendicular to APC and thus EP is parallel to BCD. Therefore P divides OD in the same ratio that E divides OB:

$$\frac{OE}{EB} = \frac{OP}{PD} = \frac{1}{2}.$$

But since O is the midpoint of AB, OD is a *median* of $\triangle ABD$. The point P, then, in trisecting median OD, is none other than the centroid of $\triangle ABD$, and hence APC is also a median, and the conclusion follows.

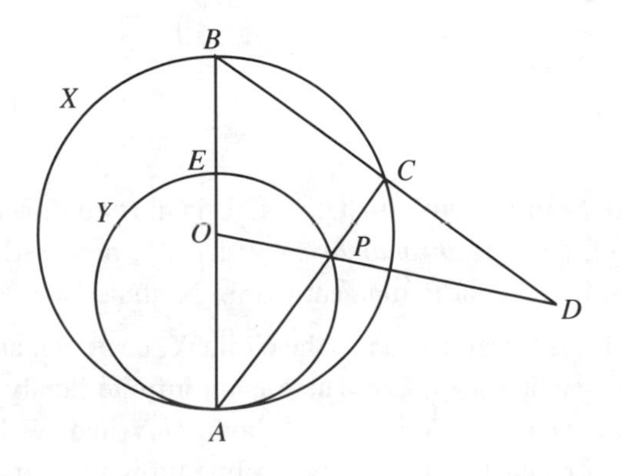

FIGURE 8

2. A Problem from the Journal $\pi\mu\epsilon$. (Problem 711, Fall, 1990; proposed by James N. Boyd, St. Christopher's School, Richmond Va; solved by Murray Klamkin, University of Alberta.)

> An n-gon P has $n-1$ of its sides of unit length. How long should the nth side be to give P the greatest possible area?

Let the length of the remaining side be w. If P is reflected in this nth side, a $(2n-2)$-gon Q is obtained which has all its sides of unit length. Thus the area of Q is twice the area of P and the maximum area of P occurs for the same w that provides the maximum area of Q. But the maximum area of an equilateral polygon is attained by the *regular* polygon of the given side length, and therefore it is *cyclic.* Hence, for maximum area, w is the *diameter* of the circumcircle of a regular $(2n-2)$-gon of unit side. Accordingly, in figure 10,

$$w = 2r = \frac{r}{\frac{1}{2}} = \csc\frac{\pi}{2n-2}.$$

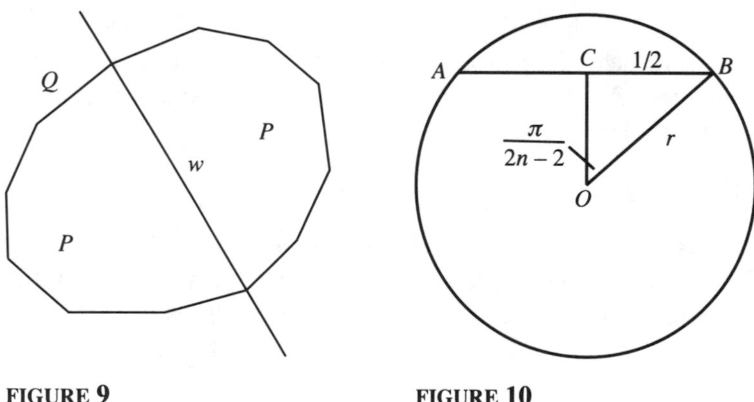

FIGURE 9 FIGURE 10

3. A Problem of Stanley Rabinowitz. (This is the two-dimensional version of Problem 1070, *Crux Mathematicorum,* 1987, 31; proposed and solved by Stanley Rabinowitz, Digital Equipment Corp., Nashua, New Hampshire.)

> In figure 11, AD is a diameter of the circle X, center O, and MN is a chord perpendicular to AD. There is an infinite family of circles Y in the segment MDN that touch both MN and X. Prove that the points of contact B and Q always line up with the point A.

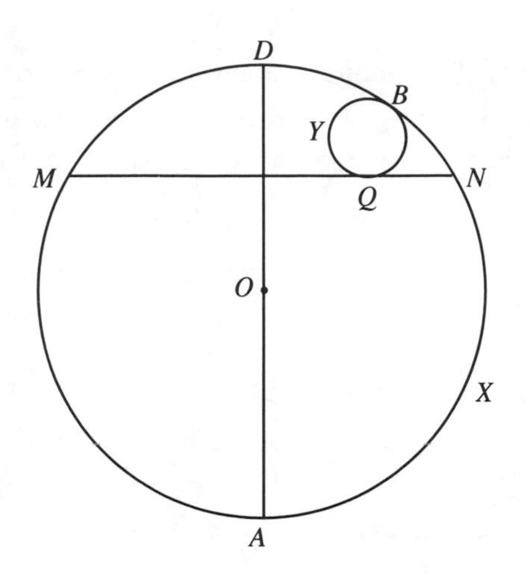

FIGURE 11

Let C be the center of Y. Then the line OC, joining the centers of touching circles, goes through the point of contact B. Now, AD is perpendicular to MN, as is the radius CQ to the point of contact with MN. Thus AO and QC are parallel and the corresponding angles AOB and QCB are equal. But triangles AOB and QCB are isosceles, and since their vertical angles are equal, so are their base angles. Therefore $\angle OBA = \angle CBQ$, (figure 12) and Q lies on BA.

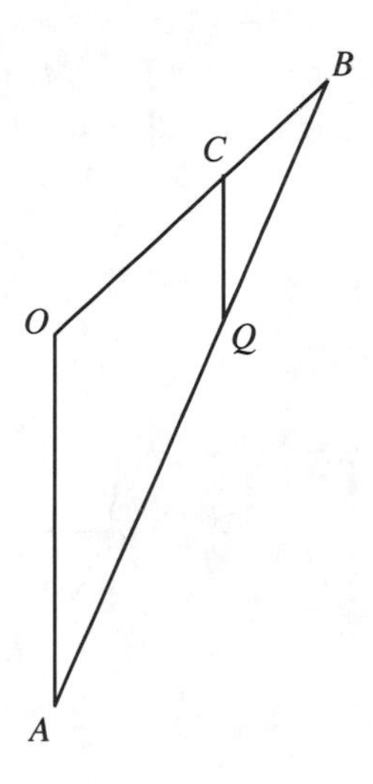

FIGURE 12

4. A Problem from the Common Room. (Source unknown to me)

If the median AM, the angle-bisector AE, and the altitude AD of $\triangle ABC$ divide angle A into four equal parts, what is the size of angle A?

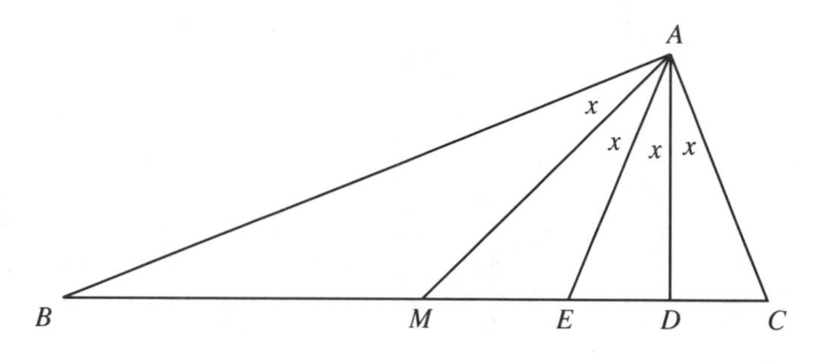

FIGURE 13

The sides of the four equal angles at A cut off four equal arcs on the circumcircle of $\triangle ABC$ (figure 14). Hence T is the midpoint of arc BTC and the equal arcs TS and TU make the chord SU parallel to BC. Since altitude ADU is perpendicular to BC, it is also perpendicular to SU, implying that AMS is a *diameter.*

Now, since M is the midpoint of chord BC and T is the midpoint of the arc BTC that is cut off by BC, the line TM must go through the center of the circle. That is to say, TM will meet every diameter in the center of the circle. Since TM meets diameter AS at M, then M must be the center of the circle, making BC a diameter and angle A a right angle.

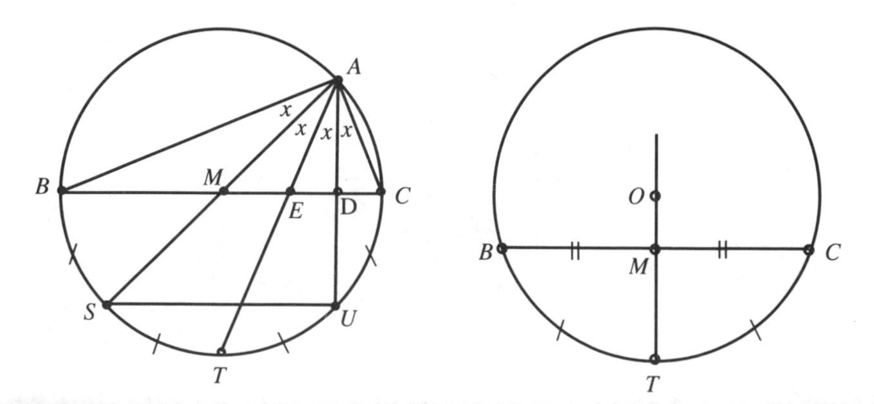

FIGURE 14

5. A Problem of Ian McGee. (University of Waterloo)

At points A and B on a circle, equal tangents AP and BQ are drawn as in figure 15. Prove that the line AB bisects PQ.

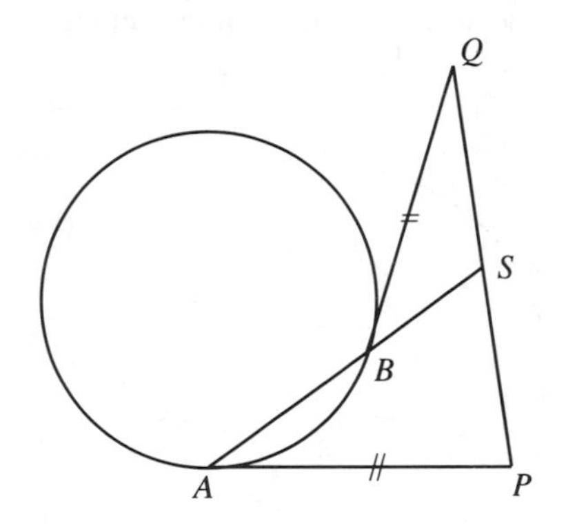

FIGURE 15

Let PA be extended its own length to R and let QB meet PA at T (figure 16). Now, tangents TA and TB are equal, and since $AR = AP = BQ$, then $TR = TQ$. Therefore triangles TAB and TRQ are isosceles triangles which have a common vertical angle at T. Consequently their base angles are equal, and the base angles at R and A imply that RQ is parallel to AB. Thus AB issues from the midpoint of side RP in $\triangle PQR$, is parallel to side RQ, and therefore bisects the third side PQ.

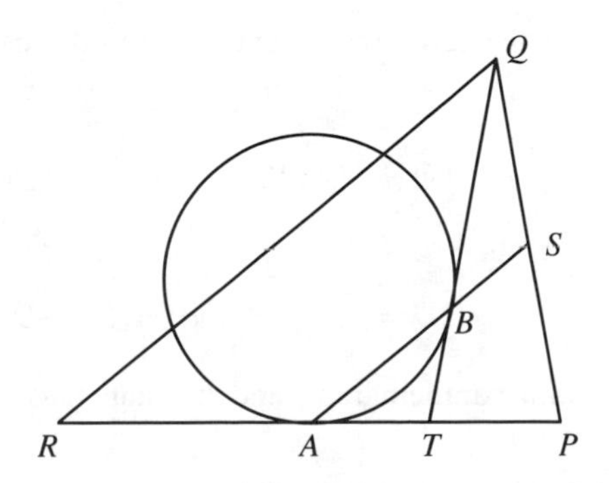

FIGURE 16

6. Another Problem from the Common Room.

The chord AB in a circle X, center C, is extended an arbitrary distance to a point D and circles Y and Z are drawn through D to touch X respectively at A and B. Prove that CD always subtends a right angle at the second point of intersection T of the circles Y and Z.

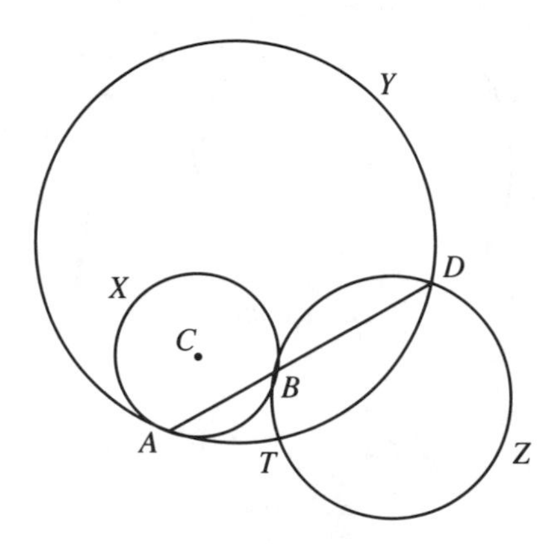

FIGURE 17

Let P and Q be the centers of Y and Z (figure 18). Since the line joining the centers of touching circles passes through the point of contact, then A, C, and P are collinear, as are C, B, and Q. Now, the four base angles in isosceles triangles ABC and QBD are all equal because of the vertically opposite angles at B. But triangle PAD is also isosceles and therefore

$$\angle PDA = \angle PAD = \alpha.$$

Hence equal alternate angles imply the opposite sides of quadrilateral $PCQD$ are parallel:

(i) $$\angle PAD = \angle ADQ = \alpha \Rightarrow PC || DQ,$$

and

(ii) $$\angle QBD = \angle BDP = \alpha \Rightarrow QC || DP.$$

Hence $PCQD$ is a parallelogram, and its diagonals bisect each other at a point E.

Now, clearly $\angle CTD$ is a right angle if and only if the circle on diameter CD goes through T. Consequently, it is only necessary to show that this

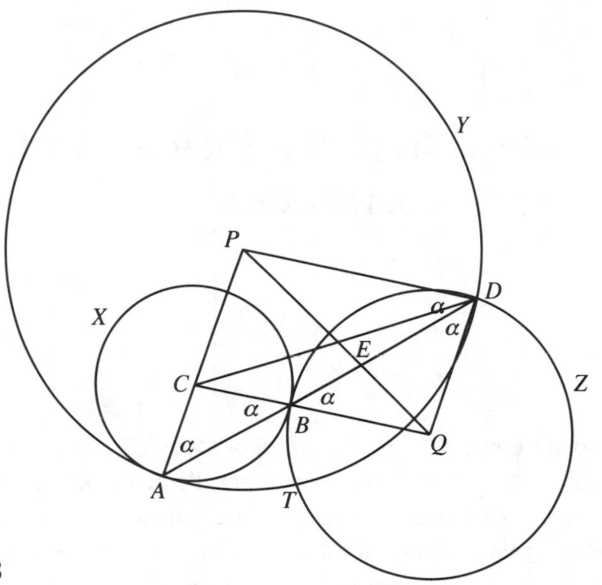

FIGURE 18

circle goes through T. Since E is already known to be the midpoint of CD, it remains only to show that $ET = ED$. But this is easy.

The line PQ, joining the centers of the intersecting circles Y and Z, is the perpendicular bisector of their common chord TD. Therefore T and D are equidistant from every point on PQ, in particular from E.

Two Problems from the 1989 Swedish Mathematical Competition[†]

1. Around a circle $4n$ points are chosen and alternately colored blue and yellow. The $2n$ blue points are divided arbitrarily into n pairs and the members of each pair are joined by a blue chord; similarly n yellow chords are drawn. If no three of the chords are concurrent, prove there are at least n blue-yellow points of intersection determined by a blue chord and a yellow chord.

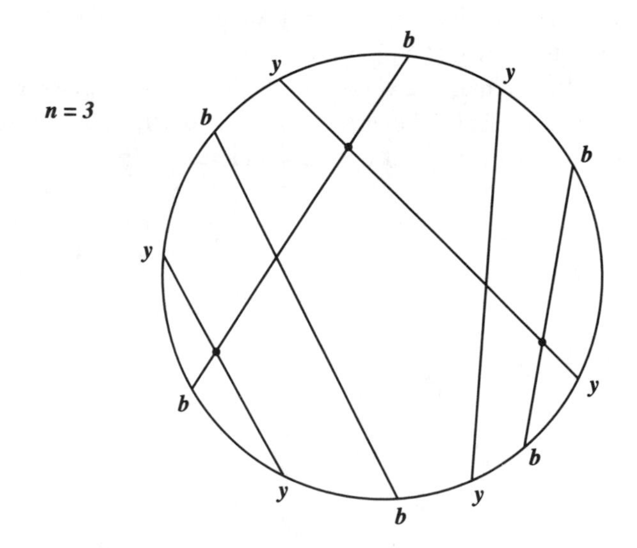

FIGURE 19

† (*Crux Mathematicorum,* 1991, 130)

The following clever solution is due to my colleague Ian Goulden.

(a) It is easy to see that each chord must be crossed by a chord of one color or the other. Consider the blue chord c in figure 20. Clearly the open arcs p and q that are cut off by c each contain one more yellow point than blue point (in traversing p or q, the first and last interior points encountered are both yellow). Therefore each of these arcs contains an *odd* number of points of one color and an even number of the other color. Now, an odd number of points of the same color cannot pair up completely among themselves—at least one of them must be paired with a point in the opposite arc, thus providing a chord that intersects c.

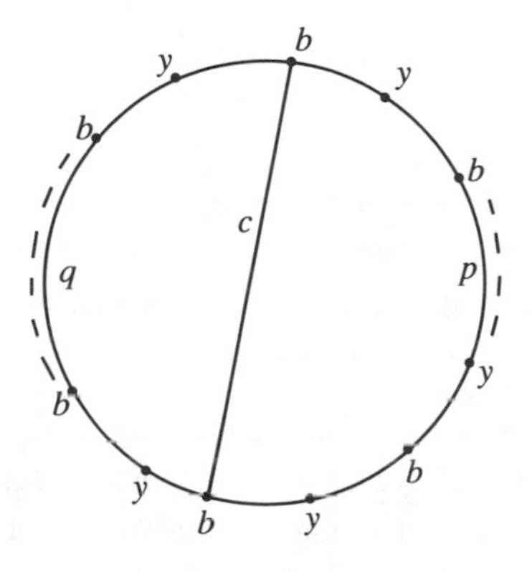

FIGURE 20

Therefore if a blue chord is not crossed by any other blue chord, it must be crossed by a yellow one. Consequently, in the event that no two blue chords intersect, each of the n blue chords must be crossed by a yellow chord, for a favorable total of at least n blue-yellow points of intersection.

(b) On the other hand, suppose there is at least one point of intersection X determined by two blue chords AB and CD (figure 21). Now, what would be the effect of undoing this blue-blue point X by pairing A with C and B with D (or the equivalent pairs (A, D) and (B, C)), leaving everything else as it is? A yellow chord like EF would continue to determine a single blue-yellow point of intersection (it would move along EF from L to M, but that is immaterial); a yellow chord like GH would continue to determine two blue-yellow points, and a chord like IJ would lose its two blue-yellow points. Of the four arcs which are determined by A, B, C, D, a yellow chord must join

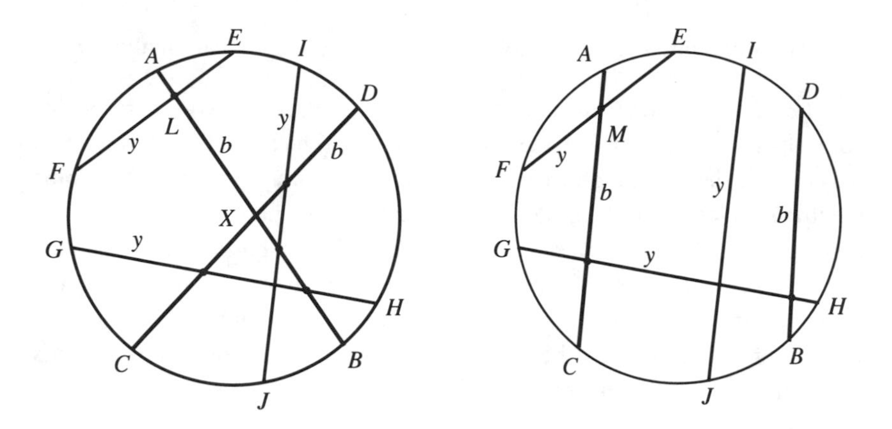

FIGURE 21

points in arcs that are either adjacent or opposite one another. Hence the three chords EF, GH, and IJ cover all the possibilities. Thus, while undoing a blue-blue point might result in a decrease in the number of blue-yellow points of intersection, it *never results in an increase.*

(c) Now, it is conceivable that, in undoing a blue-blue, new blue-blue points might be created. However, a similar analysis immediately puts this concern to rest (figure 22): undoing a blue-blue point X not only gets rid of X but, if anything else, causes other blue-blue points to vanish as well, resulting in *at least one fewer* blue-blue.

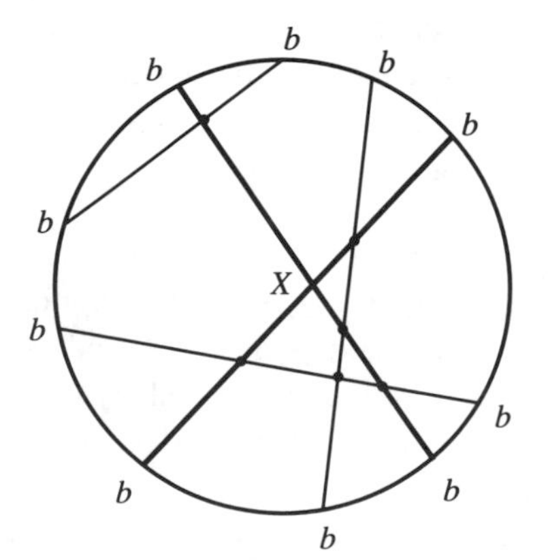

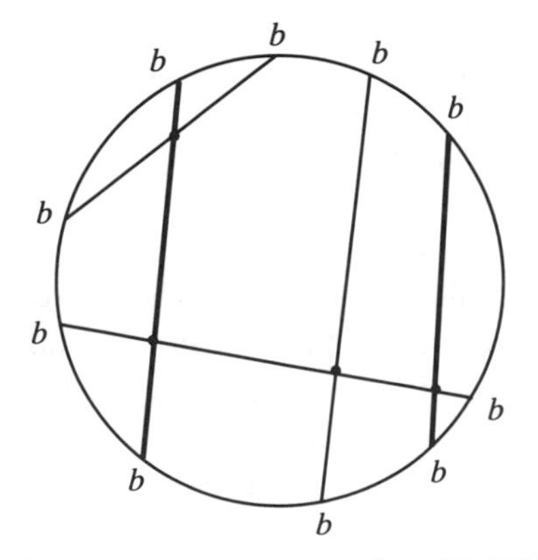

FIGURE 22

(d) Therefore, by continuing to undo blue-blue points, one ultimately obtains a figure with no blue-blue points at all. As seen above, a figure with no blue-blue points has at least n yellow-blue points of intersection, and since the unraveling process cannot increase their number, it follows that there must have been at least n of them to start with in the original configuration.

2. Find all positive integers n such that

$$f(n) = n^3 - 18n^2 + 115n - 391$$

is a perfect cube.

The first two terms of $f(n)$ ensure that it can't take a value that is very far from $(n-6)^3$:

$$(n-6)^3 = n^3 - 18n^2 + 108n - 216,$$

and so

$$f(n) = (n-6)^3 + 7n - 175.$$

Thus, when $7n - 175 = 0$, $f(n)$ is equal to $(n-6)^3$, and we have immediately that $f(25) = 19^3$, and it is clear that $n = 25$ is the *only* integer for which $f(n) = (n-6)^3$.

Since $f(n)$ is generally close to $(n-6)^3$, let us check the nearby cubes.
(i) $(n-5)^3$:

$$(n-5)^3 = n^3 - 15n^2 + 75n - 125,$$

giving

$$\begin{aligned} f(n) &= (n-5)^3 - 3n^2 + 40n - 266. \\ &= (n-5)^3 - (3n^2 - 40n + 266). \end{aligned}$$

However, the discriminant of

$$3n^2 - 40n + 266 = 1600 - 12(266),$$

which is clearly negative, implying $3n^2 - 40n + 266$ is always positive and that $f(n)$ is never as big as $(n-5)^3$.

(ii) $(n-7)^3$:

$$(n-7)^3 = n^3 - 21n^2 + 147n - 343,$$

giving

$$f(n) = (n-7)^3 + 3n^2 - 32n - 48,$$

and

$$f(n) = (n-7)^3$$

if and only if $3n^2 - 32n - 48 = 0$, that is, for $n = 12$ or $-4/3$. Thus

$$f(12) = 5^3,$$

and this is the *only* integer for which $f(n) = (n-7)^3$.

(iii) $(n-8)^3$:

$$(n-8)^3 = n^3 - 24n^2 + 192n - 512,$$

giving

$$f(n) = (n-8)^3 + 6n^2 - 77n + 121,$$

and

$$f(n) = (n-8)^3$$

if and only if $6n^2 - 77n + 121 = 0$, that is, for $n = 11$ or $11/6$. Thus

$$f(11) = 3^3,$$

and this is the *only* time $f(n) = (n-8)^3$.

However, as we shall see, this is as far as we can go. For $(n-9)$ we have

$$(n-9)^3 = n^3 - 27n^2 + 243n + 729,$$

giving

$$f(n) = (n-9)^3 + 9n^2 - 128n + 338$$

and

$$f(n) = (n-9)^3$$

if and only if $9n^2 - 128n + 338 = 0$, that is, for $n = \dfrac{128 \pm \sqrt{4216}}{18} = 10.7$ or 3.5 approximately. Hence $f(n)$ is never equal to $(n-9)^3$. Not only that, but, for n greater than the root 10.7, i.e., for $n \geq 11$, the function $9n^2 - 128n + 338$ is positive and $f(n)$ is always greater than $(n-9)^3$.

Now, differentiating $f(n)$, we get

$$f'(n) = 3n^2 - 36n + 115,$$

whose discriminant $1296 - 12 \cdot 115$ is clearly negative, implying $f'(n)$ is always positive, making $f(n)$ an increasing function. Checking the value of $f(10)$, we find

$$f(10) = 1000 - 1800 + 1150 - 391 < 0,$$

and therefore $f(n)$ is negative for all $n \leq 10$. Thus we need concern ourselves only with values of $n \geq 11$, for which values we have just observed that $f(n) > (n-9)^3$. For $n \geq 11$, then, we have

$$(n-9)^3 < f(n) < (n-5)^3,$$

implying the only times $f(n)$ can be a perfect cube are when it is equal to $(n-6)^3$, $(n-7)^3$, or $(n-8)^3$, which we have seen occur only for $n = 25$, 12, and 11. (Essentially the same solution is given in 1992, 227.)

Two Problems from the 1989 Austrian-Polish Mathematics Competition†

1. If each point of the plane is colored either red or blue, prove that some equilateral triangle has all its vertices the same color.

(a) It is easy to see that a monochromatic equilateral triangle is bound to arise if some three equally-spaced points along a straight line all have the same color. To this end, suppose A, B, and C are three such red points (figure 23). If ACZ is the equilateral triangle drawn on base AC and X and Y are the midpoints of the other two sides, then each of the four triangles formed by joining X, Y, and B is also equilateral.

Now, each of the equilateral triangles XAB, YBC, and ZAC has two vertices from the red triple $\{A, B, C\}$. Thus an all-red triangle results if any

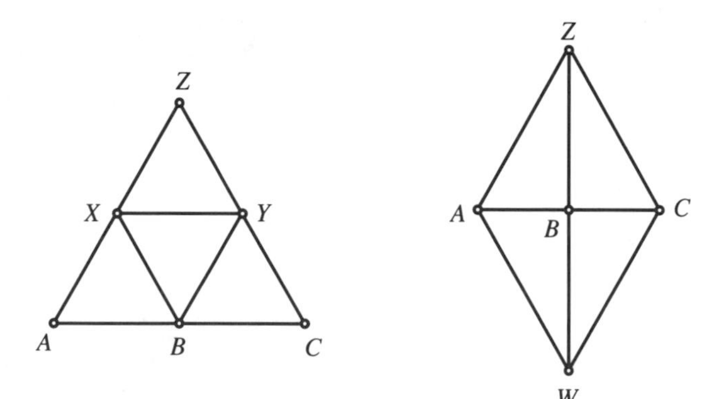

FIGURE 23

† (*Crux Mathematicorum,* 1991, 65)

of the third vertices X, Y, Z is also colored red. Consequently the only way to avoid such an all-red triangle is to make triangle XYZ itself an all-blue one.

(b) Our argument is completed by showing that, in trying to stave off a monochromatic equilateral triangle, one can't avoid coloring some three equally-spaced collinear points all the same color.

Suppose A and C are the same color, say red. Then the third vertices Z and W of equilateral triangles on either side of AC must be blue to avoid an all-red triangle with A and C (figure 23). But, since the midpoint B of AC is also the midpoint of ZW, coloring B results either in making A, B, C an equally-spaced collinear all-red triple or Z, B, W an all-blue one. (An alternative solution is given in 1992, 105.)

2. Let $S = \{a^2, b^2, c^2, \ldots\}$ be an increasing sequence of perfect squares in which the difference between each pair of consecutive terms is either a prime or the square of a prime; for example, $S = \{16, 25, 36, 49\}$, with successive differences 9, 11, 13.

Prove that all such sequences S are finite in length and find the longest possible sequence S.

Let us begin by recalling the well-known fact that the differences between consecutive perfect squares are simply the odd numbers $3, 5, 7, \ldots$:

Squares:	1,		4,		9,		16,		25,		36, ...,	
Differences:		3,		5,		7,		9,		11,		....

Now, let t^2, k^2 be consecutive terms in a sequence S. Then, for some prime p,

$$k^2 - t^2 = (k-t)(k+t) = p \quad \text{or} \quad p^2.$$

Thus $k - t$ is the smaller of two unequal factors whose product is p or p^2. It follows, then, that $k - t$ must be 1, implying that the terms of a sequence S proceed through *consecutive* perfect squares:

$$S = \{\ldots, t^2, (t+1)^2, (t+2)^2, \ldots\},$$

and that succeeding differences are *consecutive* odd numbers.

However, every third odd number is divisible by 3, and for values exceeding $3^2 = 9$, a multiple of 3 is neither a prime nor the square of a prime. Therefore, a sequence S, all of whose terms are at least 25 (beyond which the differences exceed 9) cannot extend through more than two differences, confining S to a maximum length of three terms (e.g., 64, 81, 100, with the two differences 17 and 19, at which point the next difference is 21, an unacceptable multiple of 3). A sequence longer than three terms, then, is obliged to contain squares smaller than 25 and the longest possible sequence is clearly

$$1, \quad 4, \quad 9, \quad 16, \quad 25, \quad 36, \quad 49,$$

with differences

$$3, \quad 5, \quad 7, \quad 9, \quad 11, \quad 13.$$

(An alternative solution is given in 1992, 107.)

Two Problems from the 1990 Australian Olympiad†

1. In a certain library there are n shelves, each holding at least one book. k new shelves are acquired and the books are rearranged on the $n + k$ shelves, again with at least one book on each shelf.

A book is said to be *privileged* if it is on a shelf with fewer books in the new arrangement than it was in the original arrangement.

Prove that there are at least $k + 1$ privileged books in the rearranged library.

To each book on a shelf that holds r books let a "value" of $\frac{1}{r}$ be assigned. In this way the sum of the values of the books on each shelf is 1, for a grand total of n in the initial arrangement of the library.

Now, let the library be revalued in the same way after the rearrangement. If a book is on a shelf with fewer books than before, its second value $\frac{1}{s}$ will be greater than its original value of $\frac{1}{r}$ because $s < r$. A book is privileged, then, if and only if its value *increases* in the second valuation.

The values of the books on each shelf still add up to 1 and therefore the $n+k$ shelves of the new arrangement provide a total valuation of $n+k$. That is to say, there is an increase of k units in the second assessment. However, no privileged book can provide an increase even as big as 1: clearly

$$\frac{1}{s} - \frac{1}{r} < 1 \quad \text{for } s < r \text{ and } r, s \geq 1.$$

Therefore it must take more than k books to bring about an increase of k units, implying at least $k + 1$ privileged books.

---

† (*Crux Mathematicorum,* 1991, 101)

2. Circles A and B, of radii R and r respectively, intersect in distinct points S and T, and a common tangent meets A at M (figure 24). If MS is tangent to B and ϕ is the angle between MS and the tangent SL to A as shown, prove that

$$\frac{r}{R} = \left(2\sin\frac{\phi}{2}\right)^2.$$

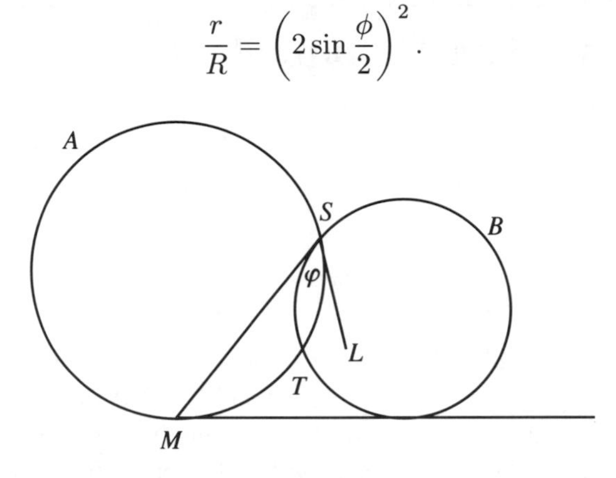

FIGURE 24

(i) If SL meets the common tangent at C (figure 25(a)), then CS and CM are equal tangents to A from C, making $\triangle CMS$ isosceles and $\angle SMC = \phi$.

(ii) Suppose the common tangent meets B at N (figure 25(b)). If O is the center of B, then OM bisects the angle between the tangents MS and

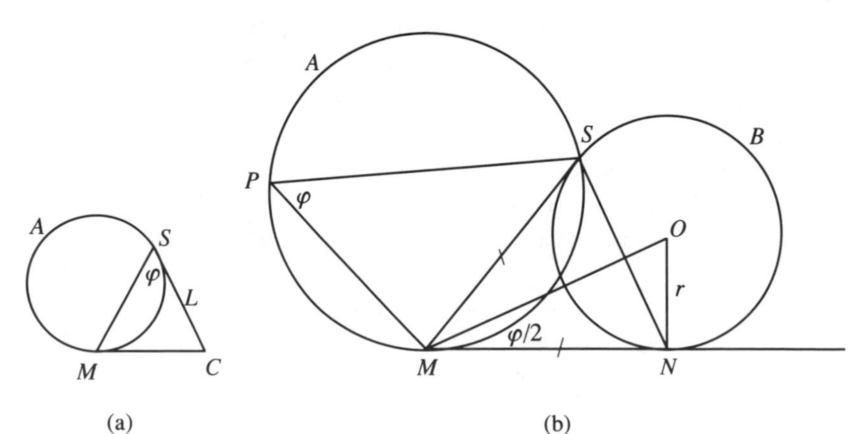

FIGURE 25

MN and $\angle OMN = \frac{1}{2}\angle SMN = \frac{\phi}{2}$; also these tangents are the same length. From $\triangle OMN$, then, we have

$$r = ON = MNtan\frac{\phi}{2}.$$

To calculate R we observe that ϕ $(= \angle SMN)$ is the angle between the tangent to A at M and the chord MS and is therefore equal to $\angle MPS$ in the segment of A on the opposite side of MS. Thus the circumradius R of $\triangle MPS$ is given by the standard formula

$$R = \frac{MS}{2\sin\phi}.$$

Recalling that $MN = MS$, then

$$\begin{aligned} \frac{r}{R} &= \frac{MN\tan\frac{\phi}{2}}{\frac{MS}{2\sin\phi}} \\ &= 2\sin\phi\tan\frac{\phi}{2} \\ &= 2\left(2\sin\frac{\phi}{2}\cos\frac{\phi}{2}\right)\cdot\frac{\sin\frac{\phi}{2}}{\cos\frac{\phi}{2}} \\ &= \left(2\sin\frac{\phi}{2}\right)^2. \end{aligned}$$

Problem 1367 from Crux Mathematicorum†

This engaging problem concerns certain towers of pennies arranged row on row. Gaps are not allowed between the pennies in any row and each penny, except those on the bottom, rests on two pennies in the row below it. Thus none of the upper rows can overhang the row below it at either end, and such towers as those in figures 26(a) and 26(b) are not allowed. However, an acceptable tower may have any number of rows, as in figure 26(c).

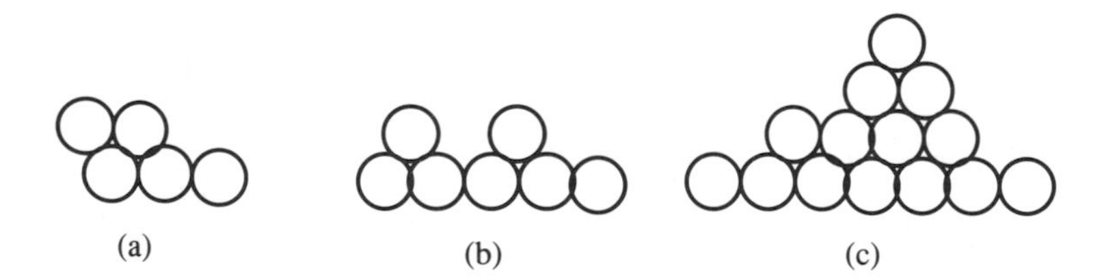

FIGURE 26

Determine the number of towers with n pennies in the bottom row. For example, there are five towers when $n = 3$ (figure 27).

FIGURE 27

† (1989, 278; proposed by Richard Guy, University of Calgary.)

The following beautiful solution is the work of Colin Springer, a student of whom we are all very proud at Waterloo.

Consider the two sequences $\{a_n\}$ and $\{b_n\}$, where a_n equals the number of acceptable towers which have *exactly* n pennies in the bottom row (i.e., the number we seek), and b_n equals the number of acceptable towers which have *at most* n pennies in the bottom row.

Since a tower having at most n pennies in the bottom row has either exactly n pennies or at most $n-1$ pennies in the bottom row, we have

$$b_n = a_n + b_{n-1}.$$

Now consider the towers counted by a_{n+1}. Since their bottom rows all have $n+1$ pennies, what distinguishes these towers are the superstructures above the bottom row. Clearly there are only two kinds of towers:

(i) those in which the first cell x of the second row is occupied (figure 28(i)),

(ii) those in which this first cell x is unoccupied (figure 28(ii)).

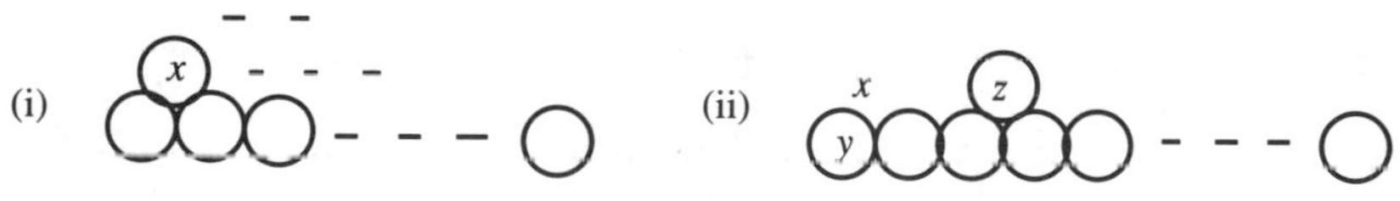

FIGURE 28

We note that, with $n+1$ pennies in the bottom row, there can't be more than n pennies in the second row.

The superstructure of a tower of type (i), then, is itself one of the towers counted by b_n (it needn't go all the way across the n places in the second row to be counted by b_n); conversely, any tower counted by b_n, placed to begin at x in the second row, yields a tower of type (i). This 1-1 correspondence implies that the number of towers of type (i) is simply b_n.

On the other hand, suppose cell x is not occupied. In this case, the first penny y in the bottom row is at least one place to the left of the first penny z in the second row (z might be some distance from the unoccupied x). Consequently, deleting y leaves a tower that is counted by a_n, for the bottom row would now contain exactly n pennies. Conversely, adding a penny (y) to the left end of a tower that is counted by a_n yields a tower of type (ii), for the bottom row would now have $n+1$ pennies and the first cell (x) in the second row would be unoccupied. This second 1-1 correspondence implies the surprising result that the number of towers of type (ii) is a_n, and we have altogether that

$$a_{n+1} = b_n + a_n.$$

Finally, consider the shuffled sequence

$$\{c_n\} = \{a_1, b_1, a_2, b_2, \ldots, b_{n-1}, a_n, b_n, a_{n+1}, \ldots\}.$$

Our recursions $a_{n+1} = b_n + a_n$ and $b_n = a_n + b_{n-1}$ show that, for $n > 2$, c_n is the sum of the two immediately preceding terms. Since a_1 and b_1 are each obviously 1 (i.e., $c_1 = c_2 = 1$), it follows that $\{c_n\}$ is none other than the Fibonacci sequence $\{f_n\}$. Hence

$$a_n = f_{2n-1} = \frac{\alpha^{2n-1} - \beta^{2n-1}}{\alpha - \beta}, \quad \text{where } \alpha = \frac{1+\sqrt{5}}{2} \text{ and } \beta = \frac{1-\sqrt{5}}{2}.$$

Three Problems from Japan

The problems in this section are taken from the report "Japanese University Entrance Examination Problems in Mathematics", edited by Ling-Erl Eileen T. Wu (MAA, 1993). Although these problems are necessarily somewhat routine and not generally of olympiad difficulty, I hope you will find them attractive and enjoyable. Their particular associations are given at the end.

1. A tetrahedron rests on a plane on one of its faces. The tetrahedron is flopped onto a different face by rolling it about one of the edges of its bottom face. If the tetrahedron is thus sent on a zig-zag course in the plane, selecting *at random* the axis of rotation from the edges of the face that is currently on the bottom, what is the probability that, after n rolls, the tetrahedron again rests on the face F that was originally on the bottom?

If the required probability is denoted by p_n, it is clear that $p_0 = 1$. Now, the one thing that would kill all chance of having the tetrahedron rest on F after n rolls is if F were the bottom face after $n-1$ rolls; in this case, the nth step would have to roll the tetrahedron off F onto a different face. Since p_{n-1} is the probability of such a disaster, the probability is $1 - p_{n-1}$ that this doesn't happen, that the nth roll could possibly lead to the desired result. Even so, there is only one chance in three that it will be the preferred face F that is rolled down in the last step, and so we obtain the neat result that

$$\begin{aligned} p_n &= (\text{the prob. } F \text{ is not on the bottom after } n-1 \text{ rolls})\times \\ &\qquad (\text{the prob. } F \text{ is the face that is rolled down in the } n\text{th step}) \\ &= (1 - p_{n-1}) \cdot \frac{1}{3}. \end{aligned}$$

The problem is essentially solved at this point, but it is interesting how the examiners completed their solution.

While I expect most of us would finish off by using standard procedures to solve this recursion, the official solution proceeds brilliantly as follows: from

$$p_n = \frac{1}{3}(1 - p_{n-1}),$$

we have

$$p_n = -\frac{1}{3}(p_{n-1} - 1),$$

and therefore

$$p_n - \frac{1}{4} = -\frac{1}{3}(p_{n-1} - 1) - \frac{1}{4}$$

(who would ever think of subtracting $\frac{1}{4}$?) giving

$$p_n - \frac{1}{4} = -\frac{1}{3}\left(p_{n-1} - \frac{1}{4}\right),$$

revealing that the sequence $\{p_n - \frac{1}{4}\}$ is a geometric progression with common ratio $-\frac{1}{3}$. With first term

$$p_0 - \frac{1}{4} = 1 - \frac{1}{4} = \frac{3}{4},$$

the $(n+1)$th term is

$$p_n - \frac{1}{4} = \frac{3}{4}\left(-\frac{1}{3}\right)^n,$$

and we have

$$p_n = \frac{3}{4}\left(-\frac{1}{3}\right)^n + \frac{1}{4}.$$

Comment. That magical move of subtracting a quarter is really part of a clever strategy based on calculating the value of x which would make

$$p_n - x = -\frac{1}{3}(p_{n-1} - x).$$

Since $p_n = -\frac{1}{3}(p_{n-1} - 1)$ from the known recursion, this would require

$$-\frac{1}{3}(p_{n-1} - 1) - x = -\frac{1}{3}(p_{n-1} - x),$$
$$\frac{1}{3} = \frac{4}{3}x,$$
$$x = \frac{1}{4}.$$

2. Obviously the roots of $x^3 - 3x - p = 0$ change as different real numbers are substituted for p. Let the function $f(p)$ be defined as follows:

(i) for values of p that provide three real roots to the equation, let $f(p)$ be the product of the *greatest and least* of the roots,

(ii) for values of p that give only one real root, let $f(p)$ be the *square* of that root.

Determine the minimum value of $f(p)$ as p varies over all real numbers.

First an algebraic solution and then the elegant geometric solution of the examiners.

Solution 1. (i) Three Real Roots. Suppose p gives an equation which has three real roots a, b, and c. In order to talk about the greatest and least of these roots, suppose also that $a \leq b \leq c$; this incurs no loss of generality and provides an algebraic representation for the intriguing function $f(p)$ in the expression ac. By equating coefficients in

$$x^3 - 3x - p = (x - a)(x - b)(x - c),$$

we obtain the standard relations

$$a + b + c = 0 \qquad \text{(i)}$$

$$ab + bc + ca = -3 \qquad \text{(ii)}$$

$$abc = p. \qquad \text{(iii)}$$

From (i), $a + c = -b$; then (ii) gives

$$ab + bc + ca = b(a + c) + ac = -b^2 + ac = -3,$$

and we have

$$f(p) = ac = b^2 - 3.$$

Since the real number $b^2 \geq 0$, then $f(p) \geq -3$, with $f(p) = -3$ *if* it is possible for b to take the value 0.

From $abc = p$, b could be 0 only when $p = 0$. Checking $p = 0$, we obtain $x^3 - 3x = 0$, whose roots are $0, \pm\sqrt{3}$. Thus $b = 0$ *is* feasible and we conclude that $f(p)$ has the minimum of -3 for values of p that provide and equation having three real roots.

(ii) One Real Root. Suppose p gives a single real root a, making $f(p) = a^2$. Since only one root is real, it must be that

$$x^3 - 3x - p = (x - a)(x^2 + qx + r),$$

where the discriminant $q^2 - 4r$ is negative. Equating the coefficients of the terms in x^2 and x yields

$$-a + q = 0, \quad \text{and} \quad r - aq = -3.$$

Thus $a = q$, and

$$f(p) = a^2 = aq = r + 3.$$

Now, $q^2 - 4r < 0$, where the real number $q^2 \geq 0$. Hence r must be positive, implying

$$f(p) = r + 3 > 3.$$

Hence, over the entire range of real numbers, the minimum $f(p) = -3$.

Solution 2. The roots of $x^3 - 3x - p = 0$, that is, of $x^3 - 3x = p$, are the x-coordinates a, b, c of the points of intersection of the graphs of

$$y = x^3 - 3x \quad \text{and} \quad y = p \qquad \text{(figure 29).}$$

It is easy to determine that the graph of $y = x^3 - 3x$ crosses the x-axis at the origin and at the points $S(-\sqrt{3}, 0)$ and $T(\sqrt{3}, 0)$. Now, from the graph it is evident that the equation has three real roots when the line $y = p$ lies in the closed strip between the tangents L_1 and L_2 which are parallel to the x-axis, and only one when $y = p$ is either above or below this strip. It behooves us, then, to determine the local maximum M and local minimum N in figure 29. Clearly,

$$\frac{dy}{dx} = 3x^2 - 3,$$

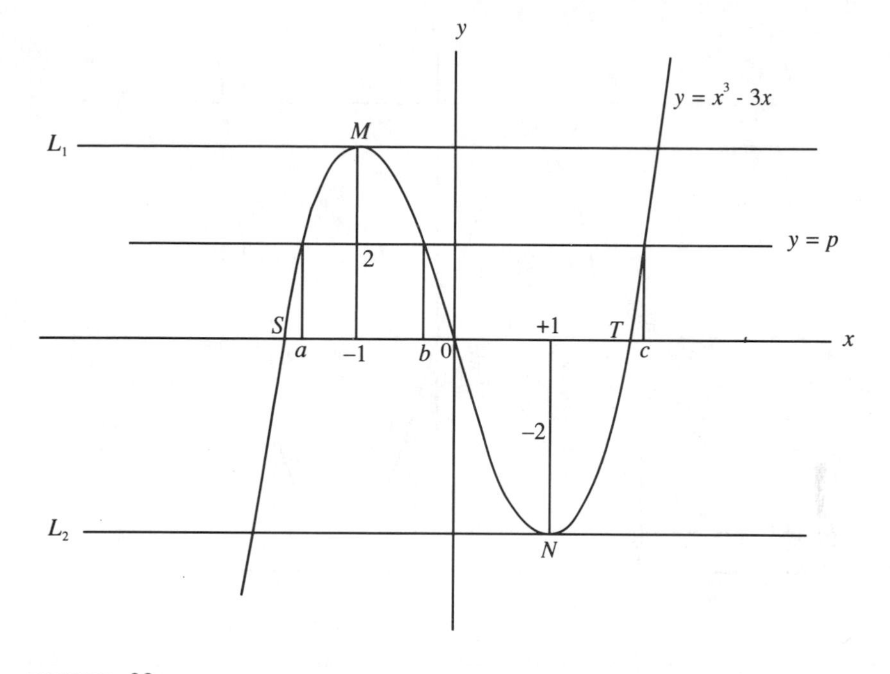

FIGURE 29

which is equal to 0 for $x = \pm 1$, making M and N the points $(-1, 2)$ and $(1, -2)$. Therefore the equation $x^3 - 3x - p = 0$ has three real roots for $|p| \leq 2$ and only one for $|p| > 2$.

(i) Three Real Roots $a \leq b \leq c$. As in solution 1, we have $abc = p$, from which

$$f(p) = ac = \frac{p}{b}.$$

Now, b is a root, and so it satisfies the equation, giving $b^3 - 3b = p$, and hence

$$f(p) = \frac{p}{b} = b^2 - 3,$$

where b is the "middle" root. But, it is also clear from the graph that the middle root must lie between the x-coordinates of M and N, which are -1 and $+1$, and that it is equal to 0 when $p = 0$. Thus b^2 lies between 0 and 1 inclusive and

$$f(p) = b^2 - 3 \geq -3,$$

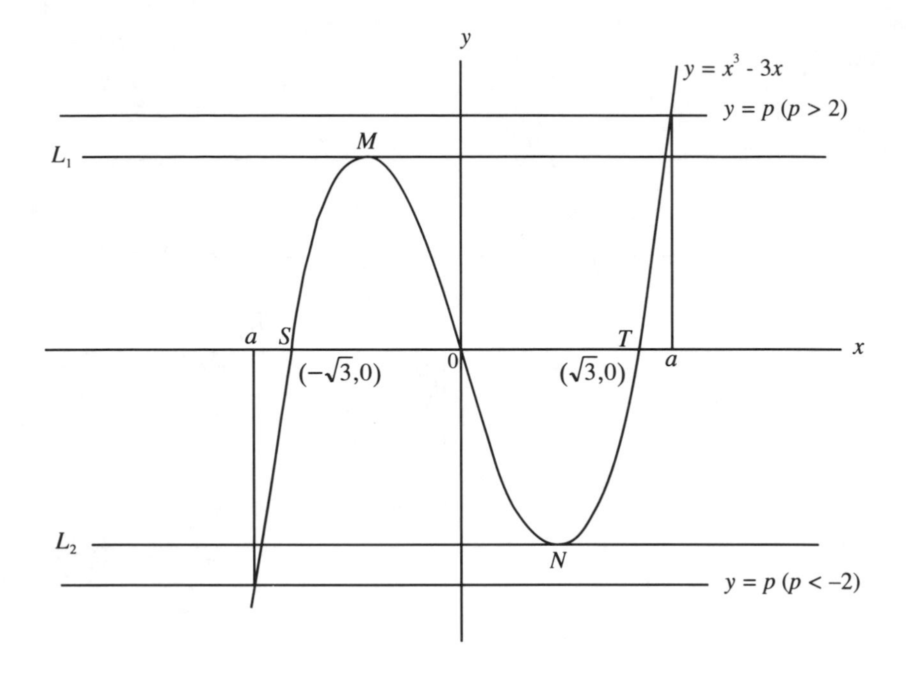

FIGURE 30

with equality for $p = 0$.

(ii) One Real Root $x = a$. In this case, either $p > 2$, making the root a considerable greater than the intercept $\sqrt{3}$ (recall T is $(\sqrt{3}, 0)$), or $p < -2$, making it considerably less than $-\sqrt{3}$ (figure 30). In either case, then,

$$f(p) = a^2 > 3,$$

and we have the final conclusion already that the minimum value of $f(p)$ over all real numbers is -3.

> **3.** In a coordinate plane, about each lattice point as center let a circle of the same radius r be drawn. Now consider all the straight lines in the plane that have slope $\frac{2}{5}$. Clearly, if r is very small, lots of these lines run across the plane without meeting any of the circles. What is the minimum radius r such that *every* straight line with slope $\frac{2}{5}$ will meet some circle in the family?

Since all parts of a lattice have the same structure, the neighborhood around the origin is representative of the whole plane. Consider, therefore,

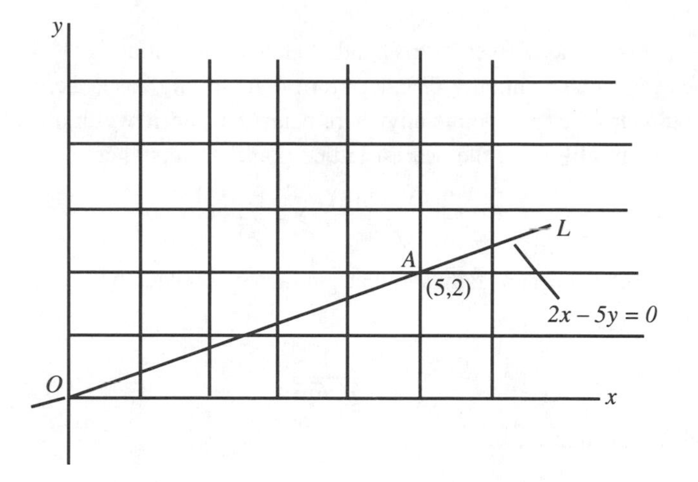

FIGURE 31

the line L with equation $2x - 5y = 0$; it has slope $\frac{2}{5}$ and goes through the origin O and the point $A(5, 2)$ (figure 31). The distribution of the lattice points in the vicinity of the segment OA is the same as that in the vicinity of any segment between consecutive lattice points on L (e.g., from $(5, 2)$ to $(10, 4)$), and since L itself is a general representative, the region around OA really tells the story of all lines through a lattice point which have slope $\frac{2}{5}$.

Going through the lattice point O, L would meet the circle centered at O no matter how small the radius might be. However, if L were translated slightly, keeping its direction constant at slope $\frac{2}{5}$, a gap would open up which would allow it to miss a small circle at O. But L couldn't be shifted very far like this without running into a circle about some nearby lattice point. In fact, the limit would be one-half the distance to the lattice point which is nearest OA. The problem, then, is simply to determine the lattice point nearest OA.

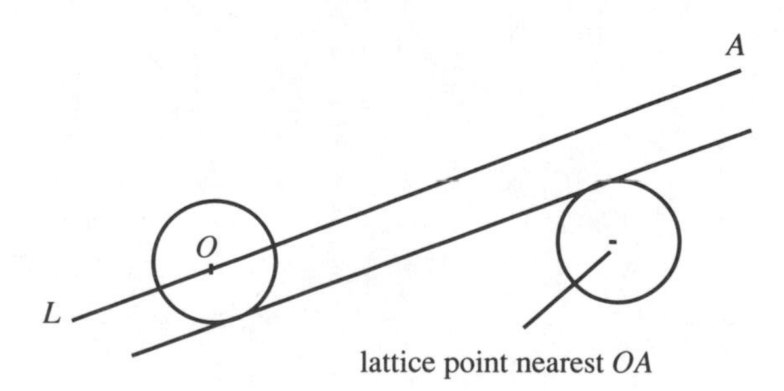

FIGURE 32

But this is easy. Even a freehand sketch reveals that $(2,1)$ and $(3,1)$ are each the same minimum distance from OA. In any case, there is only a small number of lattice points anywhere near OA and it wouldn't take long to check them all. Since the nearest lattice point is a distance

$$\frac{2(3)-5(1)}{\sqrt{29}} = \frac{1}{\sqrt{29}}$$

from OA, the minimum radius r that guarantees contact with every line of slope $\frac{2}{5}$ is

$$r = \frac{1}{2\sqrt{29}},$$

a number less than .1.

Associations

Problems 1, 2, and 3: Tokyo University; Examination A, for students in science.

Problem 4: Shiga University; Examination A, for students in the Education Division (Elementary and Secondary), and in Information Sciences.

Problem 5: Hokkaido University; Examination A, for students in Science I, Pre-medical and Pre-dental applicants.

Two Problems from the 1990 Canadian Olympiad†

1. A set of $\frac{1}{2}n(n+1)$ different numbers is arranged *at random* in a triangular tower of n rows, the kth row from the top containing k numbers, as shown. If M_k is the largest number in row k, what is the probability that

$$M_1 < M_2 < M_3 < \cdots < M_n?$$

$$\begin{array}{ccccccccc} & & & & x & & & & \\ & & & x & & x & & & \\ & & x & & x & & x & & \\ & & & & \cdots & & & & \\ x & & x & & \cdots & & x & & x \end{array}$$

This wonderful mind-boggling problem provides another opportunity to show off the tremendous capabilities of the method of recursion, for the problem is almost trivial when approached in this way.

Accordingly, let p_n denote the required probability for a tower of n rows, and let N be the biggest of the given numbers. Clearly N will be the greatest number in its row, wherever it occurs, and therefore the required $M_1 < M_2 < \cdots < M_n$ is out of the question unless N occurs in the last row, making $M_n = N$. Since the last row receives n of the $\frac{1}{2}n(n+1)$ numbers, the probability that N lies in the last row is

$$\frac{n}{\frac{1}{2}n(n+1)} = \frac{2}{n+1}.$$

---

† (*Crux Mathematicorum,* 1990, 161)

With N in the last row, it doesn't matter how the rest of the last row is filled in. In any case, the numbers remaining after completing the last row are all different, and therefore the probability of putting them in the first $n-1$ rows so that $M_1 < M_2 < \cdots < M_{n-1}$ is p_{n-1} by definition. Hence

$$\begin{aligned} p_n = &\text{(prob. } N \text{ lies in the last row)}\times \\ &\text{(prob. of suitably completing last row)} \times \\ &\text{(prob. of suitably filling the first } n-1 \text{ rows),} \end{aligned}$$

giving

$$p_n = \frac{2}{n+1} \cdot 1 \cdot p_{n-1},$$

that is,

$$(n+1)p_n = 2p_{n-1}.$$

It remains only to iterate this relation and observe that $p_1 = 1$.

$$\begin{aligned} (n+1)p_n &= 2p_{n-1} \\ np_{n-1} &= 2p_{n-2} \\ (n-1)p_{n-2} &= 2p_{n-3} \\ &\cdots \\ 3p_2 &= 2p_1 \\ 2p_1 &= 2. \end{aligned}$$

Multiplying gives

$$(n+1)!p_n p_{n-1} \cdots p_1 = 2^n p_{n-1} p_{n-2} \cdots p_1,$$

from which

$$p_n = \frac{2^n}{(n+1)!)}.$$

(The same solution is given in 1990, 198.)

Now for a lovely geometry problem.

2. The diagonals of cyclic quadrilateral $ABCD$ meet at X. Perpendiculars from X to the sides of $ABCD$ determine a second quadrilateral $PQRS$ (figure 33).

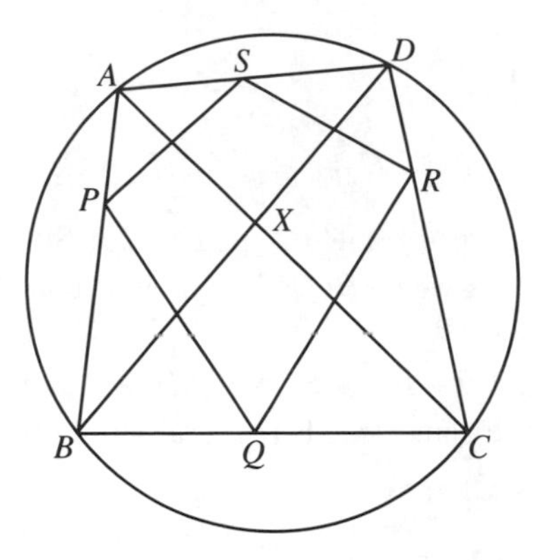

FIGURE 33

Prove that the sum of one pair of opposite sides of $PQRS$ is the same as the sum of the other pair:

$$PQ + RS = PS + QR.$$

The right angles at P and S make $APXS$ cyclic, and therefore

$$\angle XPS = \angle XAS \quad (= \alpha \text{ in figure 34) on chord } XS;$$

similarly, $PBQX$ is cyclic and

$$\angle XPQ = \angle XBQ \quad (= \beta) \text{ on chord } XQ.$$

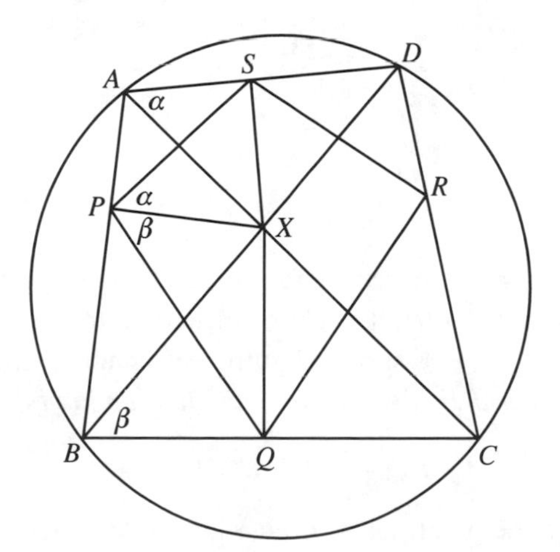

FIGURE 34

But, in the given circle,

$$\alpha = \angle CAD = \angle CBD = \beta,$$

and so XP is the bisector of angle P in $PQRS$. Similarly, XQ, XR, and XS bisect the other angles in $PQRS$, implying that X is the *incenter* of $PQRS$.

If the incircle touches the sides at K, L, M, and N, then the tangents PK and PN have the same length (a), and similarly at the other vertices (figure 35). Therefore

$$PQ + RS = (a + b) + (c + d) = (b + c) + (a + d) = QR + PS.$$

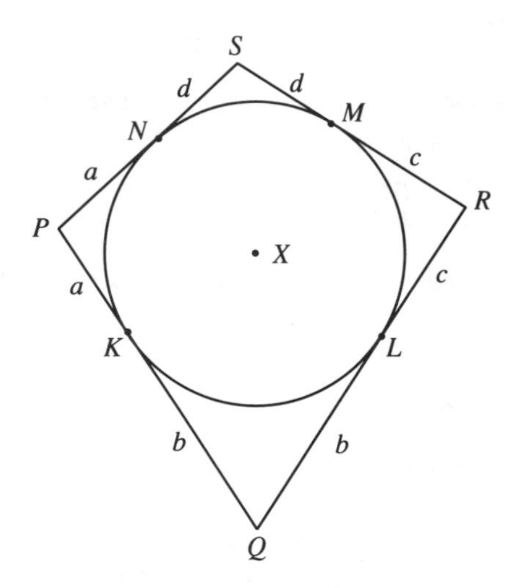

FIGURE 35

Comment. Having just observed for an inscriptible quadrilateral $PQRS$ that $PQ + RS = PS + QR$, it might be of interest to note that the converse is also true, giving a necessary and sufficient condition for inscriptibility: *a convex quadrilateral $PQRS$ has an incircle if and only if*

$$PQ + RS = PS + QR.$$

To complete the proof, let K denote the circle which touches the three sides PQ, QR, and RS (figure 36): its center O is the point of intersection of the bisectors of angles Q and R and its radius is given by the perpendicular

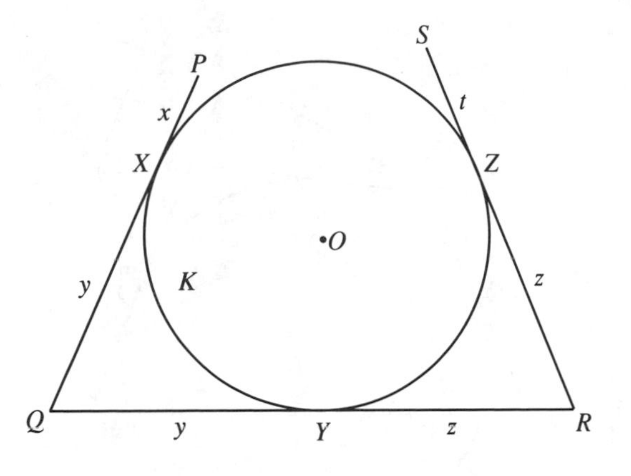

FIGURE 36

OX to PQ. Suppose K touches PQ, QR, and RS at X, Y, Z and that these points divide their sides into parts of lengths x, y, z, and t as shown in figure 36 (the tangents from Q are equal, and similarly from R). We would like to show that PS is also a tangent to K.

We observe that the given property

$$PQ + RS = PS + QR$$

gives

$$x + y + z + t = PS + y + z,$$

implying

$$PS = x + t.$$

Now suppose the tangent from P touches K at M and meets the line of RS at N. Then

$$PM = PX = x,$$

and, letting the length of the equal tangents MN and NZ be w, we have that

(i) $NS = t - w$ if N lies between S and Z (figure 37),
(ii) $NS = w - t$ if N lies on ZS extended (figure 38).

In either case, however, two sides of triangle PSN add up to the third side, implying the triangle is degenerate, in which case P, N, and S are collinear and N and S coincide, completing the proof.

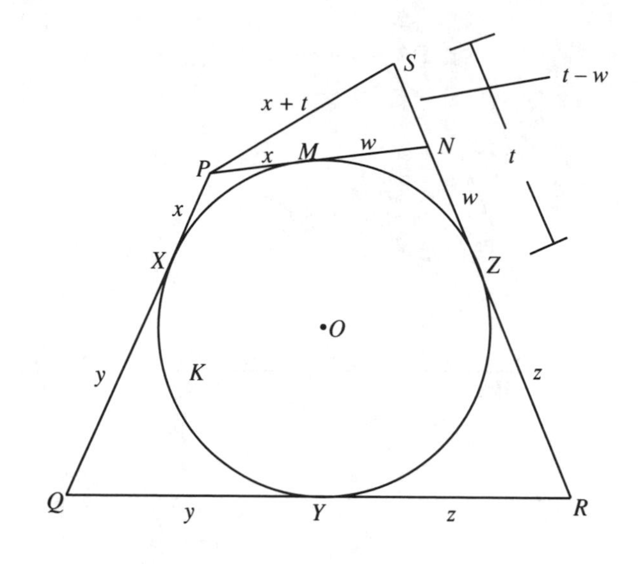

FIGURE 37

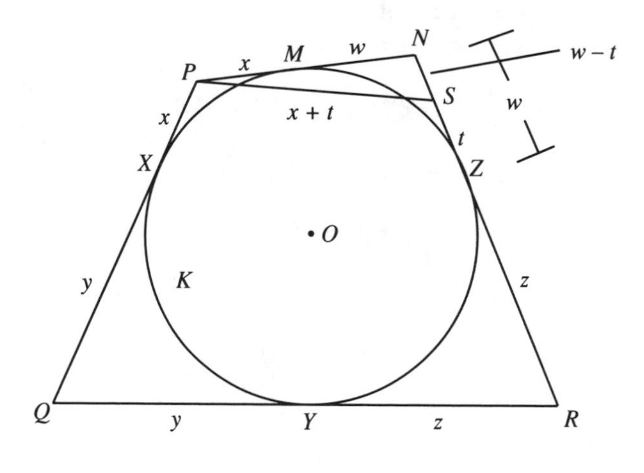

FIGURE 38

A Problem from the 1989 U.S.A. Olympiad†

If u and v are real numbers such that

$$u + u^2 + \cdots + u^8 + 10u^9 = v + v^2 + \cdots + v^{10} + 10v^{11} = 8,$$

which is bigger, u or v?

(How can they say there's no humor in mathematics?)

Our solution depends on the fact that v is positive, so let's prove this right away to avoid interrupting things later. It is simply a matter of observing that no $v \leq 0$ satisfies the defining condition on v, which may be written in the following two ways:

$$(v + v^2) + (v^3 + v^4) + \cdots + (v^9 + v^{10}) + 10v^{11} = 8, \qquad (A)$$

$$(v + v^{11}) + (v^2 + v^{11}) + \cdots + (v^{10} + v^{11}) = 8. \qquad (B)$$

Clearly v cannot be zero, and

(i) for $-1 < v < 0$, each bracket and the final term on the left side of (A) is negative because the odd powers of v are negative and each has greater magnitude than the higher power coupled with it, and therefore the left side of (A) cannot add up to 8;

(ii) for $v \leq -1$, each bracket in (B) is ≤ 0 since v^{11} is negative and is of greater magnitude than any of the other powers except when $v = -1$, when their magnitudes are equal, and so the left side of (B) never adds up to 8.

Now let us proceed indirectly by supposing $u \geq v$. In this case, simply replacing u by v gives

$$u + u^2 + \cdots + u^8 + 10u^9 \geq v + v^2 + \cdots + v^8 + 10v^9,$$

† (*Crux Mathematicorum,* 1989, 163)

that is,

$$8 \geq v + v^2 + \cdots + v^8 + 10v^9,$$

implying

$$8 - 9v^9 \geq v + v^2 + \cdots + v^8 + v^9.$$

Substituting this result in the condition on v we get

$$(v + v^2 + \cdots + v^8 + v^9) + v^{10} + 10v^{11} \leq (8 - 9v^9) + v^{10} + 10v^{11},$$

that is,

$$8 \leq 8 - 9v^9 + v^{10} + 10v^{11},$$

giving

$$10v^{11} + v^{10} - 9v^9 \geq 0,$$
$$v^9(10v^2 + v - 9) \geq 0,$$

and since v is positive,

$$10v^2 + v - 9 \geq 0.$$

Solving the equation

$$10v^2 + v - 9 = 0,$$

we have

$$(10v - 9)(v + 1) = 0,$$

and $v = .9$ and -1. Thus, in order for

$$10v^2 + v - 9 \geq 0,$$

it must be that the positive number $v \geq .9$.

However, in this case,

$$\begin{aligned} 8 &= v + v^2 + \cdots + v^{10} + 10v^{11} \\ &\geq .9 + (.9)^2 + \cdots + (.9)^{10} + 10(.9)^{11} \\ &= \frac{.9 - (.9)^{11}}{1 - (.9)} + 10(.9)^{11} \\ &= 10[.9 - (.9)^{11}] + 10(.9)^{11} \\ &= 9 - 10(.9)^{11} + 10(.9)^{11} \\ &= 9, \end{aligned}$$

a contradiction.

Hence u cannot be as big as v.

A Problem on Seating Rearrangements

The following exercise appears in my Mathematical Gems I (MAA, Dolciani Series, 1973, page 76):

> A classroom has 5 rows of 5 desks per row. The teacher requests each pupil to change his seat by going either to the seat in front of him the one behind him, the one to his left, or the one to his right (of course, not all these options are open to all the pupils). Determine whether this directive can be carried out.

It is easy to see that the request cannot be honored by imagining each desk to be on a square of a 5×5 checkerboard:

> the directions require each pupil to move to a desk on a square of the "other" color; however, 12 of the 25 desks are on squares of one color and 13 are on squares of the other color, making a complete interchange impossible.

Obviously the same situation exists for any rectangular classroom of odd dimensions. However, if either dimension is even, not only can the directive be carried out, but in almost all cases it can be done in many ways. Just how many ways depends on the size of the classroom and is generally a difficult problem. Therefore let us take the simplest possible case.

> In how many ways can the pupils in a $2 \times n$ classroom rearrange their seating in accordance with the teacher's directive?

This problem and the case of a $3 \times n$ classroom are solved in a paper [1] by Robert E. Kennedy and Curtis Cooper of the Central Missouri State University. The following discussion owes its inspiration to the ideas presented in their paper.

To begin, consider the 9 rearrangements in a 2×3 classroom (figure 39). These exhibit *endings* that may be classified to our advantage into the

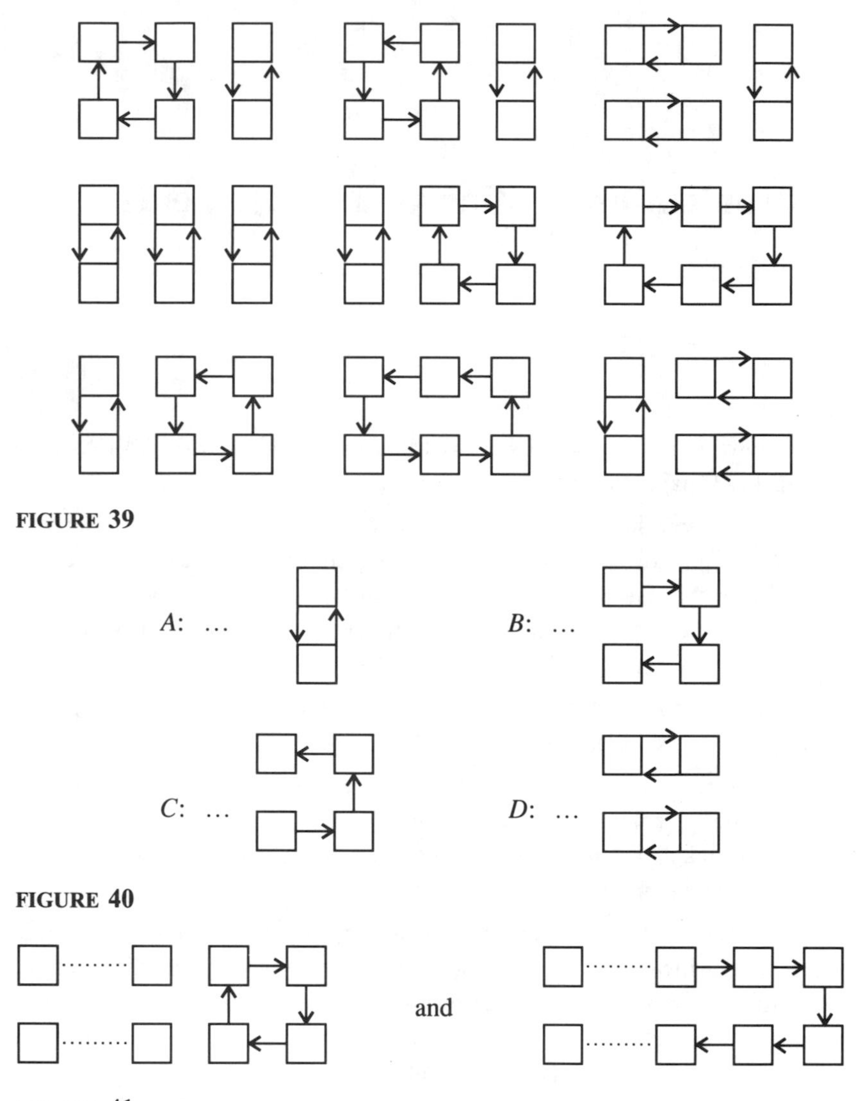

FIGURE 39

FIGURE 40

FIGURE 41

four types shown in figure 40. For long classrooms, type B may be further subdivided into the two kinds shown in figure 41; similarly for type C.

Thus, in general, the set T_n of all $2 \times n$ rearrangements may be partitioned into four subsets A_n, B_n, C_n, D_n according to the following six endings (figure 42). If we let the cardinalities of A_n, B_n, C_n, D_n, T_n, respectively, be a_n, b_n, c_n, d_n, t_n, then the number we wish to determine is

$$t_n = a_n + b_n + c_n + d_n.$$

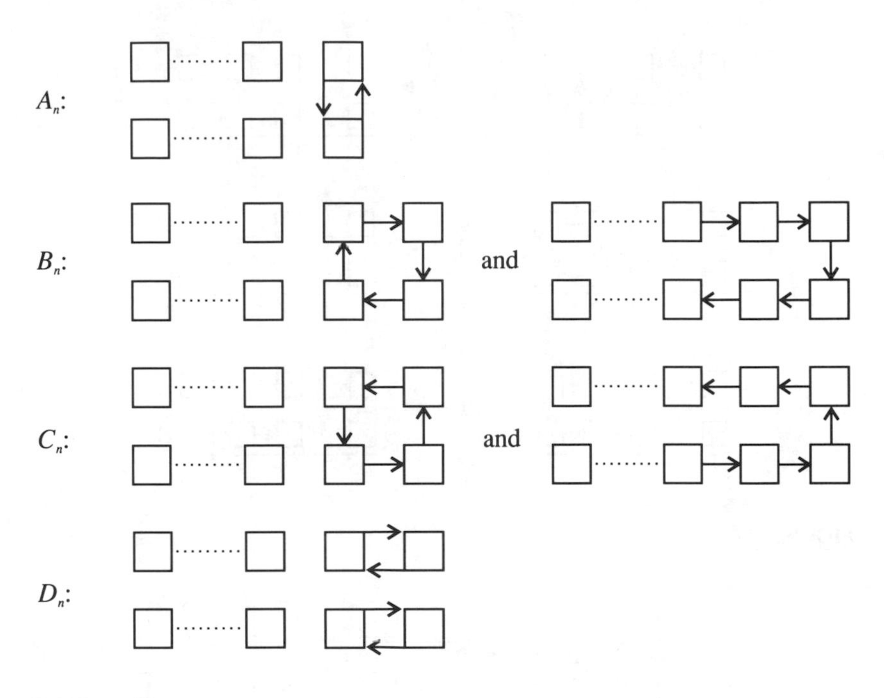

FIGURE 42

Now, as we shall see, these numbers are governed by the recursions

$$a_{n+1} = t_n = a_n + b_n + c_n + d_n \qquad \text{(i)}$$

$$b_{n+1} = a_n + b_n \qquad \text{(ii)}$$

$$c_{n+1} = a_n + c_n \qquad \text{(iii)}$$

$$d_{n+1} = a_n. \qquad \text{(iv)}$$

To see (i), we need only observe that any member of T_n becomes a member of A_{n+1} when an extra column (figure 43) is attached to the end, and conversely.

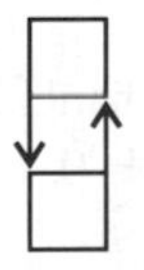

FIGURE 43

Recursion (ii) is established by the 1-1 correspondence between $A_n \cup B_n$ and B_{n+1} which is described in figure 44. A similar correspon-

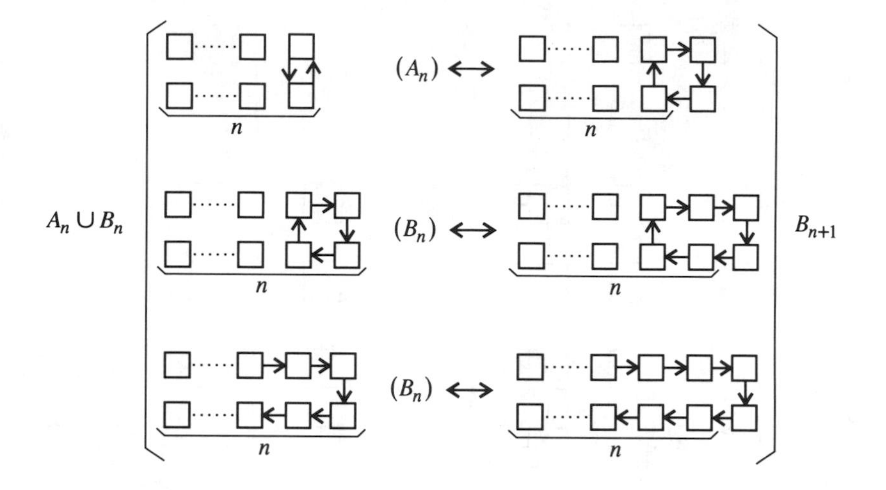

FIGURE 44

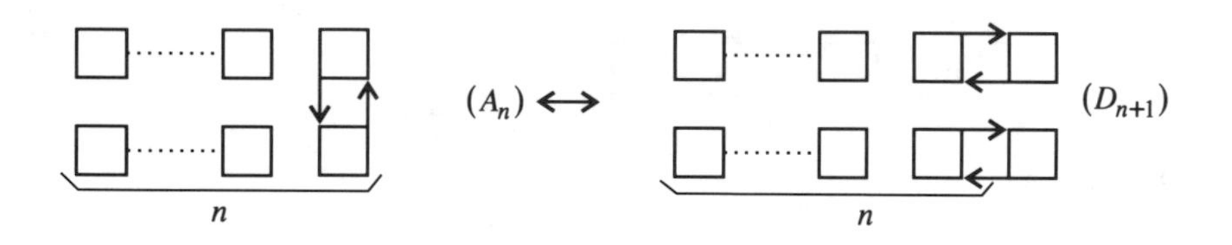

FIGURE 45

dence establishes (iii), and (iv) is given immediately by the correspondence in figure 45. The results are summarized nicely in the matrix equation

$$\begin{pmatrix} 1 & 1 & 1 & 1 \\ 1 & 1 & 0 & 0 \\ 1 & 0 & 1 & 0 \\ 1 & 0 & 0 & 0 \end{pmatrix} \begin{pmatrix} a_n \\ b_n \\ c_n \\ d_n \end{pmatrix} = \begin{pmatrix} a_{n+1} \\ b_{n+1} \\ c_{n+1} \\ d_{n+1} \end{pmatrix}.$$

From the single way of interchanging the two pupils in a 2×1 array, we have $a_1 = 1$ and $b_1 = c_1 = d_1 = 0$, and therefore

$$\begin{pmatrix} a_2 \\ b_2 \\ c_2 \\ d_2 \end{pmatrix} = \begin{pmatrix} 1 & 1 & 1 & 1 \\ 1 & 1 & 0 & 0 \\ 1 & 0 & 1 & 0 \\ 1 & 0 & 0 & 0 \end{pmatrix} \begin{pmatrix} 1 \\ 0 \\ 0 \\ 0 \end{pmatrix} = \begin{pmatrix} 1 \\ 1 \\ 1 \\ 1 \end{pmatrix},$$

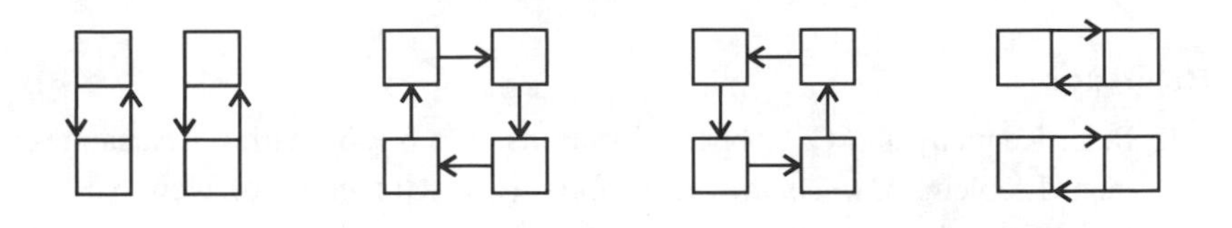

FIGURE 46

making $t_2 = 4$, which is easily verified by enumerating the cases (figure 46). Also,

$$\begin{pmatrix} a_3 \\ b_3 \\ c_3 \\ d_3 \end{pmatrix} = \begin{pmatrix} 1 & 1 & 1 & 1 \\ 1 & 1 & 0 & 0 \\ 1 & 0 & 1 & 0 \\ 1 & 0 & 0 & 0 \end{pmatrix} \begin{pmatrix} 1 \\ 1 \\ 1 \\ 1 \end{pmatrix} = \begin{pmatrix} 4 \\ 2 \\ 2 \\ 1 \end{pmatrix},$$

giving $t_3 = 9$, as seen above. So far, then, we have $t_1 = 1$, $t_2 = 4$, $t_3 = 9$, that is, 1^2, 2^2, and 3^2. It is probably too much to hope that t_n is always equal to n^2, and indeed the very next term dispels this conjecture. However, further easy calulations give $t_4 = 25$, $t_5 = 64$, and $t_6 = 169$, extending our results to $1^2, 2^2, 3^2, 5^2, 8^2$, and 13^2, from which one is not likely to miss the delightful possibility that

$$t_n = f_{n+1}^2,$$

the square of the $(n+1)$th Fibonacci number. This is indeed the case and, in fact, an almost trivial induction establishes that

$$\begin{pmatrix} a_n \\ b_n \\ c_n \\ d_n \end{pmatrix} = \begin{pmatrix} f_n^2 \\ f_n f_{n-1} \\ f_n f_{n-1} \\ f_{n-1}^2 \end{pmatrix}:$$

noting $f_0 = 0$, this is clearly true for $n = 1$, and if it holds for any $n \geq 1$, then, from the recursions, we have

$$a_{n+1} = f_n^2 + 2f_n f_{n-1} + f_{n-1}^2 = (f_n + f_{n-1})^2 = f_{n+1}^2;$$

$$b_{n+1} = a_n + b_n = f_n^2 + f_n f_{n-1} = f_n(f_n + f_{n-1}) = f_n f_{n+1};$$

similarly for c_{n+1}; and finally

$$d_{n+1} = a_n = f_n^2.$$

Accordingly,

$$t_n = a_{n+1} = f_{n+1}^2$$

for all n.

References

1. R. E. Kennedy and C. Cooper: Variations on a 5×5 Seating Rearrangement Problem, *Mathematics in College* (City University of New York), Fall-Winter 1993, 59–67.
2. Toshi Otake, Robert E. Kennedy, and Curtis Cooper: On a Seating Rearrangement Problem, *Mathematics and Informatics Quarterly,* vol. 6, 1996, 63–71.

Three Problems from the 1980 and 1981 Chinese New Year's Contests†

1. (1980) If P is a convex polygon which, due to its size or shape, is unable to cover any triangle of area $\frac{1}{4}$, prove that P itself can be covered by some triangle of area 1.

For example, a square cannot cover any triangle of area greater than one-half its own area but there is a triangle with twice its area which is capable of covering it (figure 47). In particular, a square of side .6 cannot cover any

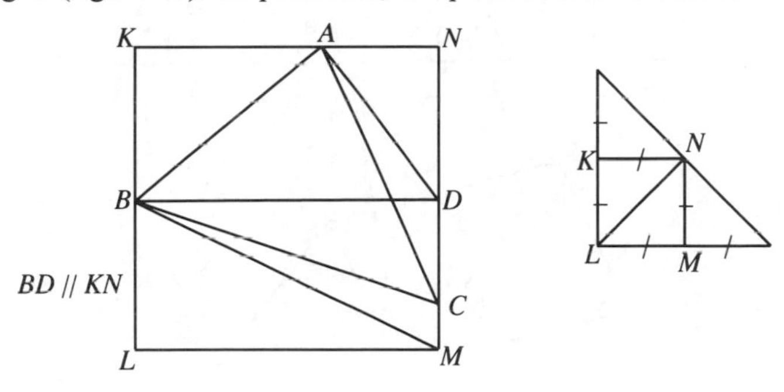

$$\begin{aligned}\triangle ABC &< BADM \\ &= \triangle BAD + \triangle BDM \\ &= \tfrac{1}{2}(KBDN) + \tfrac{1}{2}(BLMD) \\ &= \tfrac{1}{2}(KBDN + BLMD) \\ &= \tfrac{1}{2}KLMN.\end{aligned}$$

FIGURE 47

† (*Crux Mathematicorum,* 1991, 1)

triangle of area exceeding .18, let alone one as big as $\frac{1}{4}$ (although its own area is much larger, namely .36), while it can itself be covered by a triangle of area only .72 (which is far less than 1).

To say that P is unable to cover any triangle of area $\frac{1}{4}$ is just another way of saying no three points of P determine a triangle of area as big as $\frac{1}{4}$. Thus a triangle XYZ of *maximum* area inscribed in P has area $< \frac{1}{4}$.

From such a maximum inscribed triangle XYZ, let a triangle ABC be constructed by drawing lines through X, Y, Z which are respectively parallel to the opposite sides of $\triangle XYZ$ (figure 48). The parallelograms thus obtained clearly reveal that X, Y, and Z are the midpoints of the sides of $\triangle ABC$; and since a diagonal bisects the area of a parallelogram, then

$$\triangle ABC = 4(\triangle XYZ) < 4\left(\frac{1}{4}\right) = 1.$$

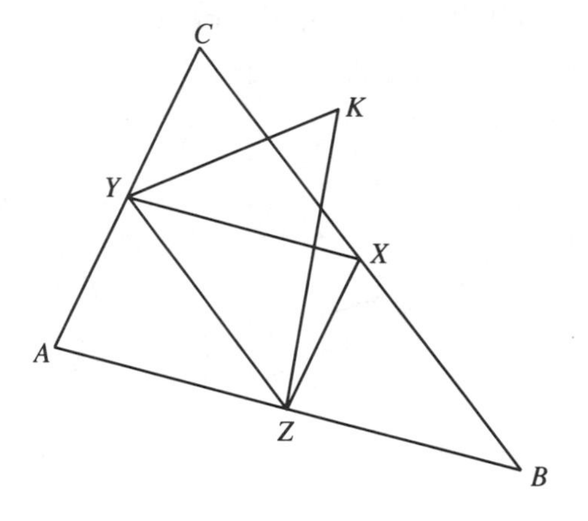

FIGURE 48

Now, it is easy to see that $\triangle ABC$ covers the entire polygon P. Suppose to the contrary that a point K of P were outside $\triangle ABC$, say on the side of BC opposite YZ (figure 48). Since P is convex, it would contain the entire triangle KYZ. But $\triangle KYZ$ would have the same base YZ as $\triangle XYZ$ and a greater altitude to YZ, giving the contradiction of a triangle inscribed in P which has a greater area than the maximum such triangle XYZ. The conclusion follows. (Essentially the same solution is given in 1992, 272.)

2. (1980) If a_n denotes the integer *nearest* $\sqrt{n}$, what is the value of

$$S = \frac{1}{a_1} + \frac{1}{a_2} + \cdots + \frac{1}{a_{1980}}?$$

Although potentially ominous, this particular nearest integer function is very well behaved and gives up its secrets almost without a fight. From a table of the first few values it is clear that the function takes the value 1 twice, 2 four times, 3 six times, and 4 eight times, strongly suggesting that it changes its value from k to $k+1$ as n goes from $k(k+1)$ to $k(k+1)+1$.

n:	1	2	3	4	5	6	7	8	9	10	11	12	13	14	15	16	17	18	19	20	21
f:	1	1	2	2	2	2	3	3	3	3	3	3	4	4	4	4	4	4	4	4	5

That is to say, it appears as if

$$\sqrt{k(k+1)} < k + \frac{1}{2}, \tag{i}$$

making the nearest integer k, while

$$\sqrt{k(k+1)+1} > k + \frac{1}{2}, \tag{ii}$$

making the nearest integer $k+1$. Squaring shows immediately that both these conjectures are correct:

$$k(k+1) = k^2 + k < k^2 + k + \frac{1}{4}, \tag{i}$$

and

$$k(k+1) + 1 = k^2 + k + 1 > k^2 + k + \frac{1}{4}. \tag{ii}$$

Thus this nearest integer function starts its run of values equal to k at the number $(k-1)k+1$ and continues them through to the integer $k(k+1)$, that is, through a total of

$$k(k+1) - (k-1)k = k^2 + k - k^2 + k = 2k$$

integers. Sincc $1980 = 44 \cdot 45$, S contains the reciprocals of the integers in all the blocks of repeated values up to the 88 values equal to 44 ($n = 1981$ starts the block of 45's), giving

$$\begin{aligned} S &= 2\left(\tfrac{1}{1}\right) + 4\left(\tfrac{1}{2}\right) + 6\left(\tfrac{1}{3}\right) + \cdots + 88\left(\tfrac{1}{44}\right) \\ &= \quad 2 \quad + \quad 2 \quad + \quad 2 \quad + \cdots + \quad 2 \quad = 88. \end{aligned}$$

(An alternative solution is given in 1992, 101.)

3. (1981) If

$$f(x) = x^{99} + x^{98} + \cdots + x + 1,$$

what is the remainder when $f(x^{100})$ is divided by $f(x)$?

This problem is the following general problem with $n = 99$: if

$$f(x) = x^n + x^n - 1 + \cdots + x + 1,$$

what is the remainder when $f(x^{n+1})$ is divided by $f(x)$? An examination of a few early cases reveals a tell-tale pattern:

$$\begin{aligned} x^2 + 1 &= (x+1)(x-1) + 2, \\ x^6 + x^3 + 1 &= (x^2 + x + 1)(x^4 - x^3 + 2x - 2) + 3, \\ x^{12} + x^8 + x^4 + 1 &= (x^3 + x^2 + x + 1)\times \\ &\qquad (x^9 - x^8 + 2x^5 - 2x^4 + 3x - 3) + 4, \end{aligned}$$

leading to the conjecture

$$\begin{aligned} &x^{9900} + x^{9800} + \cdots + x^{100} + 1 \\ &= (x^{99} + x^{98} + \cdots + 1)\times \\ &\quad (x^{9801} - x^{9800} + 2x^{9701} - 2x^{9700} + 3x^{9601} - 3x^{9600} + - \cdots + 99x - 99) \\ &\quad + 100, \end{aligned}$$

with remainder 100. It remains only to confirm this result by multiplying out the right side.

The right side is given by

$$\begin{aligned} &(x^{99} + x^{98} + \cdots + 1)\times \\ &\quad \left[x^{9800}(x-1) + 2x^{9700}(x-1) + 3x^{9600}(x-1) + \cdots + 99(x-1)\right] + 100 \\ &= (x^{99} + x^{98} + \cdots + 1)\times \\ &\quad \left[(x-1)\cdot(x^{9800} + 2x^{9700} + 3x^{9600} + \cdots + 98x^{100} + 99)\right] + 100 \\ &= (x^{100} - 1)\sum_{k=0}^{98}(99-k)x^{100k} + 100 \\ &= \sum_{k=0}^{98}(99-k)(x^{100(k+1)} - x^{100k}) + 100. \end{aligned}$$

Now, for $k = 98, 97, \ldots, 1$, the power x^{100k} appears only in the two terms

$$(99-k)(x^{100(k+1)} - x^{100k}) \quad \text{and} \quad \left[99-(k-1)\right](x^{100k} - x^{100(k-1)}),$$

and so its coefficient is

$$-(99-k)+\big[99-(k-1)\big]=1.$$

In addition, the terms x^{9900} and x^0 in this large sum are given by the single terms

$$(99-98)(x^{(98+1)})=x^{9900} \quad \text{and} \quad (99-0)(-x^0)=-99.$$

Thus the right side gives the left side precisely when a remainder of 100 is present to convert the 99 into the required absolute term of 1, and our solution is complete.

Comment. By the way, what's wrong with the following argument?

When $f(x^{100})$ is divided by $f(x)$, let the quotient be $Q(x)$ and the remainder $R(x)$. Then

$$f(x^{100})=f(x)\cdot Q(x)+R(x)$$

is an identity in which the degree of $R(x)$ does not exceed 98 (i.e., one less than the degree of the divisor $f(x)$). But there are 100 values of x which make $x^{100}=x$, namely 0 and the ninety-nine 99th-roots of unity, and for each of these values

$$f(x^{100})=f(x),$$

in which case $Q(x)=1$ and $R(x)=0$. Thus $R(x)$ has more zeros than its degree, making it identically zero. (An alternative solution is given in 1992, 102; additional comment appears in 1994, 46.)

A Problem in Arithmetic

Four consecutive even numbers are removed from the set

$$A = \{1, 2, 3, \cdots, n\}.$$

If the average of the remaining numbers is 51.5625, which four numbers were removed?

This problem and the following solution are due to my colleagues Lloyd Auckland and Ron Dunkley.

First let us note that

$$51.5625 = \frac{515625}{10000} = \frac{825}{16}.$$

Therefore, denoting by S the sum of the remaining $n - 4$ numbers, we have

$$\frac{S}{n-4} = \frac{825}{16}, \quad \text{and} \quad 16S = 825(n-4).$$

Since 16 and 825 are relatively prime, it follows that $16|n - 4$, and therefore

$$n = 16t + 4$$

for some positive integer t (t cannot be zero since n must be at least 8 in order to provide A with four even numbers).

Now, the sum of all n numbers in A is $\frac{1}{2}n(n + 1)$, implying an average value of $\frac{1}{2}(n + 1)$. With four numbers removed, the average would likely be altered, but perhaps not by very much. Thus we might expect $\frac{1}{2}(n + 1)$ to be in the vicinity of 51.5625, and hence $n + 1$ not very far from 103.125. Since the values of $16t + 4$ nearest 103.125 are 84, 100, and 116, it is very probable that $n = 100$. However, this is not certain and we had better check nearby values.

$n = 84$: Since there are 80 numbers left no matter which four are removed, the *greatest* average value of the remaining numbers is

$$\frac{(1+2+\cdots+84)-(2+4+6+8)}{80} = \frac{1}{80}\left(\frac{1}{2}\cdot 84\cdot 85 - 20\right) = 44.375;$$

Thus $n = 84$ can't muster an average up to the required 51.5625.

$n = 116$: Similarly, the *least* possible average of the remaining 112 numbers is

$$\frac{(1+2+\cdots+116)-(110+112+114+116)}{112} = \frac{1}{112}\left(\frac{1}{2}\cdot 116\cdot 117 - 452\right) = 56.55\ldots,$$

which overshoots the required 51.5625.

Thus $n = 100$, and the deleted numbers can be calculated easily as follows. Denoting them by $a-3$, $a-1$, $a+1$, and $a+3$, we have

$$\frac{(1+2+\cdots+100)-4a}{96} = \frac{825}{16},$$
$$5050 - 4a = 6\cdot 825 = 4950.$$

Thus $4a = 100$, $a = 25$, and the numbers are 22, 24, 26, and 28.

While this discussion is most enjoyable and convincing, it does not actually prove that $n = 100$. However, its ideas can be expanded into a rigorous treatment as follows.

(a) The least average of the numbers remaining in A is

$$\frac{(1+2+\cdots+n)-(n+n-2+n-4+n-6)}{n-4} = \frac{1}{n-4}\left[\frac{1}{2}n(n+1)-(4n-12)\right] = \frac{n^2-7n+24}{2(n-4)}.$$

In order for this to be as little as 51.5625, we must have

$$n^2 - 7n + 24 \le 51.5625[2(n-4)] = 103.125(n-4),$$

requiring

$$n^2 - 110.125n + 463.5 \le 0.$$

Now, this is true only for n between the roots of the corresponding equation, which are

$$\frac{110.125 \pm \sqrt{110.125^2 - 1746}}{2}, \quad \text{i. e., } 8.235\ldots \text{ and } 106.007\ldots.$$

Thus for $n \geq 116$ this average never gets as low as 51.5625.

(b) Similarly, the greatest average is

$$\frac{\frac{1}{2}n(n+1) - (2+4+6+8)}{n-4} = \frac{n^2+n-40}{2(n-4)}.$$

To be as great as 51.5625, this requires

$$n^2 + n - 40 \geq 103.125(n-4), \text{ i. e.,}$$

$$n^2 - 102.125n + 372.5 \geq 0,$$

which fails for n between the roots $7.57\ldots$ and $98.3\ldots$. For $n \leq 84$, then, the average is never as big as 51.5625.

(c) Since 100 is the only value of $16t + 4$ that is both ≤ 106 and > 98, $n = 100$ is the only value which can yield the required average of 51.5625; and since the problem promises the existence of at least one value of n, n must be 100.

A Checkerboard Problem

A *straight tromino* is a 1×3 rectangle, that is, three squares of a checkerboard in a row, and a single square is called a *monomino*. The problem is to devise a covering of an ordinary 8×8 checkerboard with 21 straight trominos and a monomino (figure 49).

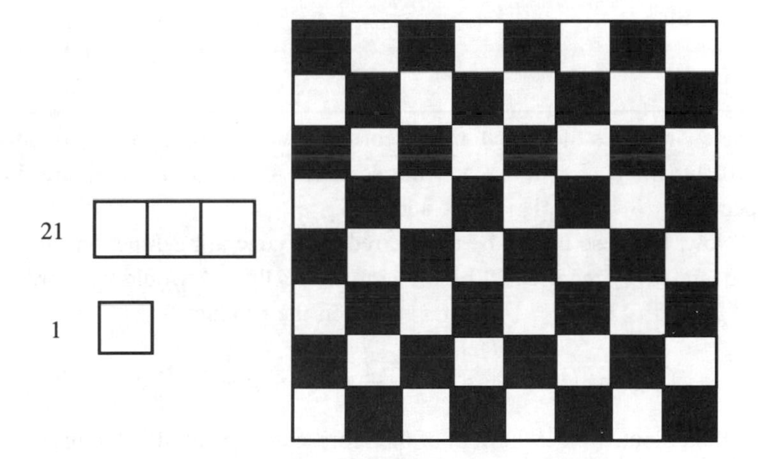

FIGURE 49

This is an old problem which is discussed in my *Mathematical Gems II,* (MAA, Dolciani Series, 1976, pages 62–63). However, in his book *Lure of the Integers* (MAA Spectrum Series, 1992), Joe Roberts gives the following beautiful solution by generating functions that was published by N. Mackinnon in the *Mathematical Gazette* (1989, pages 210–211).

Suppose the checkerboard sits in the first quadrant of a coordinate plane, up against the axes at the origin, and that each square of the board is known

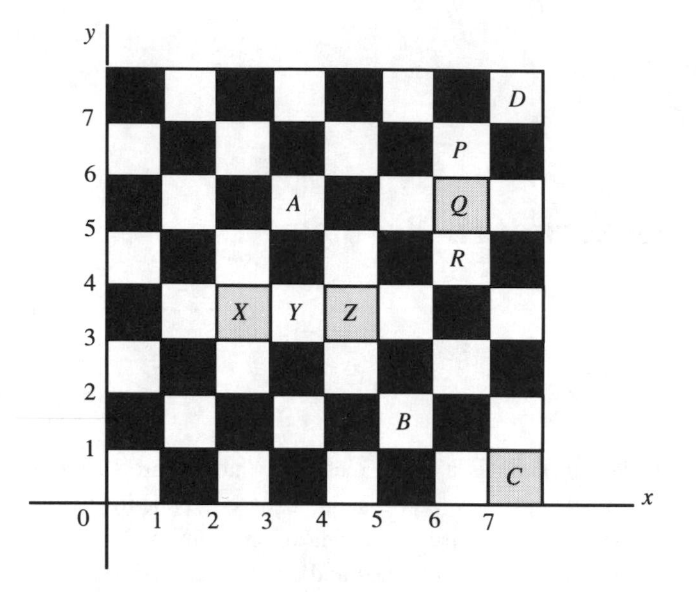

FIGURE 50

by the coordinates of its bottom left corner. Thus the first column of squares is $\{(0,0),(0,1),\ldots,(0,7)\}$ and the squares A, B, C, D in figure 50 are respectively $(3,5)$, $(5,1)$, $(7,0)$, and $(7,7)$.

Now, let these names be transferred to a generating function by having the square (a,b) represented by the term $x^a y^b$; thus A would be represented by x^3y^5 and C by x^7y^0 (or just x^7). When the product

$$f(x,y) = (1 + x + x^2 + \cdots + x^7)(1 + y + y^2 + \cdots + y^7)$$

is multplied out, we get the sum of the representatives of all 64 squares of the board; that is to say, $f(x,y)$ is the generating function of the entire board.

Clearly, a straight tromino must run across the board, like XYZ in figure 50, or up and down the board like PQR. The representations of the three squares in XYZ are x^2y^3, x^3y^3, x^4y^3, and so, in $f(x,y)$, the tromino XYZ accounts for the sum

$$x^2y^3 + x^3y^3 + x^4y^3 = x^2y^3(1 + x + x^2);$$

similarly the tromino PQR accounts for

$$x^6y^4 + x^6y^5 + x^6y^6 = x^6y^4(1 + y + y^2).$$

In the same way, no matter where a straight tromnino might be placed on the board, it will occupy three squares whose representation in $f(x, y)$ will be, for some (i, j), either

$$\text{(a)}\quad x^i y^j(1+x+x^2) \qquad \text{or} \qquad \text{(b)}\quad x^i y^j(1+y+y^2),$$

depending on whether (i, j) is its leftmost square or its lowest square. Thus 21 straight trominoes will account for 21 expressions like (a) and (b). Removing the common factors from these terms gives an expression that simplifies to

$$\begin{aligned}(1+x+x^2)(x^i y^i + x^k y^t + \cdots) + (1+y+y^2)(x^m y^n + \cdots)\\ = (1+x+x^2)\cdot g(x,y) + (1+y+y^2)\cdot h(x,y),\end{aligned}$$

for some functions g and h.

Now, this represents $21 \cdot 3 = 63$ of the squares of the board, all but some single square (a, b). Therefore this expression must be

$$f(x,y) - x^a y^b = (1+x+x^2+\cdots+x^7)(1+y+y^2+\cdots+y^7) - x^a y^b,$$

and we have

$$\begin{aligned}(1+x+x^2)\cdot g(x,y) + (1+y+y^2)\cdot h(x,y)\\ = (1+x+x^2+\cdots+x^7)(1+y+y^2+\cdots+y^7) - x^a y^b.\end{aligned} \tag{A}$$

If these products are expanded, the 63 terms $x^i y^j$ that occur on the left side of equation (A) would appear again on the right side; in other words, this is an *identity* and yields valid equations for all values we might care to substitute for x and y.

The presence of $(1+x+x^2)$ and $(1+y+y^2)$ might lead one to recall that $1+\omega+\omega^2 = 0$, where ω is either of the nonreal cube roots of unity. What better choice for x and y, then! Since

$$\omega^n + \omega^{n+1} + \omega^{n+2} = \omega^n(1+\omega+\omega^2) = 0,$$

substituting $x = y = w$ in (A) gives

$$\begin{aligned}0 &= (1+\omega)(1+\omega) - \omega^{a+b},\\ \omega^{a+b} &= 1+\omega+\omega^2+\omega-\omega,\end{aligned}$$

and since $\omega^3 = 1$, it follows that

$$a+b \equiv 1 \pmod 3.$$

Happily, when $x = \omega^2$, we again have

$$1+x+x^2 = 1+\omega^2+\omega^4 = 1+\omega^2+\omega = 0,$$

and so the substitutions $x = \omega^2$, $y = \omega$ similarly yield

$$0 = (1 + \omega^2)(1 + \omega) - \omega^{2a+b},$$

$$\omega^{2a+b} = 1 + \omega + \omega^2 + \omega^3 = 1,$$

implying

$$2a + b \equiv 0 \pmod 3.$$

These congruences immediately give $a \equiv -1 \equiv 2 \pmod 3$, from which it follows that $b \equiv 2 \pmod 3$. Thus a and b must be 2 or 5 and the single square (a, b) that is left uncovered by the 21 trominos must be either $(2, 2)$, $(2, 5)$, $(5, 2)$, or $(5, 5)$, all of which are seen to be equivalent by spinning the board through a succession of quarter-turns.

Once one knows where to put the monomino, a solution follows easily (figure 51).

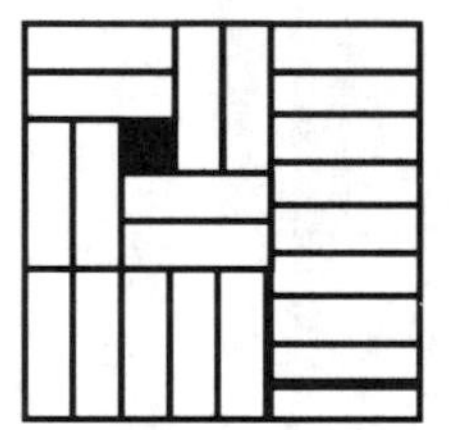

FIGURE 51

Two Problems from the 1990 Asian Pacific Olympiad†

1. Consider all the triangles ABC which have a fixed base AB and whose altitude from C is a constant h. For which of these triangles is the *product* of the altitudes a maximum?

The variable vertex C can be any point on the two lines parallel to AB which are at a distance h on either side; since these lines yield equivalent cases, we need consider only one of them, say the line L in figure 52.

Let the altitudes from A and B have lengths x and y (figure 52). Since h is constant, the product of the altitudes xyh has its maximum when xy is a maximum.

Since the base AB and the altitude to AB are constants, the area of triangle ABC is some constant k. Therefore

$$\frac{1}{2}ax = \frac{1}{2}by = \frac{1}{2}ab\sin C = k,$$

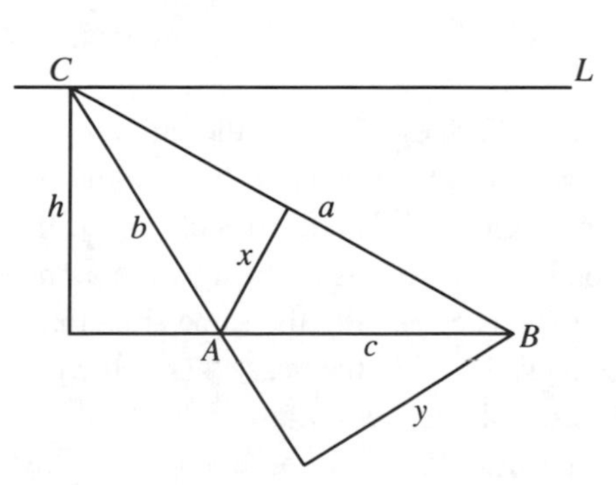

FIGURE 52

† (*Crux Mathematicorum*, 1990, 257)

giving

$$\left(\frac{1}{2}ax\right)\left(\frac{1}{2}by\right) = \frac{1}{4}abxy = k^2 = \frac{1}{4}a^2b^2\sin^2 C,$$

and

$$xy = ab\sin^2 C = 2k\sin C,$$

implying xy is greatest when $\sin C$ is greatest.

Now, the range of values of angle C depends on the relative sizes of AB and h. If h is small enough to permit L to contact the circle on diameter AB (figure 53), $\angle C$ becomes a right angle at each point of intersection, giving $\sin C$ its greatest possible value of 1 and making xy a maximum at each such point.

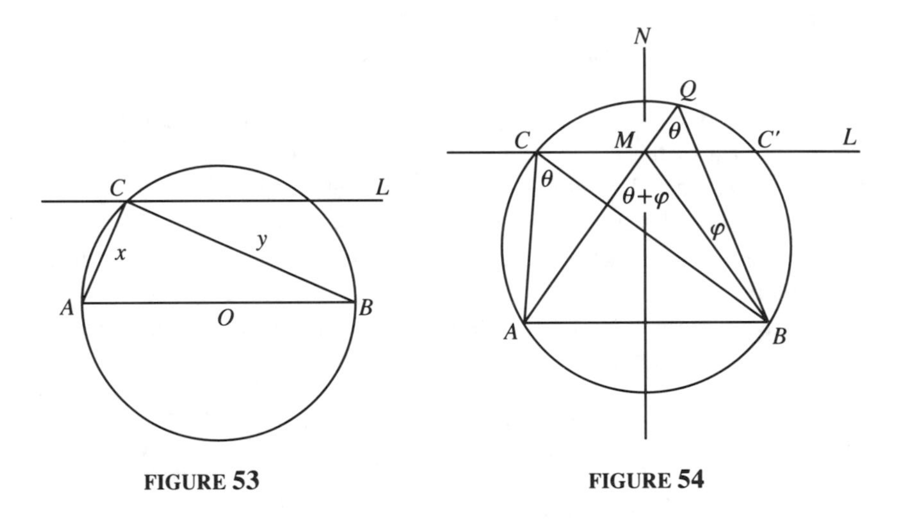

FIGURE 53 FIGURE 54

On the other hand, if L does not meet the circle on diameter AB, the angle C is never as big as a right angle and the maximum value of $\sin C$ is given by the maximum angle C. It is not difficult to see that this maximum occurs for C at the point M where L is crossed by the perpendicular bisector N of AB (figure 54). (The is essentially the same situation as in the "statue" problem of Regiomontanus, posed in the year 1471, which is discussed in my *Ingenuity in Mathematics,* MAA, New Mathematical Library Series, 1970.)

Suppose L does not meet the circle on diameter AB and that C is any point of L except M. Since N is a diameter of the circumcircle of $\triangle ABC$, reflection in N takes L and the circle into themselves, implying that the circle intercepts L in a chord CC' which has M as its midpoint (figure 54). Now it is easy to see that $\angle AMB$ is greater than $\angle ACB$. Let AM meet the circle at Q. Then, for $\triangle MQB$,

$$\begin{aligned} \text{exterior angle } AMB &> \text{interior angle at } Q \\ &= \angle ACB \quad \text{(on chord AB).} \end{aligned}$$

Thus $\angle C$ is a maximum when C is at M, providing the desired maximum of xy.

> **2.** For each integer $n \geq 6$, prove that there exists a *convex hexagon* which can be decomposed into exactly n congruent triangles; equivalently, for each integer $n \geq 6$, prove there exists a triangle t_n such that n copies of t_n can be assembled into a convex hexagon.

Since an equilateral triangle can be decomposed into two congruent triangles by an altitude, into three congruent triangles by segments from the center to the vertices, and into four congruent triangles by joining the midpoints of the sides, an equilateral triangle seems a likely candidate for the basic building block of the desired hexagon.

However, after several fruitless attempts to make use of these facts, I discovered that the 1-2-$\sqrt{5}$ right-angled triangle can be taken as t_n for all values of n, as shown in figures 56 and 57. (An alternative solution is given in 1992, 41.)

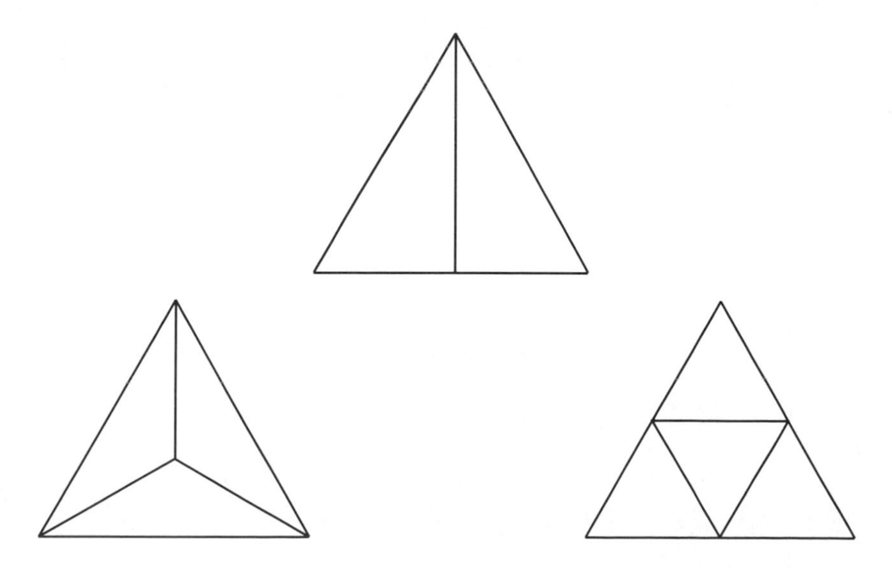

FIGURE 55

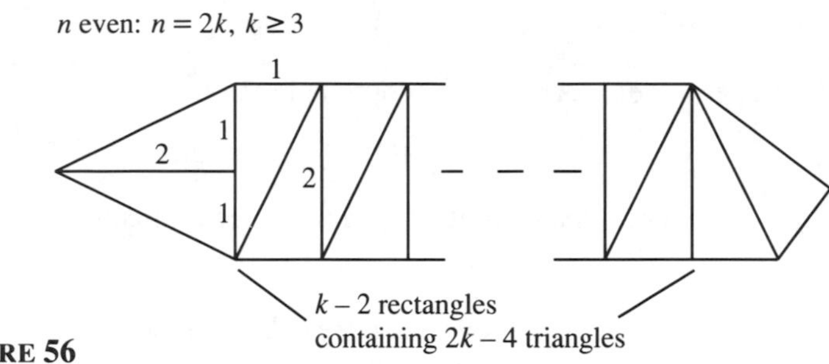

FIGURE 56

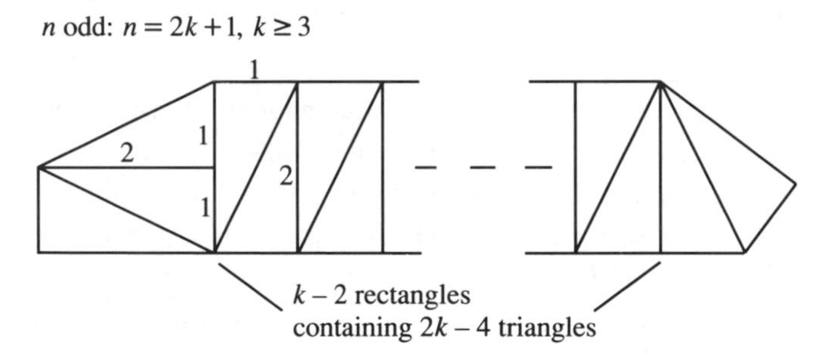

FIGURE 57

Four Problems from the 1989 AIME†

This year's American Invitational Mathematics Examination did not disappoint those who expected it would contain some wonderful problems; the following four particularly caught my eye.

> **1.** Let S be a subset of $\{1, 2, 3, \ldots, 1989\}$ in which no two members differ by *exactly* 4 or by *exactly* 7. What is the largest number of elements S can have?

At first I wondered whether the simple prescription of "take 4 and skip 7, take 4 and skip 7, ..." all the way along $\{1, 2, \ldots, 1989\}$, might yield a subset S of maximum size:

$$S = \{1, 2, 3, 4;\ 12, 13, 14, 15;\ 23, 24, 25, 26; \ldots\}.$$

Since $\{1, 2, \ldots, 1989\}$ contains 180 abutting intervals of length 11 plus the first nine integers in a 181st interval, this subset would contain $180 \cdot 4 = 720$ members plus whatever can be gleaned from the part interval at the end. But, like the other intervals, this final section can contribute its first four integers for a total of 724 integers in S. After repeated failures to prove that 724 is the maximum value of $|S|$, it dawned on me that not four but five numbers can be taken from this last part, namely

$$1981,\ 1983,\ 1984,\ 1986,\ \text{and } 1989.$$

Although this didn't help, for I couldn't prove that 725 was the maximum either, it did lead to the key idea that it might also be possible to take five integers instead of just four from each of the abutting intervals of length 11

† (*Crux Mathematicorum,* 1989, 97)

along $\{1, 2, \ldots, 1989\}$. Taking the first, third, fourth, sixth, and ninth integers in each interval gives the subset

$$T = \{1, 3, 4, 6, 9;\ 12, 14, 15, 17, 20;\ 23, 25, 26, 28, 31;\ 34, \ldots, 1896, 1899\}$$

containing $181 \cdot 5 = 905$ members. The first thing to check is whether this is indeed a legitimate subset of the type prescribed for S; that is, is any integer among those selected from one interval incompatible with the integers selected from its neighboring intervals? It suffices to consider the integers 12, 14, 15, 17, 20 from the second interval of length 11:

$$\left.\begin{array}{l} \text{12 is incompatible with 5, 8, 16, and 19} \\ \text{14 is incompatible with 7, 10, 18, and 21} \\ \text{15 is incompatible with 8, 11, 19, and 22} \\ \text{17 is incompatible with 10, 13, 21, and 24} \\ \text{20 is incompatible with 13, 16, 24, and 27.} \end{array}\right\} \textbf{ none} \text{ of which is in } T.$$

Hence T is legitimate, and we have $|S| \geq 905$. While it seems unlikely that S could claim as many as six integers from an interval of length 11, this possibility needs to be considered.

Let's try to take six integers from $\{1, 2, \ldots, 11\}$ without taking two that differ by 4 or by 7. Consider the section $\{1, 2, \ldots, 8\}$. These integers go together into four pairs, $(1, 5)$, $(2, 6)$, $(3, 7)$, and $(4, 8)$, in which the two integers differ by 4. By the pigeonhole principle, then, any selection of five integers from $\{1, 2, \ldots, 8\}$ would include both members of one of these pairs and therefore contain a pair of integers that differ by 4. Consequently, we must not put more than four numbers from $\{1, 2, \ldots, 8\}$ into the proposed subset S, implying that at least two integers must come from $\{9, 10, 11\}$. There are four possibilities:

(i) $\{9, 10\}$: Since only two integers come from $\{9, 10, 11\}$, we must take the full complement of four integers from $\{1, 2, \ldots, 8\}$. Now, 9 eliminates 2 and 5, while 10 eliminates 3 and 6, forcing the four integers to be $\{1, 4, 7, 8\}$, thus introducing the forbidden pair $\{1, 8\}$ into S.

(ii) $\{9, 11\}$: Similarly, 9 and 11 eliminate $\{2, 5, 4, 7\}$ from $\{1, 2, \ldots, 8\}$, forcing $\{1, 3, 6, 8\}$ into S, including $\{1, 8\}$ again.

(iii) $\{10, 11\}$: 10 and 11 eliminate $\{3, 6, 4, 7\}$, forcing $\{1, 2, 5, 8\}$ into S with the forbidden $\{1, 8\}$.

(iv) $\{9, 10, 11\}$: In this case only three integers are to be taken from $\{1, 2, \ldots, 8\}$. But 9, 10, and 11 eliminate $\{2, 5, 3, 6, 4, 7\}$, leaving only two integers (and, as if that wasn't enough, the two are the spoilers 1 and 8).

Thus S can always take five integers from each of the 181 abutting intervals, but never six, and therefore

$$\max|S| = 181 \cdot 5 = 905.$$

2. If $x_1, x_2, \ldots, x_7$ are real numbers such that

$$\begin{aligned} x_1 + 4x_2 + 9x_3 + 16x_4 + 25x_5 + 36x_6 + 49x_7 &= 1, \\ 4x_1 + 9x_2 + 16x_3 + 25x_4 + 36x_5 + 49x_6 + 64x_7 &= 12, \\ 9x_1 + 16x_2 + 25x_3 + 36x_4 + 49x_5 + 64x_6 + 81x_7 &= 123, \end{aligned}$$

what is the value of

$$S = 16x_1 + 25x_2 + 36x_3 + 49x_4 + 64x_5 + 81x_6 + 100x_7?$$

This promises to be really messy unless S is some combination of the left sides of the given equations. Our fervent hope is that constants a, b, and c exist such that

$$an^2 + b(n+1)^2 + c(n+2)^2 = (n+3)^2$$

holds for positive integers $n \leq 7$. Thus we seek a solution of the equation

$$(a+b+c)n^2 + (2b+4c)n + (b+4c) = n^2 + 6n + 9,$$

i.e., of the system

$$\begin{aligned} a + b + c &= 1, \\ 2b + 4c &= 6, \\ b + 4c &= 9. \end{aligned}$$

But this immediately gives $b = -3$, $c = 3$, and $a = 1$ (for all positive integers n). Therefore, from the right sides of the given equations, it follows that

$$\begin{aligned} S &= a(1) + b(12) + c(123) \\ &= 1 - 36 + 369 = 334. \end{aligned}$$

3. One of Euler's conjectures was disproved in the 1960's by three American mathematicians who showed that it *is* possible for the sum of four perfect fifth powers to be another perfect fifth power. They found that

$$S = 133^5 + 110^5 + 84^5 + 27^5 = n^5.$$

What is the value of n?

The last digits of the powers in S are respectively 3, 0, 4, and 7, implying that n^5 ends in the digit 4. Now, fifth powers are special in that n and n^5 always end in the same digit; to see this observe that, modulo 10, we have $m \equiv 0, 1, 2, 3, 4, 5, 6, 7, 8, 9$, giving, respectively,

$$\begin{aligned} m^2 &\equiv 0, 1, 4, 9, 6, 5, 6, 9, 4, 1, \\ m^3 &\equiv 0, 1, 8, 7, 4, 5, 6, 3, 2, 9, \quad \text{and} \\ m^5 &\equiv 0, 1, 2, 3, 4, 5, 6, 7, 8, 9. \end{aligned}$$

Thus n also ends in 4.

Since a modulo 10 analysis provided some useful information, let's see if a similar investigation modulo 9 is helpful. Mod 9, we have the table

$$\begin{aligned} m &\equiv 0, 1, 2, 3, 4, 5, 6, 7, 8, \\ m^2 &\equiv 0, 1, 4, 0, 7, 7, 1, 4, 1, \\ m^3 &\equiv 0, 1, 8, 0, 1, 8, 6, 1, 8, \\ m^5 &\equiv 0, 1, 5, 0, 7, 2, 6, 4, 8. \end{aligned}$$

Since 133, 110, 84, and 27 are respectively congruent (mod 9) to 7, 2, 3, and 0, then

$$\begin{aligned} S = n^5 &\equiv 7^5 + 2^5 + 3^5 + 0^5 \\ &\equiv 4 + 5 + 0 + 0 \qquad \text{(from the table)} \\ &\equiv 0, \end{aligned}$$

and therefore, again from the table, $n \equiv 0$ or $3 \pmod 9$.

Now, n^5 is clearly bigger than 133^5, and so, ending in 4, n must be one of the integers $\{134, 144, 154, 164, 174, 184, \ldots\}$. Being congruent to 0 or 3 (mod 9), n must also belong either to

$$\{135, \mathbf{144}, 153, 162, 171, 180, \ldots\} \text{ or to } \{138, 147, 156, 165, \mathbf{174}, 183, \ldots\}.$$

Therefore the two smallest candidates for n are 144 and 174. Let's consider the feasibility of $n = 174$.

By the binomial theorem, we have

$$\begin{aligned} 174^5 &= (133 + 41)^5 \\ &= 133^5 + 5 \cdot 133^4 \cdot 41 + 10 \cdot 133^3 \cdot 41^2 \\ &\quad + 10 \cdot 133^2 \cdot 41^3 + 5 \cdot 133 \cdot 41^4 + 41^5. \end{aligned}$$

Comparing terms with three of the powers in S, we have

$$133^5 = 133^5, \quad 5 \cdot 133^4 \cdot 41 > 133^5 > 110^5, \quad \text{and} \quad 41^5 > 27^5,$$

and it is beginning to appear that 174 might be greater than n. To confirm this we need to show that the remaining terms exceed 84^5: i.e.,

$$10 \cdot 133^3 \cdot 41^2 + 10 \cdot 133^2 \cdot 41^3 + 5 \cdot 133 \cdot 41^4 > 84^5.$$

It turns out that the first two terms suffice, for

$$\begin{aligned} 10 \cdot 133^3 \cdot 41^2 + 10 \cdot 133^2 \cdot 41^3 &= 10 \cdot 133^2 \cdot 41^2(133 + 41) \\ &> 10 \cdot 100^2 \cdot 1000 \cdot 100 \\ &= 10^{10} = 100^5 > 84^5. \end{aligned}$$

Hence $n < 174$, and we conclude that n must be 144.

4. Referring to figure 58, in $\triangle ABC$, AD, BE, and CF intersect at P such that

$$AP = PD = 6, \; EP = 3, \; PB = 9, \; \text{and} \; CF = 20.$$

What is the area of $\triangle ABC$?

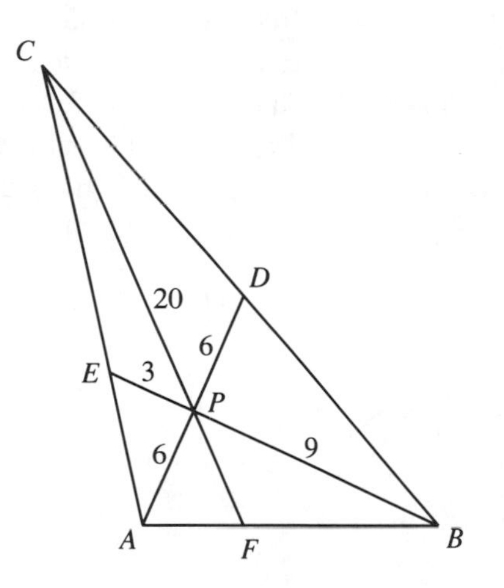

FIGURE 58

If Q is the midpoint of BE, then $PQ = 3$, and since AD and EQ bisect each other, $AQDE$ is a parallelogram (figure 59). Now, the diagonals partition

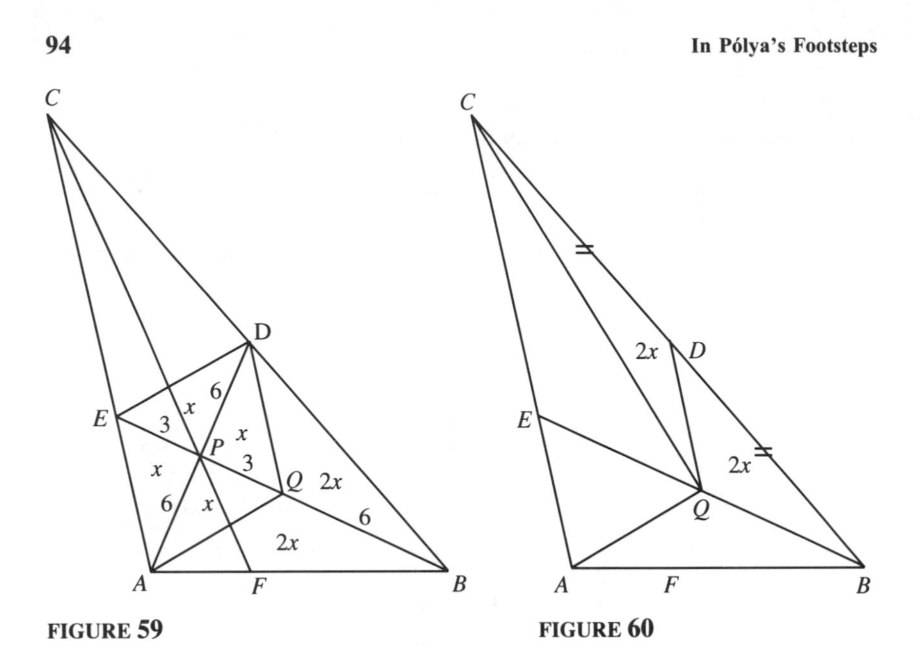

FIGURE 59

FIGURE 60

a parallelogram into four triangles of equal area and so, letting $AQDE = 4x$, we have each of the triangular quarters equal to x; and since the median DQ bisects $\triangle EDB$, it follows that $\triangle DQB = 2x$; similarly $\triangle AQB = 2x$.

Now, in $\triangle EBC$, QD goes from the midpoint Q of EB, is parallel to EC, and therefore it bisects BC. Hence D is the midpoint of BC and QD is a median of $\triangle QBC$ (figure 60). Thus $\triangle QBC = 2\triangle QBD = 4x$. But there are other medians bisecting triangles as well: CQ bisects $\triangle CEB$, implying $\triangle CEQ = 4x$, in which (figure 61) median CP gives $\triangle CEP = \triangle CPQ = 2x$. We note in passing that the desired area of $\triangle ABC$ is therefore $12x$.

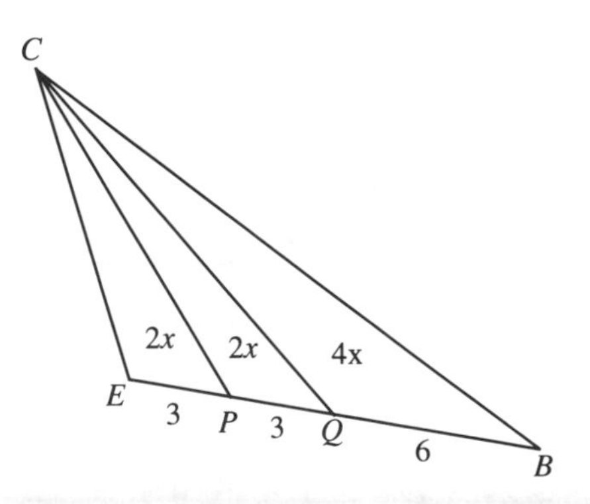

FIGURE 61

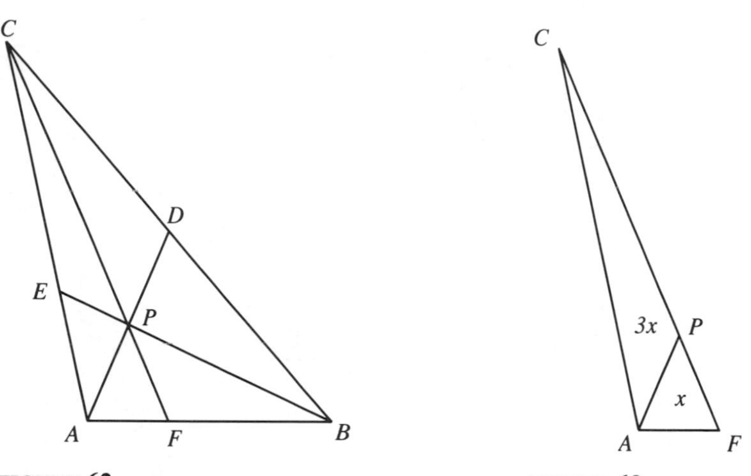

FIGURE 62

FIGURE 63

Observing that $\triangle ECB = 8x$, while $\triangle AEB = 4x$, only half as much, it follows that their bases CE and AE bear the same ratio and we have

$$\frac{CE}{EA} = \frac{2}{1}.$$

Now, Ceva's theorem asserts that

$$\frac{CE}{EA} \cdot \frac{AF}{FB} \cdot \frac{BD}{DC} = 1 \quad \text{(figure 62)},$$

and therefore

$$\frac{2}{1} \cdot \frac{AF}{FB} \cdot \frac{1}{1} = 1,$$

yielding $FB = 2 \cdot AF$. Thus E and F *trisect* the sides AC and AB while D *bisects* BC.

Consequently, $\triangle ACF = \frac{1}{3} \cdot \triangle ABC = 4x$. Now, $\triangle ECP = 2x$ (figure 61) and $\triangle AEP = x$ (= one-quarter of the parallelogram); hence $\triangle ACP = 3x$ (figure 63), implying

$$\triangle APF = \triangle ACF - \triangle ACP = 4x - 3x = x.$$

That is to say, $\triangle ACP = 3 \cdot \triangle APF$, and therefore

$$CP = \frac{3}{4}CF = \frac{3}{4} \cdot 20 = 15 \quad \text{and} \quad PF = 5.$$

Also, since $FB = 2 \cdot AF$, it follows that $\triangle FPB = 2 \cdot \triangle APF = 2x$.

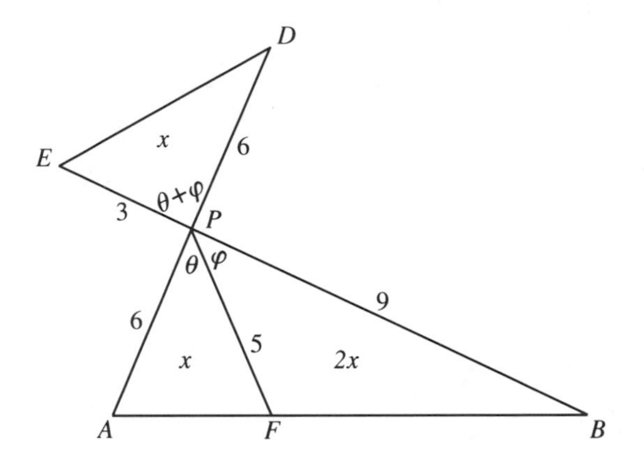

FIGURE 64

Finally, then, consider figure 64. Let $\angle APF = \theta$ and $\angle FPB = \phi$, making $\angle EPD = \angle APB = \theta + \phi$. From $\triangle APF$, we have

$$x = \frac{1}{2} \cdot 6 \cdot 5 \sin\theta = 15 \sin\theta,$$

and from $\triangle FPB$,

$$2x = \frac{1}{2} \cdot 5 \cdot 9 \sin\phi = \frac{45}{2} \sin\phi.$$

Hence

$$\sin\theta = \frac{x}{15} \quad \text{and} \quad \sin\phi = \frac{4x}{15}.$$

Therefore

$$\cos\theta = \sqrt{1 - \frac{x^2}{15^2}} = \frac{1}{15}\sqrt{15^2 - x^2},$$

and

$$\cos\phi = \sqrt{1 - \frac{(4x)^2}{45^2}} = \frac{1}{15}\sqrt{45^2 - (4x)^2}.$$

Consequently,

$$\begin{aligned} \triangle EPD = x &= \frac{1}{2} \cdot 3 \cdot 6 \sin(\theta + \phi) \\ &= 9(\sin\theta\cos\phi + \cos\theta\sin\phi), \end{aligned}$$

that is,

$$x = 9 \cdot \frac{x}{15} \cdot \frac{1}{45}\sqrt{45^2 - (4x)^2} + 9 \cdot \frac{1}{15}\sqrt{15^2 - x^2} \cdot \frac{4x}{15}$$

giving

$$1 = \frac{1}{75}\sqrt{45^2 - (4x)^2} + \frac{4}{75}\sqrt{15^2 - x^2},$$

and

$$75 = \sqrt{45^2 - (4x)^2} + 4\sqrt{15^2 - x^2}.$$

Now, I'd hate to go to all the trouble of solving this equation if I didn't have to, and as it happens it is not really necessary. Up to this point there has been no reason to alert you to the fact that everybody who writes the AIME knows that the answers are all 3-digit nonnegative integers: no solutions are required; one's reply consists entirely of a 3-digit integer from 000 to 999. Accordingly, the area of $\triangle ABC$, namely $12x$, must be an integer, and so x is at least *rational.* In view of the radical $\sqrt{15^2 - x^2}$ in our last equation, there is a reasonable chance that $15^2 - x^2$ is itself a perfect square, y^2. If this is the case, then $x^2 + y^2 = 15^2$, bringing to mind the Pythagorean triple $(9, 12, 15)$ and suggesting that x might be either 9 or 12. Since $x = 12$ would make $45^2 - (4x)^2$ negative and $\cos\phi$ nonreal, $x = 9$ must carry all our hopes. Checking $x = 9$, we have

$$\begin{aligned}
&\sqrt{45^2 - 36^2} + 4\sqrt{15^2 - 9^2} \\
&\qquad = 27 + 4 \cdot 12 \quad \text{(recall } (27, 36, 45) \text{ is a Pythagorean triple)}, \\
&\qquad = 75,
\end{aligned}$$

as desired. Hence x is indeed equal to 9 and

$$\triangle ABC = 12x = 108.$$

We note that $\triangle EPD = x$ gives

$$\frac{1}{2} \cdot 3 \cdot 6 \sin \angle EPD = 9,$$

$$\sin \angle EPD = 1,$$

implying $\angle EPD$ is a right angle. Thus AD and BE are in fact perpendicular, and if this could somehow be established independently, the solution would follow quickly and easily; the difficulty is how to take $CF = 20$ into account.

Five Unused Problems from the 1989 International Olympiad

1. Proposed by Romania. (*Crux Mathematicorum,* 1989, 261)

Find the smallest positive integer n which makes

$$m^n - 1 \text{ divisible by } 2^{1989}$$

no matter what odd positive integer greater than 1 might be substituted for m.

By a standard factoring we have

$$m^n - 1 = (m-1)(m^{n-1} + m^{n-2} + \cdots + m + 1).$$

Now, if n is odd, the n terms in the large bracket, each of which is also odd, would add up to an odd total. In this case, the only way of obtaining factors equal to 2 would be from the first factor $(m-1)$, which might contain many of them but which cannot be counted upon to have more than just one: for example, when $m = 3$ (and, in general, whenever $m = 4k + 3$, which makes $m - 1 = 4k + 2 = 2(2k + 1)$). Thus n *must be even.*

Now, every positive integer n can be expressed in the form $2^r q$, where q is odd (simply factor all the 2's out of it). Thus, repeatedly factoring the differences of squares, we have, for an even integer $n = 2^r q$, that $r \geq 1$ and hence

$$\begin{aligned}
m^n - 1 &= m^{2^r q} - 1 \\
&= \left(m^{2^{r-1}q} + 1\right)\left(m^{2^{r-1}q} - 1\right) \\
&= \left(m^{2^{r-1}q} + 1\right)\left(m^{2^{r-2}q} + 1\right)\left(m^{2^{r-2}q} - 1\right) \\
&\vdots \qquad \vdots \qquad\qquad \vdots \qquad\qquad \vdots \\
&= \left(m^{2^{r-1}q} + 1\right)\left(m^{2^{r-2}q} + 1\right)\cdots\left(m^{2q} + 1\right)\left(m^{2q} - 1\right) \\
&= \left(m^{2^{r-1}q} + 1\right)\left(m^{2^{r-2}q} + 1\right)\cdots\left(m^{2q} + 1\right)\left(m^{q} + 1\right)\left(m^{q} - 1\right).
\end{aligned}$$

Since m is odd, each of these $r+1$ final factors is even, thus contributing at least one factor 2 to the prime decomposition of $m^n - 1$. However, being odd, $m \equiv \pm 1 \pmod 4$, and so $m^2 \equiv 1 \pmod 4$, making $m^{2k} + 1 \equiv 2 \pmod 4$ (i.e., of the form $4t + 2 = 2(2t + 1)$), showing that regardless of the value of q, the first $r - 1$ of these factors (i.e., all except the last two) contribute only a single factor 2.

Since m and q are odd, for some positive integer k we have (mod 4) that

$$m^q = m^{2k+1} \equiv (m^2)^k m \equiv m \equiv \pm 1,$$

making one of the factors m^q+1, m^q-1 a multiple of 4. Since the other factor is also even, these final two factors must contribute *at least three* factors 2. However, depending on the value of m, they might contribute more than three: for example, for $m = 2^k - 1$, and n a power of 2, which makes $q = 1$, we have $m^q + 1 = m + 1 = 2^k$ itself. But it seems unlikely that there are *always* more than three factors 2 in the final two factors, suggesting that there might be cases when they need to be preceded by a full line of 1986 factors of the type containing only a single 2. This would call for a value of $r - 1 = 1986$ in the makeup of n $(= 2^r q)$, and since the smallest n is required, $q = 1$ leads us to consider the value $n = 2^{1987}$.

This value certainly works for all m, since it always provides an initial string of 1986 factors each containing a single 2 which, with the 3 or more from the last two factors, guarantees at least the required 1989 factors 2. However, it is conceivable that some $q > 1$ might always provide more than three 2's in the final two factors, thus allowing a smaller value of r and an n that is possibly smaller than 2^{1987}. It would be nice therefore if we could show that there exists a value of m whose final two factors never yield more than three factors 2 no matter what odd number q might be, thus forcing r to be at least 1987 and implying $n = 2^{1987}$ is indeed the smallest possible n.

It is a relief to discover that $m = 3$ is such an integer. We conclude, then, by showing that $(3^q + 1)(3^q - 1)$ is never divisible by $2^4 = 16$ for any odd positive integer q. We have

$$\begin{aligned}
(3^q + 1)(3^q - 1) &= 3^{2q} - 1 = 9^q - 1 \\
&= 9^{2k+1} - 1 \quad \text{for some positive integer } k, \\
&= (81)^k \cdot 9 - 1 \\
&\equiv (1)^k \cdot 9 - 1 \pmod{16} \\
&\equiv 8 \pmod{16},
\end{aligned}$$

and never congruent to 0.

2. Proposed by India. (*Crux Mathematicorum,* 1989, 226)

On Bicentric Quadrilaterals. A bicentric quadrilateral is one which is both inscriptible and circumscriptible, that is, it has both an incircle and a circumcircle.

Prove that the line joining the centers O and I of the circumscribed and inscribed circles of a bicentric quadrilateral $ABCD$ always passes through the point of intersection E of its diagonals.

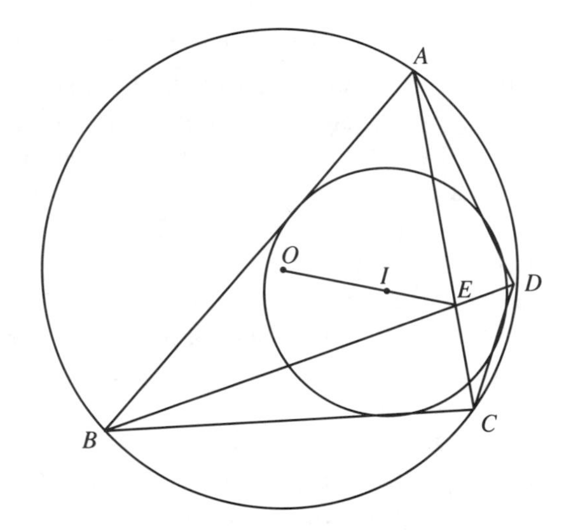

FIGURE 65

(a) I found this to be a very difficult problem. I have no idea how the proposers expected a contestant to approach the problem and I was able to solve it only with essential help from one of my all-time favorite books—Heinrich Dorrie's *100 Great Problems of Elementary Mathematics* (pages 188–193 in the 1965 Dover edition).

Problems which deal with quadrilaterals that are either inscriptible or circumscriptible seem to arise much more often than ones concerning bicentric quadrilaterals and I dare say that the suspicion lurks in the minds of many of us that there just don't exist as many bicentric quadrilaterals as either of their semi-endowed relatives. However, this is not the case, for our first step will be to show that the tangents to a circle at the ends of any pair of perpendicular chords generate a bicentric quadrilateral $ABCD$.

Let perpendicular chords PR and QS cross at E. Since the angles in each of the quadrilaterals $APES$ and $EQCR$ add up to 360 degrees, we

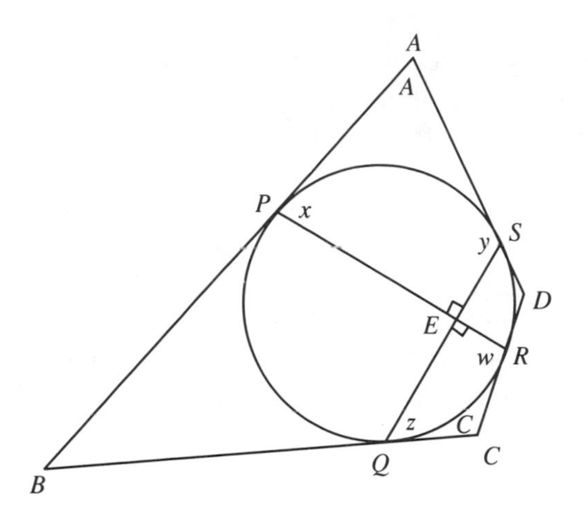

FIGURE 66

have in figure 66 that

$$(A + x + 90^\circ + y) + (90^\circ + z + C + w) = 720^\circ. \qquad (1)$$

Since the angle between a tangent and a chord is equal to the angle in the segment on the opposite side of the chord, then

$x =$ the angle in segment PQR (on one side of chord PR),

while

$w =$ the angle in segment PSR (on the other side of PR),

and we have $x + w = 180^\circ$ (the angles in the segments on both sides of a chord are opposite angles in a cyclic quadrilateral and thus add up to 180°). Similarly, $y + z = 180^\circ$, and it follows from equation (1) that $A + C = 180^\circ$, implying $ABCD$ is cyclic (i.e., circumscriptible).

Actually, what we want is the converse of this result. But clearly, if $ABCD$ *is* cyclic, then $A + C = 180^\circ$, in which case equation (1) implies the opposite angles at E are supplementary; since these angles are vertically opposite, they are equal and hence each must be a right angle.

I expect you are wondering what this has to do with the problem at hand. The fact is that these perpendicular chords PR and QS also pass through the point of intersection of the diagonals, thus allowing us to discuss things in terms of their point of intersection instead of the intersection of the diagonals. To prove this fact, we appeal to the famous theorem of Brianchon. The natural setting for this great theorem is an inscriptible hexagon $ABCDEF$

(in its full generality the inscribed figure may be any conic, not just a circle), for which it claims the concurrency of the three major diagonals AD, BE, and CF. Remarkably, the theorem also holds for the degenerate cases of inscriptible pentagons, quadrilaterals, and triangles, in which cases one simply supplements the vertices of the polygon with any of the points of tangency to raise the total number of points to six (figures 67 and 68).

Thus, in the problem at hand the major diagonals of $BQCDSA$, namely BD, QS, and CA, are concurrent (figure 69), as are the major diagonals of $APBCRD$, which are AC, PR, and BD. With diagonals AC and BD common to these triples, it follows that all of these diagonals pass through a common point (E).

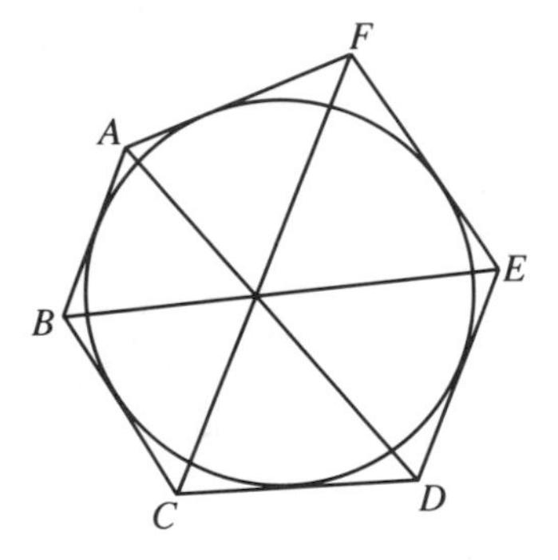

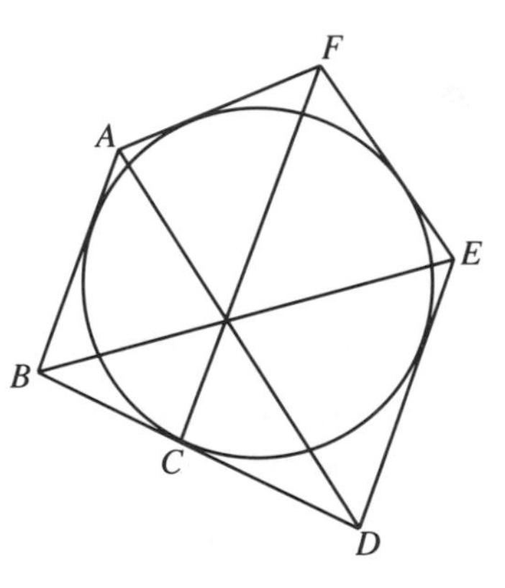

FIGURE 67

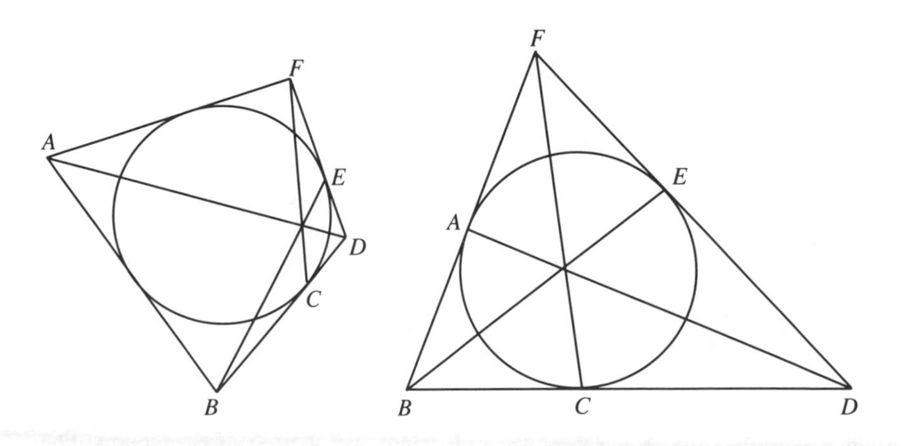

FIGURE 68

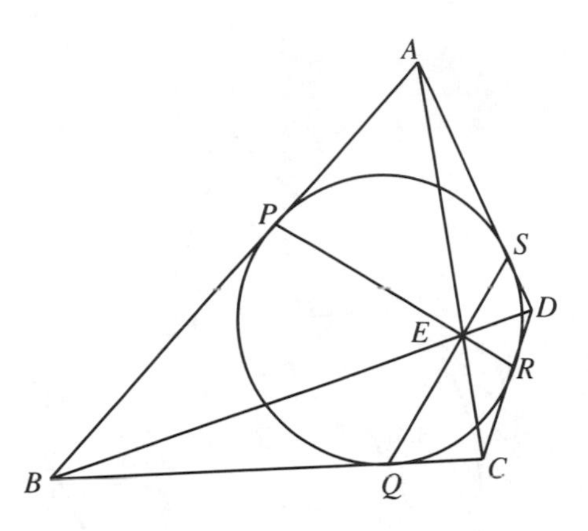

FIGURE 69

(b) At last we are ready for the main argument. Let the incenter I and the point of intersection E of the diagonals be determined for the given bicentric quadrilateral $ABCD$, but *not* its circumcenter O (figure 70). Also, let AI cross PS at T, drop perpendiculars EU and EV respectively to IT and TS, and let $IA = p$, $IE = e$, $\angle EIA = \phi$, $IU = x$ $(= e\cos\phi)$, and $UE = y$. Finally, let EI be extended a distance $z = \dfrac{r^2 e}{r^2 - e^2}$ to determine a point K. Our goal is to show that K is really the circumcenter O of $ABCD$.

Clearly the equal tangents at A and the equal radii at I make AI the perpendicular bisector of PS. Thus T is the midpoint of PS and the angles at T are right angles, and so there are two right-triangles, AIS and PES, in which the altitude is drawn to the hypotenuse (altitudes ST and EV). Accordingly, we have the mean proportions

$$EV^2 = PV \cdot VS \qquad (2),$$

and

$$IS^2 = IA \cdot IT, \quad \text{i.e.,} \quad r^2 = p \cdot IT \qquad (3).$$

Since T bisects PS and $TUEV$ is a rectangle, we have

$$EV = TU = IT - IU = IT - x,$$

$$PV = PT + TV = TS + TV, \quad \text{and} \quad VS = TS - TV.$$

Hence, from equation (2), we get

$$EV^2 = (IT - x)^2 = (TS + TV)(TS - TV),$$

$$IT^2 - 2x \cdot IT + x^2 = TS^2 - TV^2 = TS^2 - UE^2 = TS^2 - y^2,$$

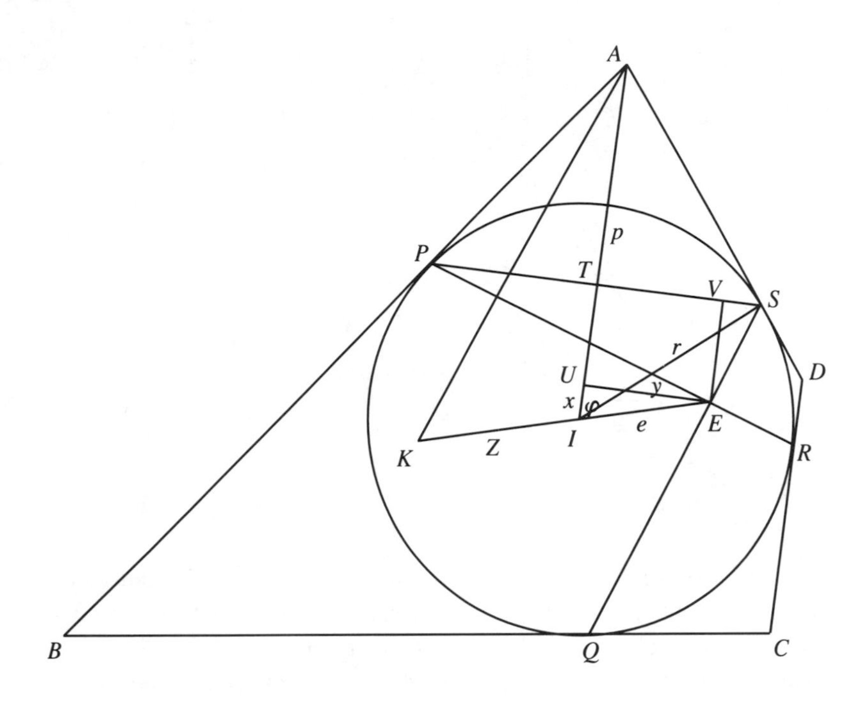

FIGURE 70

giving

$$IT^2 - 2x \cdot IT + (x^2 + y^2) = TS^2. \tag{4}$$

Now, $x^2 + y^2 = e^2$, $TS^2 = r^2 - IT^2$, and $x = e\cos\phi$, and so equation (4) yields

$$IT^2 - 2x \cdot IT + e^2 = r^2 - IT^2, \quad \text{and} \quad 2 \cdot IT^2 - 2 \cdot IT \cdot e\cos\phi + e^2 = r^2.$$

Since $IT = r^2/p$ from (3),

$$\frac{2r^4}{p^2} - \frac{2r^2}{p} e\cos\phi + e^2 = r^2,$$

$$2r^4 - 2r^2 pe\cos\phi + p^2e^2 = r^2p^2,$$

$$2r^4 = p^2(r^2 - e^2) + 2r^2 pe\cos\phi,$$

and

$$\frac{2r^4}{r^2 - e^2} = p^2 + \frac{2r^2 e}{r^2 - e^2} \cdot (p\cos\phi). \tag{5}$$

Now, applying the law of cosines to triangle KIA, we obtain

$$KA^2 = z^2 + p^2 - 2zp\cos(180° - \phi) = z^2 + p^2 + 2zp\cos\phi.$$

We have just seen in (5) that

$$p^2 + 2\frac{r^2e}{r^2 - e^2} \cdot p\cos\phi = \frac{2r^4}{r^2 - e^2},$$

from which it is clear that, when z is assigned the value, $\dfrac{r^2e}{r^2 - e^2}$ (which it is), we have

$$KA^2 = z^2 + p^2 + 2zp\cos\phi = z^2 + \frac{2r^4}{r^2 - e^2}.$$

Since this quantity is independent of the vertex A, the same expression would be obtained if we were to calculate KB^2, KC^2, and KD^2, implying that K is equidistant from A, B, C, and D, and is indeed the circumcenter O of $ABCD$, completing the argument.

3. Proposed by Hungary. (revised; *Crux Mathematicorum,* 1989, 260)

At an arbitrary selection of n different points $s_1, s_2, \ldots, s_n$ around a circular track, there are respectively n cars $c_1, c_2, \ldots, c_n$, ready to start. The cars are not about to race, for they all go at the same speed, namely one circuit of the track per hour. All the cars start off at the same moment, at which time each driver independently selects a direction and proceeds around the track. These cars are like bumper-cars, for whenever two cars meet, both of them instantly reverse direction and proceed without loss of speed.

Prove that, at some moment in the future, the original configuration will be recaptured exactly, that is, each car will be at its own starting point going in its original direction.

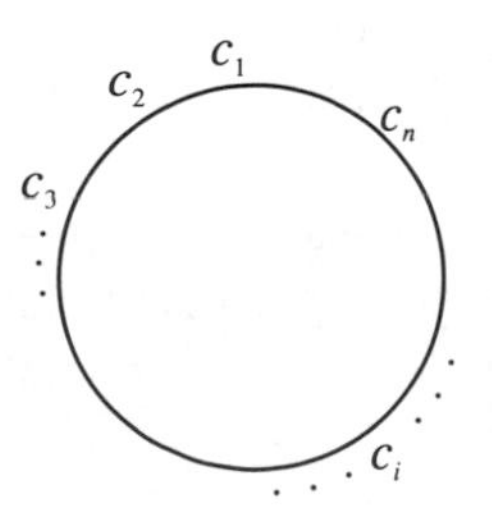

FIGURE 71

When all the cars start off in the same direction, the equal speeds maintain their mutual distances and the cavalcade simply rotates rigidly around the course to the original configuration in exactly one hour.

Generally, however, there will be a mind-boggling succession of ricochets occurring at various times and places around the track, like a glorified pinball machine. Despite all this rebounding, it turns out that, at the end of an hour, the n cars will again occupy the n starting positions; that is to say, after an hour some car will be at each of the starting positions s_i and, moreover, the car at s_i will be going in c_i's original direction. Unfortunately, the car at s_i might not be c_i, and so the result might not be the original configuration exactly. To see all this, consider the course taken by a passenger who starts in the beginning at s_i with car c_i and who transfers to the "other" car at each meeting with another car. Since the ricochets are instantaneous and without loss of speed, our traveller is thus given a perfectly smooth ride all the way around the track; an hour later, then, he is in some car that is passing through s_i going in the same direction that he left it.

Now, because no two cars ever pass each other, the cyclic order of the cars around the track never changes. Therefore if c_i is at s_{i+j} after one hour, it must be c_{i+1} that is at s_{i+j+1}, and so on. Hence the result of letting the cars run for an hour is to advance each car around the track to the jth starting point past its original position, for some $j \in \{1, 2, \ldots, n\}$. Since the car at s_i is also going in c_i's original direction, the configuration at the beginning of the second hour is essentially the same as that at the very beginning and therefore a second hour's running will have the same effect as the first hour, namely to advance each car j starting positions around the track. Clearly it is the same hour after hour. Thus c_i, beginning at s_i, is moved successively to positions $s_{i+j}, s_{i+2j}, \ldots$, to wind up after n hours at s_{i+nj}, which modulo n is just s_i itself. Since the car at s_i is always going in c_i's original direction, after n hours the original configuration is recaptured exactly. (An alternative solution is given in 1991, 204.)

4. Proposed by the United States. (*Crux Mathematicorum,* 1989, 262)

If the vertex A of acute-angled triangle ABC lies on the perpendicular bisector of the segment HO, joining the orthocenter H to the circumcenter O, determine the size of angle A (figure 72)?

Since $\triangle ABC$ is acute-angled, its circumcenter must lie inside the triangle, as shown.

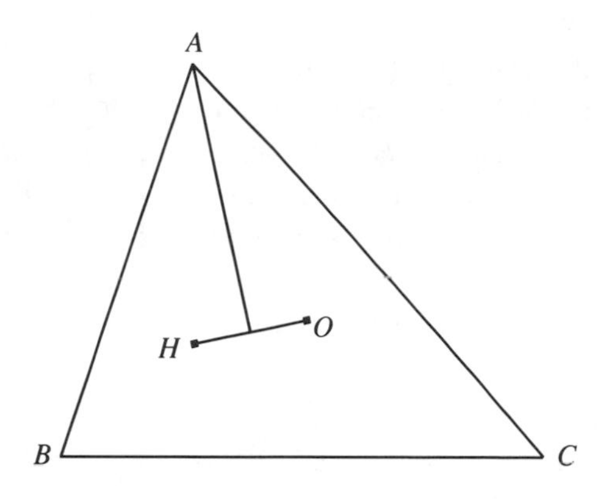

FIGURE 72

Now, the distance along an altitude from a vertex to the orthocenter is twice the distance from the circumcenter to the opposite side of the triangle. This is not difficult to prove and we shall do so shortly. In figure 73, then, $AH = 2 \cdot OD$.

Since O is the circumcenter, each of OA, OB, and OC has length R, the circumradius. Lying on the perpendicular bisector of HO, A is equidistant

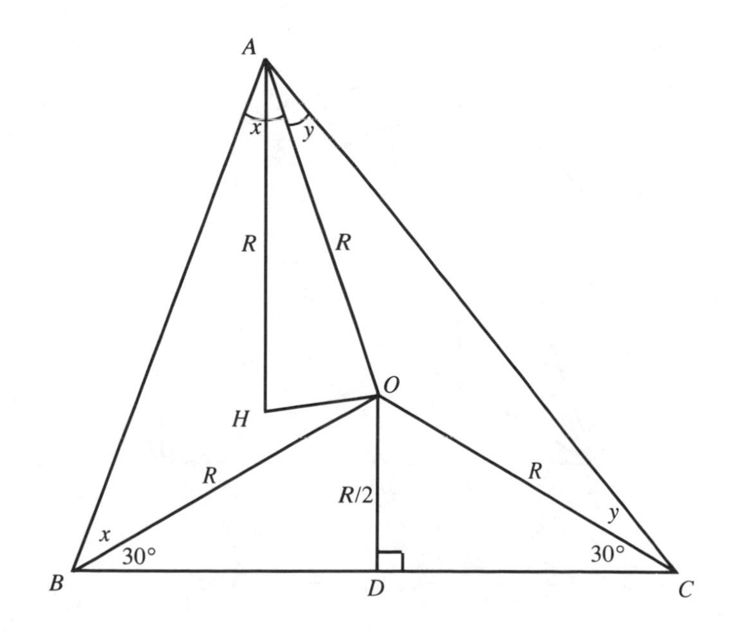

FIGURE 73

from H and O, also making $AH = R$. Thus $OD = \frac{1}{2}AH = \frac{1}{2}R$, implying that right-triangle ODC is a 30-60-90 degree triangle. Therefore $\angle OCD = 30°$, and in isosceles triangle OBC, the other base angle OBC is also 30°.

Letting the base angles in isosceles triangles AOB and AOC be x and y, respectively, the sum of the angles in $\triangle ABC$ is

$$(x + y) + (x + 30°) + (y + 30°) = 2(x + y) + 60° = 180°,$$

and we have

$$\angle A = x + y = 60°.$$

To show that $AH = 2 \cdot OD$, let AE and CF be the altitudes from A and C, and let BO be extended to meet the circumcircle at G (figure 74).

Now, D is actually the midpoint of BC, and therefore OD joins the midpoints of two side of $\triangle BCG$, implying that CG is parallel to OD and twice as long. Since OD is perpendicular to BC, as is altitude AE, it follows that AH and CG are parallel. But so are AG and CH parallel since both are perpendicular to AB (CF is an altitude, and $\angle BAG = 90°$ because BG is a diameter). Thus the pairs of opposite sides of quadrilateral $AHCG$ are parallel, making it a parallelogram. Its opposite sides AH and CG are therefore equal, and since CG is twice OD, so is AH.

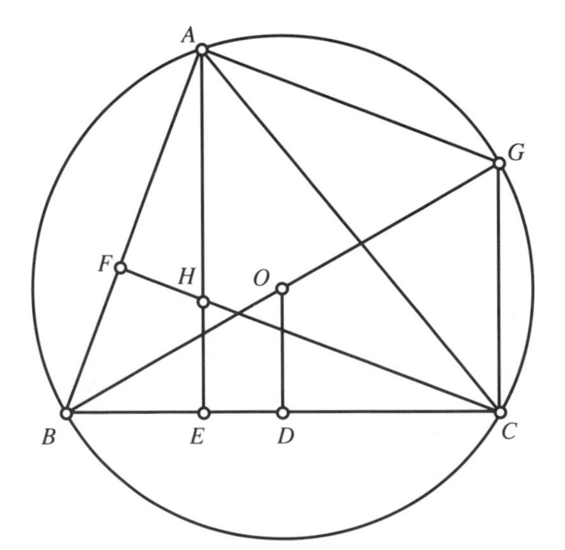

FIGURE 74

5. Proposed by Mongolia. (*Crux Mathematicorum,* 1989, 260)

$A = A_1A_2\ldots A_n$ is a plane convex polygon of area S and M is an arbitrary point in the plane of A (figure 75). M is rotated in turn about each of the vertices A_i through a fixed angle x to yield an image polygon $P = M_1M_2\ldots M_n$. Determine the area of P.

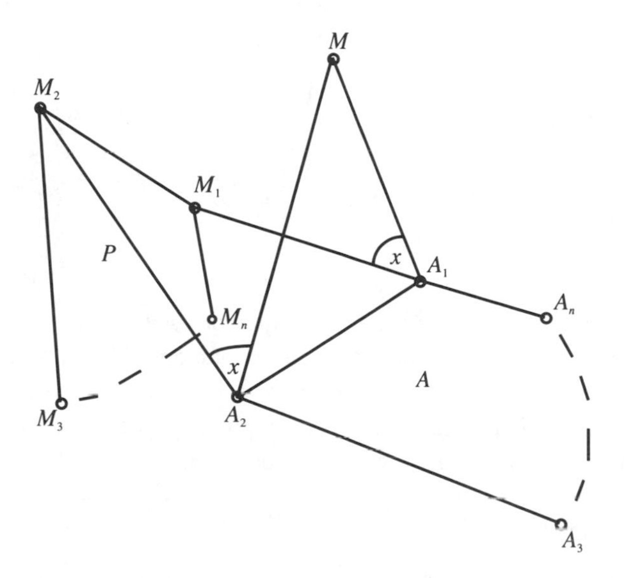

FIGURE 75

I expect this lovely problem was a hot contender for a place on this olympiad.

Clearly the prescribed rotation about the center A_i carries M to its image M_i so that $\triangle MA_iM_i$ is an isosceles triangle whose vertical angle is always the given angle x (figure 76). Thus the base angles in these isosceles triangles are always the same, namely $y = 90° - \frac{x}{2}$. Because of this, the image point M_i may be alternatively constructed by first rotating A_i about the center M (rather than rotating M about A_i) through the angle $-y$, to carry it to an intermediate position B_i, after which the dilatation with center M and magnification ratio $r = \frac{MM_i}{MB_i}$ would take B_i the rest of the way to M_i. Since $MB_i = MA_i$ in this rotation, this ratio r is just the base to an arm of the isosceles triangle. By the law of sines, then, we have

$$r = \frac{\text{base } MM_i}{\text{arm } MA_i} = \frac{\sin x}{\sin y} = \frac{\sin x}{\sin(90° - \frac{x}{2})} = \frac{\sin x}{\cos \frac{x}{2}} = 2\sin\frac{x}{2}.$$

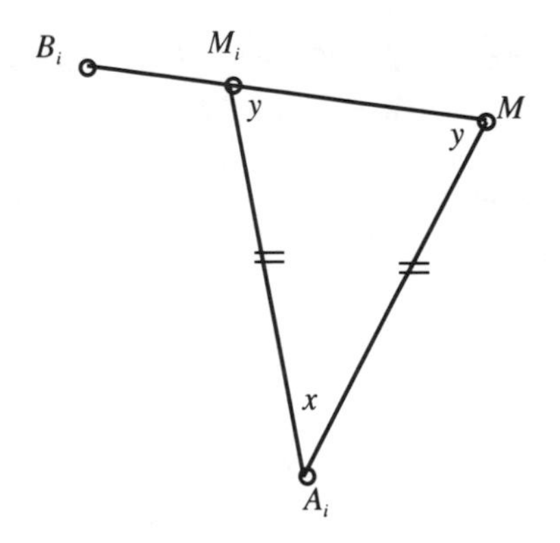

FIGURE 76

That is to say, the compound transformation of rotating the entire polygon A about M through $-y$ followed by the dilatation $M(2\sin\frac{x}{2})$ carries the vertices $A_1, A_2, \ldots, A_n$ into their respective images *all at once*. Thus the resulting polygon P is similar to A and the ratio of their areas is simply r^2, the square of the linear magnification factor. Hence

$$s' = \text{ the area of } P = \left(2\sin\frac{x}{2}\right)^2 \cdot S = 2\left(2\sin^2\frac{x}{2}\right) \cdot S = 2(1-\cos x)S.$$

We observe that S' exceeds, is equal to, or is less than S as x exceeds, is equal to, or is less than $60°$ and, surprisingly, *S' is independent of the position of M in the plane.*

Four Geometry Problems†

1. 1991 U. S. A. Olympiad. There are many triangles ABC in which angle A is twice angle B and angle C is obtuse. Determine the *minimum perimeter* of such a triangle if the lengths of its sides are *integers*.

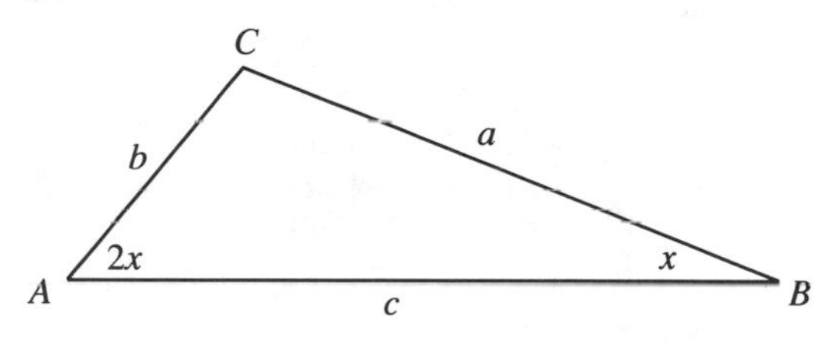

FIGURE 77

If $\angle B = x$, then $\angle A = 2x$, and since $\angle C$ is obtuse, we have

$$\angle A + \angle B = 3x < 90°,$$

implying $x < 30°$. Let AD bisect angle A and DE be perpendicular to AB (figure 78). Then AD divides CB in the ratio of the sides about A, i.e.,

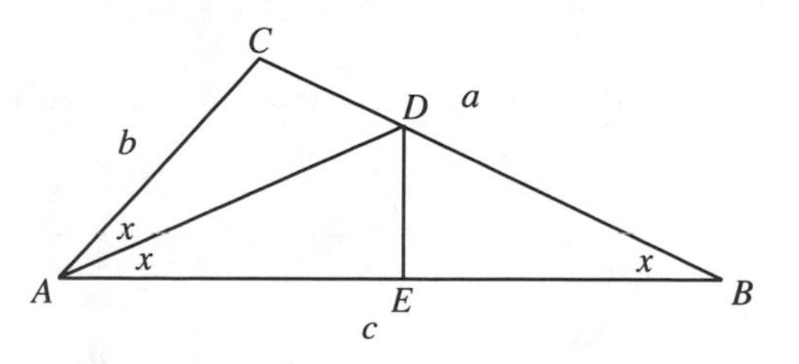

FIGURE 78

† (*Crux Mathematicorum,* 1991, 162–163)

$\frac{CD}{DB} = \frac{b}{c}$; also, $\triangle ADB$ is isosceles, and so E is the midpoint of AB, and we have

$$EB = DB\cos x = \frac{c}{2}, \quad \text{giving} \quad DB = \frac{c}{2\cos x}.$$

Then

$$CD = a - DB = a - \frac{c}{2\cos x} = \frac{2a\cos x - c}{2\cos x},$$

and then $\dfrac{CD}{DB} = \dfrac{b}{c}$ gives

$$\frac{\frac{2a\cos x - c}{2\cos x}}{\frac{c}{2\cos x}} = \frac{2a\cos x - c}{c} = \frac{b}{c},$$

from which $2a\cos x - c = b$, and $\cos x = \dfrac{b+c}{2a}$.

Now, by the law of sines,

$$\frac{a}{\sin A} = \frac{b}{\sin B},$$
$$b\sin 2x = a\sin x$$

$$b(2\sin x\cos x) = a\sin x$$

and

$$\cos x = \frac{a}{2b}.$$

Therefore

$$\cos x = \frac{b+c}{2a} = \frac{a}{2b},$$

giving

$$a^2 = b(b+c).$$

It follows that b divides a^2 and from $a^2 = b^2 + bc$, we get $c = \dfrac{a^2 - b^2}{b}$.

Now, by the triangle inequality, we have $a + b > c$. Hence

$$c = \frac{a^2 - b^2}{b} = \frac{(a+b)(a-b)}{b} > \frac{c(a-b)}{b},$$

and, cancelling c's, we get

$$1 > \frac{a-b}{b}, \quad b > a - b, \quad \text{and} \quad a < 2b.$$

On the other hand, from $x < 30^\circ$, we have

$$\frac{a}{2b} = \cos x > \cos 30^\circ = \frac{\sqrt{3}}{2},$$

giving $a > b\sqrt{3}$. Altogether, then,

$$b\sqrt{3} < a < 2b.$$

Since b divides a^2, each prime divisor of b is also a prime divisor of a. Thus, if $b = pqr\ldots$, the product of different primes, each to the first degree, then each of the primes also divides a, implying b divides a, making a a multiple of b. But since

$$b\sqrt{3} < a < 2b,$$

a cannot be a multiple of b. Therefore b cannot be the product of primes to the first degree, and we conclude that b must have a *square factor.* Thus b is divisible by a perfect square and must belong to the set

$$\{1, 4, 8, 9, 12, 16, 18, 20, 24, 25, 27, 28, \ldots\}.$$

For $b = 1$ we have $\sqrt{3} < a < 2$, which is impossible for an integer a, and so b must be at least 4. We observe that, when b is even, a must also be even since b divides a^2. For the first few values of b, then, we have

(i) $b = 4$: This gives $4\sqrt{3} = 6.8\ldots < a < 8$, making a the *odd* integer 7, a contradiction.

(ii) $b = 8$: This time $8\sqrt{3} = 13.8\ldots < a < 16$, making $a = 14$ (since a must be even for b even), and then b fails to divide a^2.

(iii) $b = 9$: Here we have $9\sqrt{3} = 15.5\ldots < a < 18$, making $a = 16$, and again b fails to divide a^2.

(iv) $b = 12$: We have $12\sqrt{3} = 20.7\ldots < a < 24$, making $a = 22$, for which b continues to fail to divide a^2.

Although this doesn't seem to be working out, let's try one more case.

(v) $b = 16$: This gives $16\sqrt{3} = 27.7\ldots < a < 32$, making $a = 28$ or 30. Now, clearly 16 does not divide 30^2, but finally our persistence has brought us to our reward, for 16 *does* divide 28^2. Thus $b = 16$ implies $a = 28$, and since $a^2 = b(b + c)$, we have $c = 33$:

$$28^2 = 16(16 + c), \quad 7^2 = 16 + c, \quad \text{and} \quad c = 33.$$

We have every hope, then, that the minimum perimeter is

$$a + b + c = 28 + 16 + 33 = 77.$$

However, it is conceivable that a larger value of b might be offset by a smaller value of $a + c$ to yield a smaller perimeter, and so we still need to show that $b > 16$ always gives perimeters ≥ 77. But this is easy.

If $b > 16$, then $b \geq 18$, and we have $18\sqrt{3} = 31.1\ldots < a$, implying $a \geq 32$. But $\angle C$ is obtuse, and so c is even bigger than a, making $c \geq 33$, and we have

$$a + b + c \geq 18 + 32 + 33 = 83 > 77,$$

as desired.

2. 1991 U. S. A. Olympiad. D is any point on side AB of triangle ABC, and E is the point inside $\triangle ABC$ where AD is intersected by the external common tangent to the incircles of triangles CAD and CDB. As D assumes all points between A and B, prove that E traces an arc of a circle (figure 79).

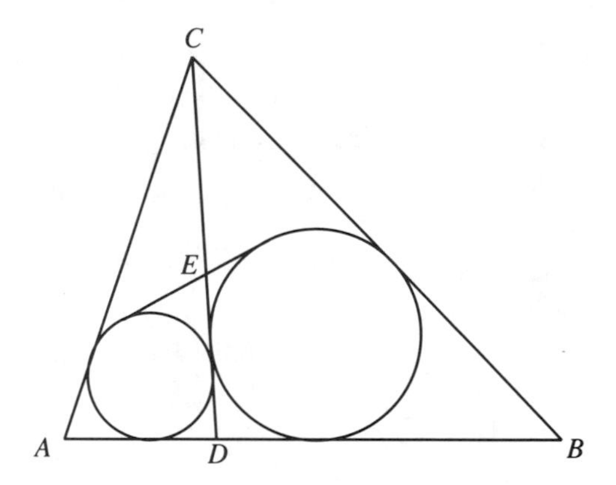

FIGURE 79

(i) As usual, let a, b, c denote the sides of $\triangle ABC$ and s its semiperimeter; also let s_1 and s_2 denote the semiperimeters of triangles CAD and CDB respectively. Finally, let the points of contact of the circles with the three common tangents be (M, N), (P, Q), and (U, V) (figure 80).

Now, as D approaches A, $\triangle CAD$ collapses and $\triangle CDB$ approaches $\triangle ABC$ itself, with the result that E approaches the point of contact X of the incircle of $\triangle ABC$ on AC (figure 81). Similarly, E approaches Y as D approaches B.

Since the tangents CX and CY are equal, it is a small step to conjecture that CE is a fixed length. Accordingly, let $CE = d$ and let us try to show that d is a constant.

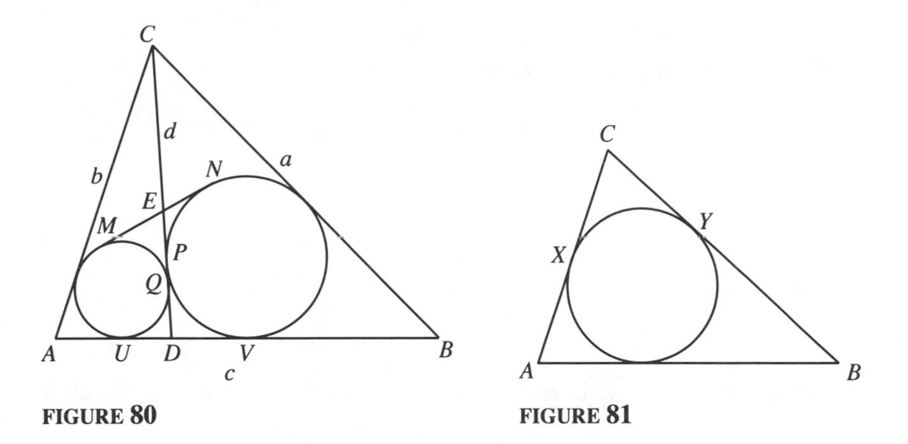

FIGURE 80

FIGURE 81

(ii) The basis of our solution is the remarkable fact that EP is always the same length as QD (figure 80). We begin with the easy proof of this result.

The symmetry of the figure implies the common tangents MN and UV have the same length. Now, clearly $EM = EQ$ and $EN = EP$, and so

$$MN = EM + EN = EQ + EP = 2EP + PQ,$$

and similarly

$$UV = UD + DV = DQ + DP = 2DQ + PQ.$$

Since $MN = UV$, then $2EP = 2DQ$ and $EP = QD$.

(iii) It is well known that the length of the tangent from vertex C of $\triangle ABC$ to the incircle is $CT = s - c$ (figure 82). Therefore, in triangles

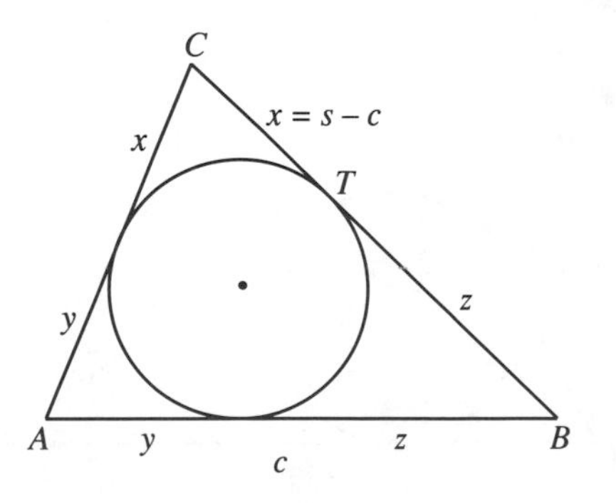

FIGURE 82

CAD and CDB (figure 80), we have

$$CQ = d + EQ = s_1 - AD$$

and

$$CP = d + EP = s_2 - DB.$$

Adding we get

$$2d + EQ + EP = s_1 + s_2 - AB,$$

and since $EP = QD$, we have $EQ + EP = EQ + QD = ED$, and therefore

$$2d + EQ + EP = d + (d + ED) = s_1 + s_2 - c,$$

i.e.,

$$d + CD = s_1 + s_2 - c.$$

Now,

$$s_1 = \frac{1}{2}(b + AD + CD) \quad \text{and} \quad s_2 = \frac{1}{2}(a + DB + CD).$$

Hence

$$s_1 + s_2 - c = \frac{1}{2}(a + b + AB + 2CD) - c = \frac{1}{2}(a + b + c) + CD - c,$$

and then

$$d + CD = s_1 + s_2 - c = \frac{1}{2}(a + b + c) + CD - c = s + CD - c,$$

giving

$$d = s - c, \quad \text{a constant.}$$

Thus, as conjectured, E lies on the circle with center C and radius $s - c$.

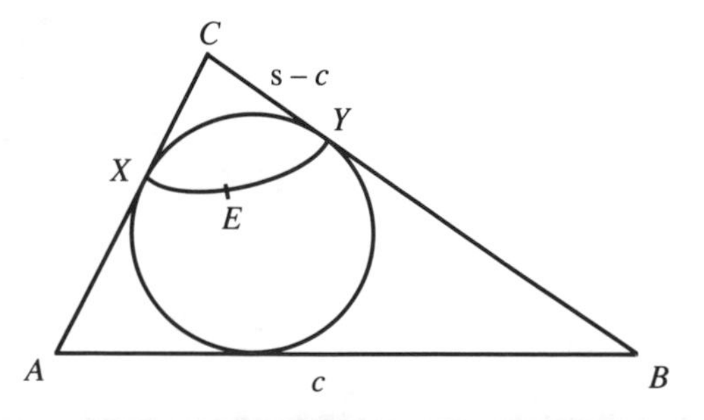

FIGURE 83

3. 15th Russian All-Union Grade Ten Olympiad. In $\triangle ABC$, $\angle C$ is a right angle and D and E are arbitrary points on sides AC and BC respectively. Perpendiculars are dropped from C to DE, AE, AB, and BD. Prove that the four feet of these perpendiculars always lie on a circle (figure 84).

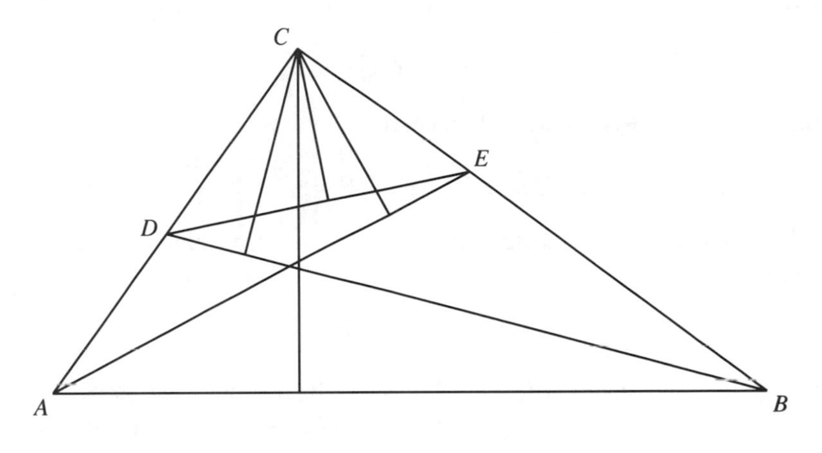

FIGURE 84

Let the feet be P, Q, R, and S (figure 85). We shall show that the angles of quadrilateral $PQRS$ at P and R are supplementary.

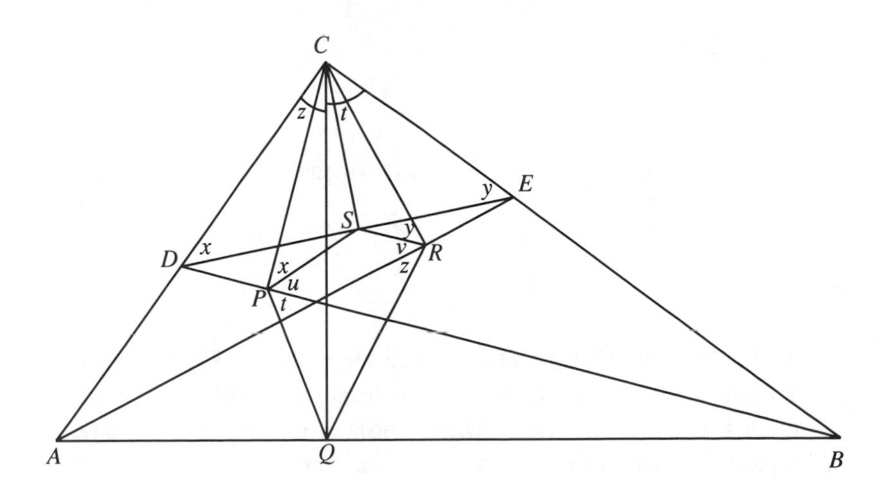

FIGURE 85

Let the angles at D and E be x and y as marked and let $\angle ACQ = z$ and $\angle BCQ = t$. Then, since $\angle C$ is a right angle, we have in $\triangle CDE$ that

$$z + t = 90^\circ = x + y.$$

Now, since CP and CS are perpendiculars to DB and DE, CD subtends a right angle at P and at S, making $CDPS$ cyclic, and on chord CS,

$$\angle CPS = \angle CDS = x.$$

Similarly, $CERS$ is cyclic, and on chord CS

$$\angle CRS = \angle CES = y.$$

Again, BC subtends right angles at P and Q, making $CPQB$ cyclic and on chord QB

$$\angle QPB = \angle QCB = t;$$

finally, the right angles at Q and R make $CAQR$ cyclic, and on chord AQ we have

$$\angle ARQ = \angle ACQ = z.$$

Now, if $\angle SPB$ and $\angle SRA$ are called u and v respectively, the right angles at P and R give

$$x + u + y + v = 180^\circ,$$

and since $x + y = 90^\circ$, then

$$u + v = 90^\circ.$$

Hence in $PQRS$, the four angles at P and R add up to

$$(u + v) + (z + t) = 90^\circ + 90^\circ = 180^\circ,$$

as claimed.

4. 15th Russian All-Union Grade Ten Olympiad. AB, CD, and EF are chords of a circle that are concurrent at K and are inclined to each other at 60° angles (figure 86). Prove that the alternate triples of segments from K have the same sum:

$$KA + KD + KE = KB + KC + KF.$$

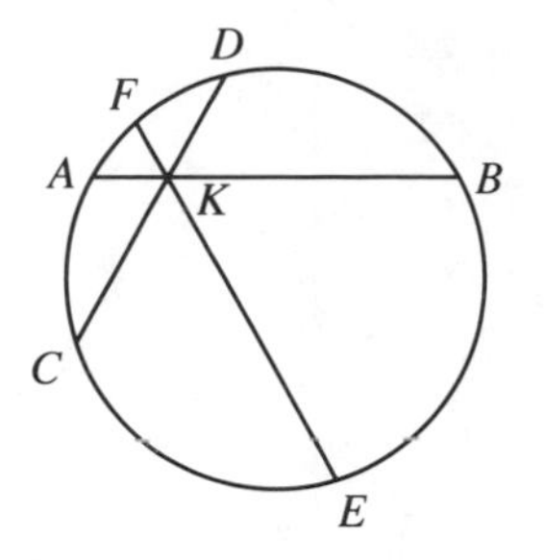

FIGURE 86

Let O be the center of the circle and let the segments be labeled a, b, c, d, e, and f (figure 87).

Since the perpendicular from the center of a circle meets a chord at its midpoint, the circle on diameter KO meets the given chords at their midpoints L, M, and N. Because the chords meet at $60°$ angles, triangle LMN is equilateral:

in $\triangle LMN$, $\angle L = \angle MKN = 60°$, and $\angle M = \angle NKL = 60°$.
Thus $LM = MN = LN$.

Applying the law of cosines to $\triangle LMK$ we get

$$LM^2 = LK^2 + MK^2 - 2LK \cdot MK \cos 120°.$$

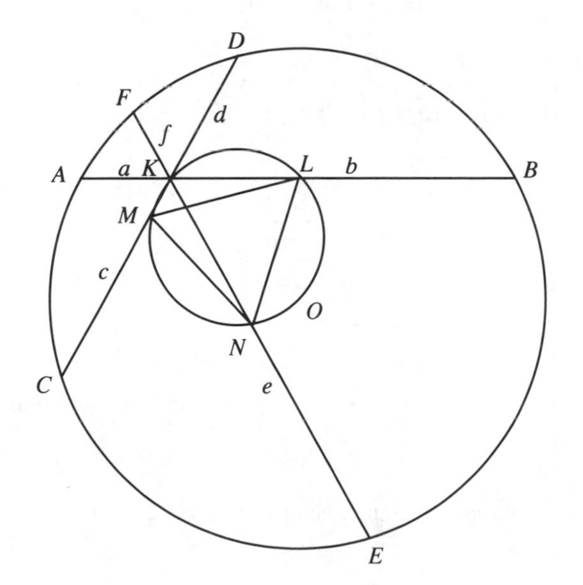

FIGURE 87

Now,

$$LK = AL - AK = \frac{a+b}{2} - a = \frac{b-a}{2},$$

and

$$MK = MD - KD = \frac{c+d}{2} - d = \frac{c-d}{2}.$$

Hence

$$LM^2 = \frac{1}{4}(b-a)^2 + \frac{1}{4}(c-d)^2 - 2 \cdot \frac{b-a}{2} \cdot \frac{c-d}{2}\left(-\frac{1}{2}\right),$$

giving

$$4LM^2 = (b-a)^2 + (c-d)^2 + (b-a)(c-d).$$

Similarly, from $\triangle MKN$, we have

$$NK = FN - FK = \frac{e+f}{2} - f = \frac{e-f}{2},$$

and

$$\begin{aligned} MN^2 &= MK^2 + NK^2 - 2MK \cdot NK \cos 60^\circ \\ &= \frac{1}{4}(c-d)^2 + \frac{1}{4}(e-f)^2 - 2 \cdot \frac{c-d}{2} \cdot \frac{e-f}{2}\left(\frac{1}{2}\right) \\ 4MN^2 &= (c-d)^2 + (e-f)^2 - (c-d)(e-f). \end{aligned}$$

Since $LM = MN$, then, we have

$$(b-a)^2 + (c-d)^2 + (b-a)(c-d) = (c-d)^2 + (e-f)^2 - (c-d)(e-f).$$

$$(b-a)[b-a+c-d] = (e-f)[e-f-c+d],$$

from which

$$\frac{b-a}{e-f} = \frac{e-f-c+d}{b-a+c-d}.$$

Adding 1 to each side gives

$$\frac{b-a+e-f}{e-f} = \frac{e-f+b-a}{b-a+c-d},$$

in which the numerators are the same. Hence, from the denominators we get

$$e-f = b-a+c-d, \quad \text{and} \quad a+d+e = b+c+f,$$

as desired. (An alternative solution is given in 1992, 234.)

Five Problems from the 1980 All-Union Russian Olympiad†

1. A payload, packed into containers, is to be delivered to the orbiting satellite "Salyut". There are at least 35 containers and the payload weighs a total of 18 tons. Seven "Progress" cargo ships are available, each of which can deliver a 3-ton load.

It seems remarkable that the ability to hold *any 35* of the containers should somehow endow the fleet with the capacity to hold another 10 or 20 containers. Prove, however, that this is so, that in fact this ability would enable the fleet to deliver the entire payload in one trip no matter how many containers there might be.

Suppose the number of containers is $35 + n$; then n is at least zero and the claim is certainly valid for $n = 0$. Proceeding by induction, suppose $n \geq 1$ and the fleet has the capacity to deliver the entire load no matter how it might be packed into $34 + n$ containers.

Clearly, not all containers can weigh more than average and so

$$\begin{aligned}\text{the weight of the lightest container } L &\leq \text{ average}\\ &= \frac{18}{35+n} \leq \frac{1}{2}, \qquad \text{since } n \geq 1.\end{aligned}$$

Now, while no container can weigh more than the 3-ton capacity of a cargo ship, with a payload of only 18 tons, fewer than 8 of the containers could even weigh as much as $2\frac{1}{2}$ tons. Therefore it is always possible to combine into a single package the lightest container L and any container C whose weight does not exceed $2\frac{1}{2}$ tons, to yield a packing of the payload into $34 + n$ containers, none of which exceeds the 3-ton limit. The entire payload can then be packed into the ships as prescribed by the induction hypothesis, at the

† (*Crux Mathematicorum,* 1990, 33 and 70)

end of which L and C can be separated and each repackaged to demonstrate how to load the $35 + n$ original containers, and the conclusion follows by induction. (An alternative solution is given in 1991, 267.)

2. E is a point on a diameter AC of a circle. Determine the chord BD through E which yields the quadrilateral $ABCD$ of greatest area.

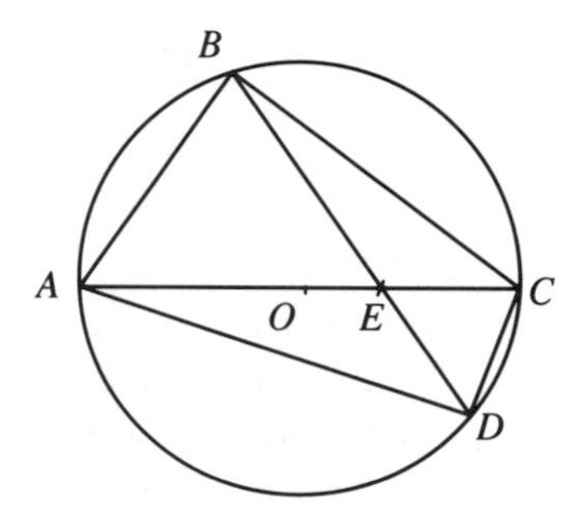

FIGURE 88

For definiteness, suppose E occurs on radius OC rather than OA. Clearly, in figure 89,

$$\begin{aligned} ABCD &= \triangle ABC + \triangle ADC \\ &= \tfrac{1}{2}AC \cdot BX + \tfrac{1}{2}AC \cdot DY \\ &= \tfrac{1}{2}AC(BX + DY), \end{aligned}$$

and the area of $ABCD$ is greatest when the sum $S = BX + DY$ is a maximum.

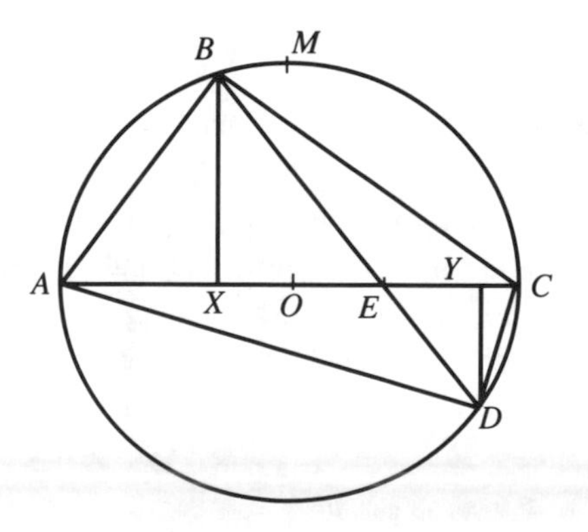

FIGURE 89

Now, as B traces the "upper" semicircular arc from A to C, the variable chord BED passes through all its positions. As B goes from A to M, the midpoint of the arc, both BX and DY steadily increase, implying that their maximum sum over the entire arc must occur at some position of B in MC, the second half of the arc, perhaps at M itself. However, as B gets past M, BX starts to decrease; but in compensation DY continues to increase. Thus their sum will continue to increase only if the increase in DY is enough to offset the decrease in BX. As we shall see, this is indeed the case until BED becomes the perpendicular PEQ to AC, after which things reverse.

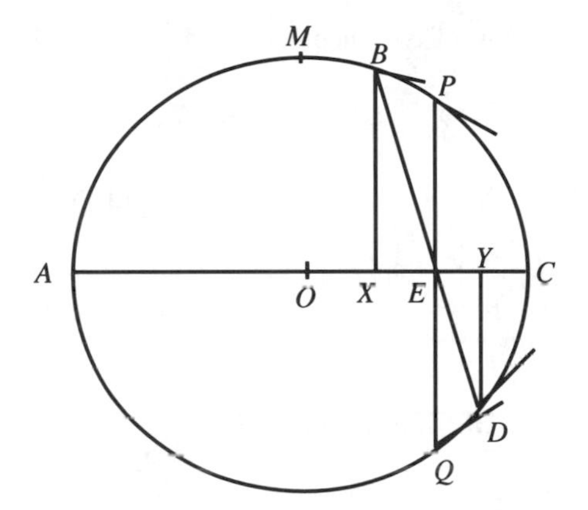

FIGURE 90

If AC is placed along the x-axis of a coordinate plane with the center O at the origin, the magnitudes of the rates of change of BX and DY are given by the magnitudes of the slopes of the tangents at B and D, which we denote by $|T_B|$ and $|T_D|$ (figure 90). Because a diameter is an axis of symmetry of a circle, and E lies on the x-axis, $|T_B|$ and $|T_D|$ have the same magnitude m when B is at P and D at Q. But, as B goes from M to P, it is evident that $|T_B|$ increases to m while $|T_D|$ decreases to it. Therefore, throughout this stage, $|T_D|$ is always greater than $|T_B|$, implying that DY is lengthening faster than BX is shortening, with a consequent increase in S until B reaches P.

As B goes beyond P, this situation obviously reverses and BX decreases more quickly than DY increases, giving a corresponding decrease in S. Hence S, and the area of $ABCD$, attains its maximum when BD is perpendicular to AC. (An alternative solution is given in 1991, 291.)

3. The set M consists of integers, the smallest of which is 1 and the greatest 100. Each member of M, except 1, is the sum of two (possibly identical) numbers in M. Of all such sets, find one with the smallest possible number of elements.

Let the members of M be $a_1 = 1 < a_2 < a_3 < \cdots < a_n = 100$. It is easy to see that $a_k \leq 2^{k-1}$. Clearly

$$a_1 = 1 \leq 2^0;$$

now, since a_k is the sum of two members of M, it can't be bigger than

$$a_{k-1} + a_{k-1} = 2a_{k-1};$$

thus, if $a_{k-1} \leq 2^{k-2}$, then $a_k \leq 2^{k-1}$, and the result follows by induction.

Hence $a_7 \leq 64 < 100$, implying that M must have at least 8 members. However, we can see that 8 members are not enough as follows:

(a) since $a_k \leq 2^{k-1}$,

the sum of two *different* members in $\{a_1, a_2, \ldots, a_7\}$

$$\leq a_6 + a_7 \leq 2^{6-1} + 2^{7-1} = 32 + 64 = 96 < 100.$$

Hence if a_8 were to be the greatest member, 100, it could not be the sum of two *different* members of M and would therefore have to be $50+50$, implying some member of M is 50. Since $a_i \leq 32$ for $i \leq 6$, it could only be $a_7 = 50$.

(b) In this case, it similarly follows, since $a_5 + a_6 \leq 16 + 32 = 48 < 50$, that $a_7 = 50 = 25 + 25$ and, since $a_i \leq 16$ for $i \leq 5$, only a_6 could be 25.

(c) But 25 is *odd,* and since $a_4 + a_5 \leq 8 + 16 = 24$ is too small to give a_6, we have that a_6 can only be $2a_5$, in view of $2a_4 \leq 2(8) = 16$. This requires $a_5 = 12\frac{1}{2}$, contradicting its integral character.

Thus the 100 can't be a_8 and we conclude that M must have at least nine members.

Once this is established, it is easy to construct an acceptable set; for example,

$$\{1, 2, 3, 6, 12, 24, 25, 50, 100\}, \quad \text{or} \quad \{1, 2, 3, 6, 12, 13, 25, 50, 100\}.$$

(Essentially the same solution is given in 1991, 297.)

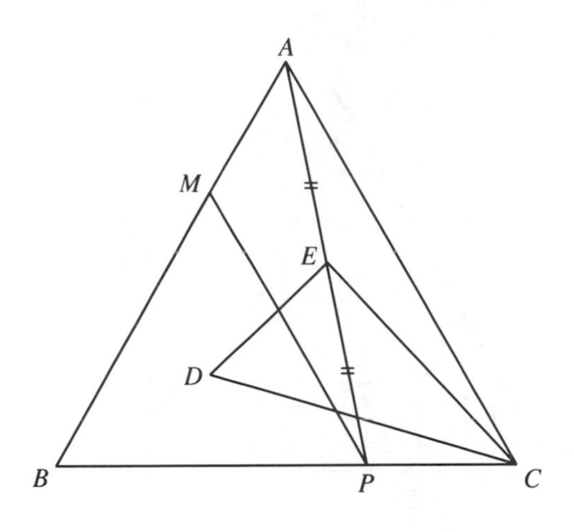

FIGURE 91

4. A line parallel to the side AC of equilateral triangle ABC intersects AB at M and BC at P, thus making $\triangle BMP$ equilateral as well (figure 91). D is the center of $\triangle BMP$ (incenter, orthocenter, ...) and E is the midpoint of AP. Determine the angles of $\triangle CDE$.

While the complete specification of D inside $\triangle BMP$ is well known, with E situated at the midpoint of AP, the triangle CDE is cast obliquely in $\triangle ABC$, and with medians not meeting the sides of a triangle at distinguished angles, it is doubtful that comparing the directions of the sides of $\triangle CDE$ with those of $\triangle ABC$ will uncover much useful information. Of course, there is always the chance that there exists an elusive construction line which will make everything clear. Finding Euclidean geometry so attractive, I am always reluctant to throw in the towel on the synthetic approach; however, after several looks at the problem from this point of view, I turned to the analytic approach.

It seems pretty obvious that *guessing* is seated in the very heart of mathematical discovery. With so many 30-60-90–degree triangles associated with an equilateral triangle, one might reasonably expect such a triangle to be a candidate in the present case; and with the visual support of even a rough figure, one is certainly justified in checking this possibility as a first step in his investigations.

In figure 92, let P be $(2a, 0)$ and C be $(2b, 0)$. The coordinates of A, M and D are found from the fact that the altitude of an equilateral triangle is $\frac{\sqrt{3}}{2}$ times the length of a side, and that the center trisects an altitude.

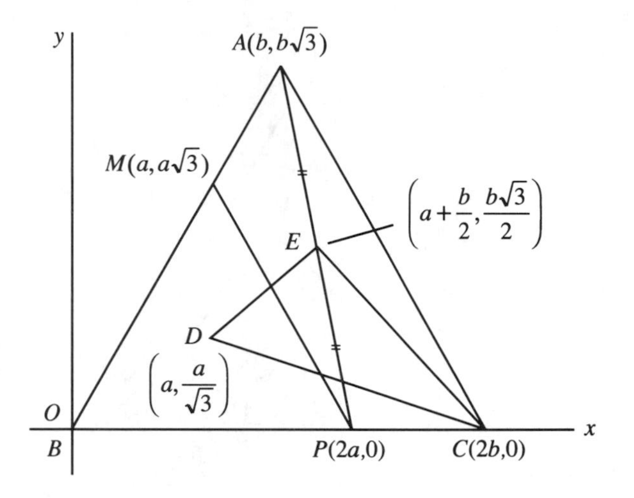

FIGURE 92

Now let's determine the slopes of DE and EC and the lengths of DE and DC. We have

$$\text{slope } DE = \frac{\frac{b\sqrt{3}}{2} - \frac{a}{\sqrt{3}}}{\frac{b}{2}} = \frac{3b-2a}{b\sqrt{3}};$$

$$\text{slope } EC = \frac{-\frac{b\sqrt{3}}{2}}{\frac{3b-2a}{2}} = -\frac{b\sqrt{3}}{3b-2a} = -\frac{1}{\text{slope } DE},$$

showing $\angle DEC$ is a right angle. Also,

$$DC^2 = (a-2b)^2 + \frac{a^2}{3} = \frac{4}{3}a^2 - 4ab + 4b^2,$$

while

$$DE^2 = \frac{b^2}{4} + \left(\frac{a}{\sqrt{3}} - \frac{b\sqrt{3}}{2}\right)^2 = \frac{b^2}{4} + \frac{a^2}{3} - ab + \frac{3b^2}{4}$$
$$= \frac{a^2}{3} - ab + b^2 = \frac{1}{4}DC^2.$$

Thus $DC = 2 \cdot DE$, making $\angle CDE = 60°$ and CDE a 30°-60°-90° triangle, as conjectured.

5. Now for the highlight of this olympiad.

Let $p(n)$ denote the product of the (decimal) digits of the positive integer n. Consider the sequences, beginning at an arbitrary positive integer, in which succeeding terms are obtained by adding to the previous term the product of its digits:

$$n_0 = n, \quad \text{and for} \quad r \geq 0, \; n_{r+1} = n_r + p(n_r).$$

For example, three such sequences proceed

- $11,\ 12,\ 14,\ 18,\ 26,\ 38,\ 62,\ 74,\ 102,\ 102, \ldots,$
- $37,\ 58,\ 98,\ 170,\ 170, \ldots,$
- $324,\ 348,\ 444,\ 508,\ 508, \ldots.$

Obviously, a digit equal to zero anywhere in n_r results in $p(n_r) = 0$, leading to $n_{r+1} = n_r$ and to n_r repeating indefinitely; otherwise, $p(n_r)$ is positive and the sequence increases. The question is "Does every such sequence contain a zero in some term and thus forever mark time at that value or is there an initial integer n for which the sequence continues to increase indefinitely?"

The thing that is so difficult to come to grips with, of course, is the *product* of the digits in an integer. I pondered this problem on and off for a couple of weeks and never did get it before my colleague Paul Schellenberg showed me how to do it.

Paul noticed that, as one might expect, the strings of consecutive 0-bearing integers at the beginning of the set of integers which contain a given number of digits gets longer as the number of digits increases; for example,

the 3-digit integers begin with the 10 integers $\{100, 101, \ldots, 109\}$,
the 4-digit integers begin with the 100 integers $\{1000, 1001, \ldots, 1099\}$.

In general, for $k \geq 3$, the k-digit integers begin with 10^{k-2} consecutive 0-bearing integers (actually there are more than this but we do not need greater accuracy). Clearly, if $p(n_r)$ were to increase n_r to any value in this initial segment of 0-bearing k-digit integers, the sequence would never get any bigger.

For small values of k, the values of $p(n_r)$ can vault n_r right over this initial quicksand. For example, $n_r = 88$ leads to $p(n_r) = 64$, carrying the sequence to 152, well beyond the interval $[100, 109]$; $n_r = 956$ goes to $n_{r+1} = 956 + 270 = 1226$, skipping the string $[1000, 1099]$. For these values of k, $p(n_r)$ is easily sufficient to extend many sequences into viable k-digit values. However, for a $(k-1)$-digit term n_r, $p(n_r)$ cannot exceed 9^{k-1}. Now, as we have noted, the initial segment of 0-bearing k-digit integers is of length at least 10^{k-2}, and if it should ever happen that this segment were too great for $p(n_r)$ to get over, that is, if ever $10^{k-2} > 9^{k-1}$, then all sequences with terms containing $k-1$ or fewer digits could never muster the increment necessary to reach any k-digit integer that didn't contain a zero—those that managed to survive to the k-digit level would simply be mired in its initial 0-bearing segment. Taking logarithms we see such a k would exist if

$$k - 2 > (k-1) \cdot \log 9,$$

$$k > \frac{2 - \log 9}{1 - \log 9} = 22.85\ldots,$$

i.e., if $k = 23, 24, 25, \ldots$.

Thus all hopes of an ever-increasing sequence are dashed. No sequence with a term containing 22 or fewer digits can get into the 23-digit integers beyond the net cast by the first 10^{21} of them, and sequences that begin at m-digit numbers, where $m \geq 23$, are similarly doomed to expire among the first 10^{m-1} $(m+1)$-digit integers if not sooner.

The Fundamental Theorem of 3-Bar Motion

Suppose a linkage is made of 15 rods connected at 10 swivel-joints as in figure 93, where three rods lie along each side of triangle ABC and the other six rods are each parallel to a side of $\triangle ABC$; this makes each of the little shaded triangles similar to $\triangle ABC$ and the unshaded quadrilaterals parallelograms. Clearly the linkage can be deformed into any number of configurations, such as the one illustrated in figure 94.

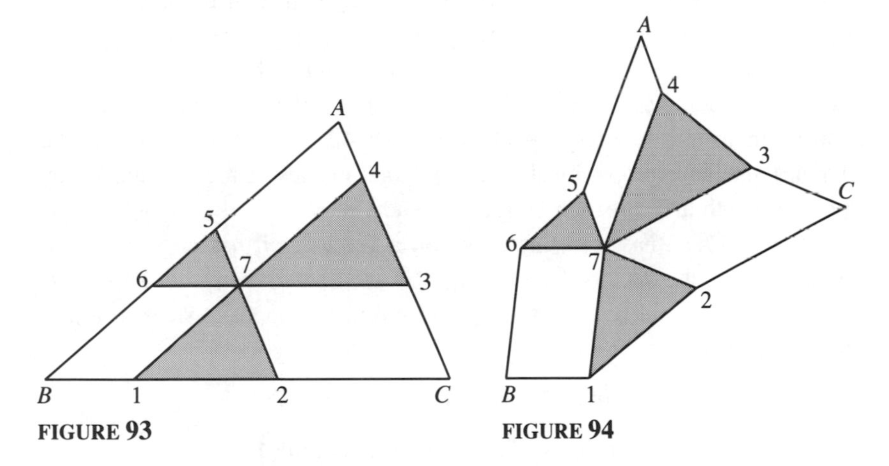

FIGURE 93

FIGURE 94

Sincc a triangle is determined by the lengths of its sides, a triangular part of the linkage, like 127, which is always bounded by the same three rods, cannot change in size or shape during a deformation, but must slide around as if it were a rigid plate. Also, since the quadrilaterals always have opposite sides equal, they remain parallelograms throughout all stages, even though their shapes vary widely. As we shall see, these characteristics lead to the basic property, the so-called fundamental theorem of 3-bar motion, due

to the nineteenth-century British mathematicians William Kingdom Clifford and Arthur Cayley:
no matter how the linkage may be deformed, ***the shape of*** $\triangle ABC$ ***never changes****—it remains similar to each of the little triangles like* 127.

Now, it is clear that if the linkage is stretched right out as in figure 93 and the points A and B are fastened down in place, the linkage would be unable to move at all; however, fixing A and B closer together would allow for continuous deformation over a limited range. From the invariant shape of $\triangle ABC$, then, we get the following surprising corollary:
if A and B are fixed in place so that the linkage can be deformed, everything else will move around except the point C, which must remain completely stationary lest $\triangle ABC$ change its shape.

Referring to figure 94, let $S(\theta, \mu)$ be the spiral similarity that transforms rod 56 into rod 57; that is to say, let θ be the angle 657 and μ the ratio (the length of 57):(the length of 56); then $S(\theta, \mu)$ is the compound transformation that first rotates rod 56 into the direction of 57 and then expands or shrinks it to make it the same length as 57. Now, a rotation $O(\theta)$ of the plane through an angle θ changes the direction of every straight line in the plane by the angle θ no matter what point O might be the center of the rotation. Thus, for any center S, $S(\theta, \mu)$ spins rods 47 and 71 through the angle θ, which is equal to the angles 743 and 172 in the triangles that are similar to $\triangle 567$, to bring them into line respectively with rods 43 and 72; and since similar triangles also have proportional sides, the final images of rods 47 and 71 are given in length and direction by rods 43 and 72.

Now, $S(\theta, \mu)$ has the same effect on a *vector* as it has on a rod. Thus under $S(\theta, \mu)$, the vectors $\mathbf{56}$, $\mathbf{47}$, and $\mathbf{71}$ are transformed into the vectors $\mathbf{57}$, $\mathbf{43}$, and $\mathbf{72}$, and because appropriately directed opposite sides of a parallelogram represent the same vector, we have that

$\mathbf{56}$ is transformed into $\mathbf{57}$, which is the same as $\mathbf{A4}$,

$\mathbf{A5}$, which is the same as $\mathbf{47}$, is transformed into $\mathbf{43}$,

$\mathbf{6B}$, which is the same as $\mathbf{71}$, is transformed into $\mathbf{72}$, which is $\mathbf{3C}$.

Since a sum of vectors is not altered by rearrangements in their order,

$$\begin{aligned}\text{the image of } \mathbf{AB} &= \text{ the image of } (\mathbf{A5} + \mathbf{56} + \mathbf{6B}) \\ &= \mathbf{43} + \mathbf{A4} + \mathbf{3C} \\ &= \mathbf{A4} + \mathbf{43} + \mathbf{3C} \\ &= \mathbf{AC}.\end{aligned}$$

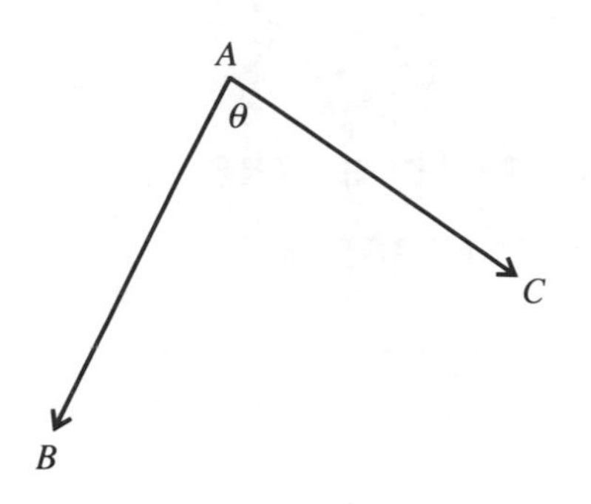

FIGURE 95

Accordingly, the angle BAC must be equal to θ and the ratio of $AC : AB$ must be μ (= ratio of the lengths of rods 57 and 56), implying that $\triangle ABC$ is *similar* to $\triangle 567$. But $\triangle 567$ changes only its position as the linkage is deformed, not its size or shape, and therefore $\triangle ABC$ must also keep the same shape throughout.

It is a pleasure to thank my good friend and colleague Mike McKiernan for bringing this gem to my attention.

Three Problems from the 1989 Austrian Olympiad†

1. (Beginner's Level) If a and b are nonnegative real numbers such that $a^2 + b^2 = 4$, show that

$$\frac{ab}{a+b+2} \le \sqrt{2} - 1.$$

(i) From

$$(a+b)^2 = a^2 + 2ab + b^2 = 4 + 2ab$$

we have

$$2ab = (a+b)^2 - 4 = (a+b+2)(a+b-2),$$

from which

$$\frac{ab}{a+b+2} = \frac{a+b-2}{2} = \frac{a+b}{2} - 1.$$

Thus the desired inequality is established upon showing that

$$\frac{a+b}{2} \le \sqrt{2}, \quad \text{i.e.,} \quad a+b \le 2\sqrt{2}.$$

(ii) Now, if both a and b exceed $\sqrt{2}$, we would immediately have the contradiction

$$a^2 + b^2 > 2 + 2 = 4;$$

similarly, if both a and b were to be less than $\sqrt{2}$, we would have

$$a^2 + b^2 < 2 + 2 = 4.$$

Therefore, either $a = b = \sqrt{2}$, in which case $a + b = 2\sqrt{2}$ as desired, or a and b *straddle* the value $\sqrt{2}$.

---

† (*Crux Mathematicorum,* 1990, 289)

In the latter case, suppose $a = \sqrt{2} + x$ and $b = \sqrt{2} - y$, where $x, y > 0$. Then

$$a^2 + b^2 = 2 + 2\sqrt{2}x + x^2 + 2 - 2\sqrt{2}y + y^2 = 4$$

giving

$$2\sqrt{2}(y - x) = x^2 + y^2 > 0,$$

implying $y > x$. Finally, then,

$$a + b = 2\sqrt{2} + x - y < 2\sqrt{2},$$

completing the solution. (A similar solution is given in 1992, 72.)

2. (Beginner's Level) If $a \le b \le c \le d$ are positive integers that add up to 30, what is the maximum value of *abcd*?

The arithmetic mean-geometric mean inequality gives

$$\sqrt[4]{abcd} \le \frac{a+b+c+d}{4},$$

implying

$$abcd \le \left(\frac{a+b+c+d}{4}\right)^4,$$

with equality if and only if $a = b = c = d$ $(= \frac{a+b+c+d}{4})$.

There isn't much doubt, then, that the maximum product occurs when a, b, c, and d are as equal as their integral characters and sum permit, i.e., for the values 7, 7, 8, 8.

For two numbers x and y, the A. M.-G. M. inequality gives

$$xy \le \left(\frac{x+y}{2}\right)^2,$$

with equality if and only if $x = y$ $(= \frac{x+y}{2})$.

Now, this can be applied to the pairs (a, b) and (c, d) as follows. Since $a \le b \le c \le d$, we have $a + b \le c + d$, and if $a + b$ were to exceed 15, then so would $c + d$, and we would have the contradiction $a + b + c + d > 30$. Hence $a + b \le 15$, and therefore, for some $t \le 0$, we have $a + b = 15 - t$; and since $a + b + c + d = 30$, then $c + d = 15 + t$.

From the inequality, then, we have that

$$ab \le \left(\frac{15-t}{2}\right)^2, \quad \text{with equality iff } a = b = \frac{15-t}{2}.$$

Similarly,

$$cd \le \left(\frac{15+t}{2}\right)^2, \quad \text{with equality iff } c = d = \frac{15+t}{2}.$$

Therefore

$$abcd \le \left(\frac{15-t}{2}\right)^2 \left(\frac{15+t}{2}\right)^2 = \left(\frac{15^2-t^2}{4}\right)^2,$$

where equality is attainable if and only if $\frac{15-t}{2}$ and $\frac{15+t}{2}$ are feasible values for (a, b) and (c, d). Since a, b, c, d are integers, this would require t to be an odd integer.

Now, for $t \ge 1$, $\left(\frac{15^2-t^2}{4}\right)^2$ takes its greatest value when $t = 1$, namely

$$\left(\frac{15^2-1}{4}\right)^2 = \left(\frac{224}{4}\right)^2 = 56^2 = 3136,$$

and therefore $abcd$ is certainly never bigger than 3136. However, if $t = 1$ yields feasible values for a, b, c, d, this upper bound is actually attainable. Checking, we have

$$a = b = \frac{15-1}{2} = 7, \qquad c = d = \frac{15+1}{2} = 8,$$

which *are* acceptable values, and we conclude that

$$\max abcd = 7 \cdot 7 \cdot 8 \cdot 8 = 56^2 = 3136,$$

as conjectured. (An alternative solution is given in 1992, 72, with a comment in 1992, 268.)

3. (Advanced Level) The tangents to a circle K from an external point P meet K at A and B. Determine the position of C on the minor arc AB such that the tangent at C cuts from the figure a triangle PQR of *maximum area* (figure 96).

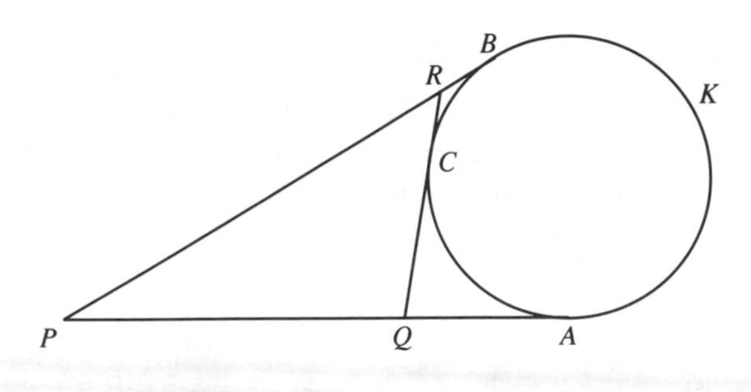

FIGURE 96

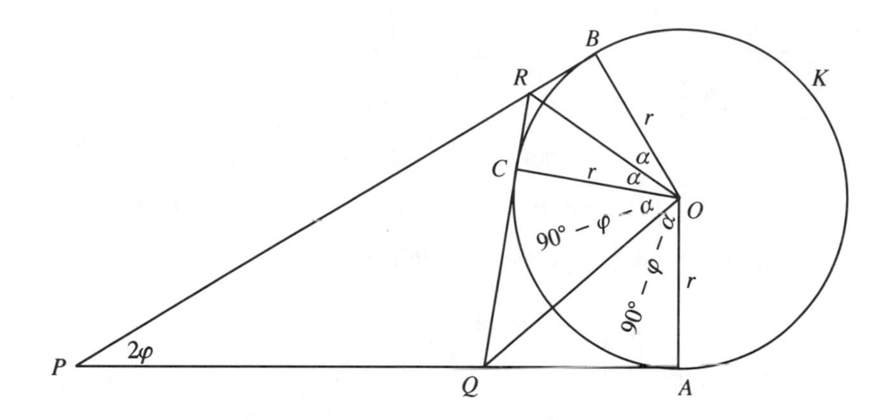

FIGURE 97

Let O be the center of K and r its radius, $\angle P = 2\phi$, and $\angle BOR = \alpha$ (figure 97). Now, the right angles at A and B imply $\angle P + \angle BOA = 180°$, giving

$$\angle BOA = 180° - 2\phi,$$

and because of the equal tangents from R and Q we have

$$\angle ROC = \angle ROB = \alpha$$

and

$$\angle COQ = \angle QOA = \frac{1}{2}\angle COA = \frac{1}{2}\left[(180° - 2\phi) - 2\alpha\right] = 90° - \phi - \alpha.$$

Therefore

$$\triangle COR \equiv \triangle ROB, \quad \text{and} \quad \triangle\, COQ \equiv \triangle QOA,$$

implying

$$\text{pentagon } AOBRQ = 2\,\triangle\, QOR = 2\left(\frac{1}{2}QR \cdot r\right) = QR \cdot r.$$

Since

$$\triangle PQR = PAOB - AOBRQ = PAOB - QR \cdot r,$$

where $PAOB$ and r are constants, $\triangle PQR$ is a maximum when QR is a *minimum.*

Now,

$$\begin{aligned} QR &= RC + CQ \\ &= r\tan\alpha + r\tan(90^\circ - \phi - \alpha) \\ &= r\big[\tan\alpha + \tan(90^\circ - \phi - \alpha)\big], \end{aligned}$$

and therefore QR takes its minimum with the function

$$f(\alpha) = \tan\alpha + \tan(90^\circ - \phi - \alpha).$$

Differentiating gives

$$\begin{aligned} f'(\alpha) &= \sec^2\alpha + \big[\sec^2(90^\circ - \phi - \alpha)\big](-1) \\ &= \sec^2\alpha - \sec^2(90^\circ - \phi - \alpha), \end{aligned}$$

and hence $f'(\alpha) = 0$ for

$$\sec\alpha = \pm\sec(90^\circ - \phi - \alpha),$$

that is, for

(i)
$$\begin{aligned} \alpha &= 90^\circ - \phi - \alpha, \\ 2\alpha &= 90^\circ - \phi = \frac{1}{2}(180^\circ - 2\phi) = \frac{1}{2}\angle BOA, \end{aligned}$$

i.e., for $\alpha = \frac{1}{4}\angle BOA$, which places C at the midpoint of the arc AB, and for

(ii)
$$\begin{aligned} \alpha + (90^\circ - \phi - \alpha) &= 180^\circ, \\ -\phi &= 90^\circ, \end{aligned}$$

making $\angle P = 2\phi = -180^\circ$, which requires P to be a point on the circle itself, a contradiction.

Thus $f(\alpha)$ takes an extreme value when C is the midpoint of the arc AB, and we need to check the second derivative $f''(\alpha)$ to confirm it is a minimum. Since

$$f'(\alpha) = \sec^2\alpha - \big[\sec^2(90^\circ - \phi - \alpha)\big],$$

then

$$\begin{aligned} f''(\alpha) &= 2\sec\alpha(\sec\alpha\tan\alpha) \\ &\quad - 2\sec(90^\circ - \phi - \alpha)\big[\sec(90^\circ - \phi - \alpha)\tan(90^\circ - \phi - \alpha)\big](-1) \\ &= 2\tan\alpha\sec^2\alpha + 2\tan(90^\circ - \phi - \alpha)\big[\sec^2(90^\circ - \phi - \alpha)\big], \end{aligned}$$

which is positive for the positive acute angle $\alpha = \frac{1}{4}\angle BOA = \frac{1}{2}(90^\circ - \phi)$, since, as seen above, this value of α also makes the angle

$$(90^\circ - \phi - \alpha) = \frac{1}{2}(90^\circ - \phi) = \alpha.$$

Thus, as expected, the unique maximum area of $\triangle PQR$ occurs when C bisects arc AB.

Three Problems from the Tournament of the Towns Competitions†

1. (1980) Consider an $n \times n$ array of numbers in which no two rows are the same, that is, completely identical: for any pair of rows, either there is a number in the one row that is not in the other, or, in the case of rows composed from the same collection of numbers, they are arranged in different orders so that, when compared component by component along the rows, there is at least one column in which the rows do not have the same entry. For example,

$$\begin{array}{ccc} 1 & 2 & 3 \\ 1 & 1 & 2 \\ 1 & 2 & 1. \end{array}$$

Prove that in any such array there is always some column whose deletion leaves an $n \times (n-1)$ array whose rows are still all different.

The following beautiful solution is due to the brilliant Bulgarian mathematician Jordan Tabov.

Proceeding indirectly, suppose no column can be deleted without leaving two rows that are identical. In this case, let us delete the columns in turn and note a resulting pair of identical rows for each column; if a deleted column yields more than one pair of identical rows, select any one of them.

Next, construct a graph G as follows (figure 98): on a set of n vertices, $r_1, r_2, \ldots, r_n$, representing the rows, let an edge c_k join r_i and r_j if and only if rows i and j are the selected identical pair that result when column k is

---

† (*Tournament of the Towns 1980–1984,* Peter Taylor (ed), Australian Mathematics Trust, 1993)

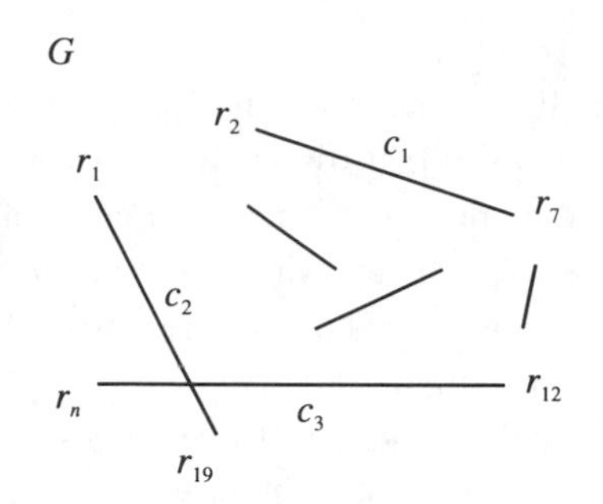

FIGURE 98

deleted; in figure 98, the deletion of column 1 makes rows 2 and 7 identical, and when column 2 is deleted, rows 1 and 19 become identical,... .

Now, as the columns are deleted in turn and G is constructed, each column must give rise to a *new* edge; that is, no existing edge $c_k = (r_i, r_j)$ can also serve as the edge c_t for a different column t. If c_k were to serve in both capacities, there would be no escaping the contradiction that rows i and j must be completely identical: $c_k = (r_i, r_j)$ implies rows i and j are *identical in all places except column k,* and for $k \neq t$, $c_t = (r_i, r_j)$, in implying rows i and j to be the same in all places except column t, would make them *the same in column k, too,* revealing the contradiction that they are the same everywhere.

As a result, our graph G contains n edges, not fewer. Now, it is a well known result in graph theory that it is impossible to apply n edges to n vertices without creating a cycle. Suppose, then, that G contains a cycle $C = (r_a, r_b, r_c, \ldots, r_{m-1}, r_m)$. Let its edges, in order, be $c_i, c_j, \ldots, c_k, c_t$ (figure 99). Since each column has its own edge, it follows that column t is a

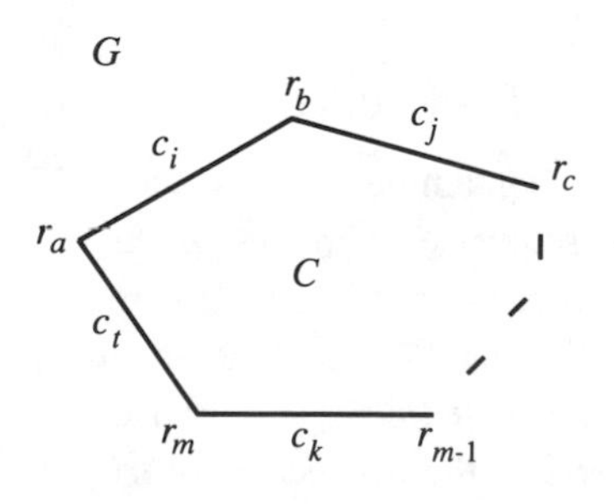

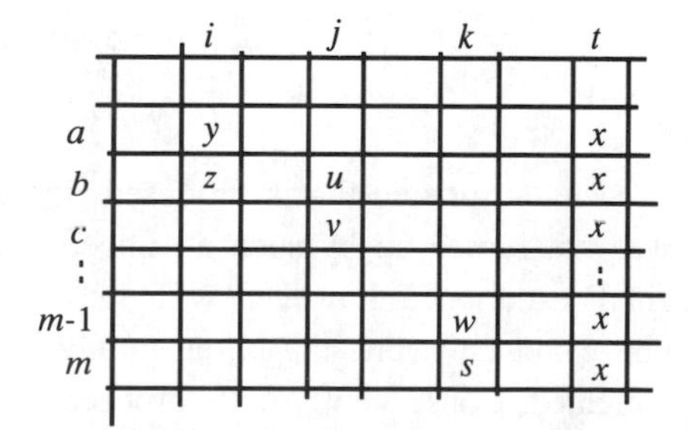

rows a and m are the same in column t

FIGURE 99

different column from all the other columns $i, j, \ldots, k$ whose edges are in C. The edge $c_i = (r_a, r_b)$ implies that rows a and b are identical in all columns except i and therefore they must have the same entry in column t. Similarly, the edge $c_j = (r_b, r_c)$ implies that rows b and c differ only in column j and have the same entry in column t. Thus all three of the rows a, b, and c have the same entry in column t (figure 99).

Continuing to apply this argument to the edges around C to the vertex r_m, we arrive at the conclusion that all the rows $a, b, c, \ldots, m$ have the same entry in column t; in particular, *rows a and m are the same in column t.* But the edge $c_t = (r_m, r_a)$ implies that *rows a and m are the same everywhere else, too,* and we obtain the contradiction that rows a and m are absolutely identical.

Our initial assumption, then, must be false, and we conclude that there is indeed some column whose deletion does not result in any pair of identical rows.

2. On the island of Camelot there are 45 chameleons. At one time 13 of them were grey, 15 brown, and 17 crimson. However, whenever two chameleons of different colors meet, they both change color to the third color. Thus, for example, if a grey and a brown chameleon were to be the first to meet, the count would change to 12 grey, 14 brown, and 19 crimson.

Is it possible to arrange a succession of meetings that would result in all the chameleons displaying the same color?

Keeping track of the count with the ordered triple (g, b, c) for grey, brown, and crimson, it is easy to see that, beginning with $(13, 15, 17)$, repeated pairings of the grey and brown chameleons yields the following progression of states:

$$(13, 15, 17) \to (12, 14, 19) \to (11, 13, 21) \to \cdots$$

$$\to (0, 2, 43) \to (2, 1, 42) \to (1, 0, 44).$$

It is not surprising that a proposed final triple containing two 0's and a 45 was not turned up by such an unconsidered sequence of moves. Even this small experience, however, with its apparent dead-end of $(1, 0, 44)$, makes one wonder whether the initial numbers 13, 15, and 17 aren't incompatibly weighted. Consequently, let's proceed by trying to settle the question in the negative, that is, by trying to find some property of the final states $(45, 0, 0)$, $(0, 45, 0)$, and $(0, 0, 45)$ that cannot be realized from the initial position of $(13, 15, 17)$.

There are only three ways a meeting of two chameleons can change the count:

$$(g, b, c) \to \begin{cases} (g-1, b-1, c+2) \\ (g-1, b+2, c-1) \\ (g+2, b-1, c-1). \end{cases}$$

For any two colors, then, their difference in number *before* the change either remains the same *after* the change or it is altered by 3; for example, before the meeting the difference between the numbers of grey and brown chameleons is $g - b$; after the meeting this difference is either

$$(g-1) - (b-1) = g - b,$$

$$(g-1) - (b+2) = (g-b) - 3, \quad \text{or}$$

$$(g+2) - (b-1) = (g-b) + 3.$$

Thus we have the key result that, for the numbers of any two colors,

the difference before the change $\equiv$ the difference after (mod 3).

At the beginning these differences are 2, 2, and 4, and therefore they must remain congruent to 2, 2, and 4 (mod 3) thereafter. Thus the difference is never congruent to 0, implying that no two colors could ever claim the *same* number of chameleons (an actual difference of 0 is $\equiv 0$). Hence none of $(45, 0, 0)$, $(0, 45, 0)$, and $(0, 0, 45)$ is attainable and the conclusion follows.

3. The altitude CE of acute-angled triangle ABC crosses the circle on diameter AB at M and N and the altitude BD crosses the circle on diameter AC at P and Q. Prove that M, N, P, and Q are concyclic.

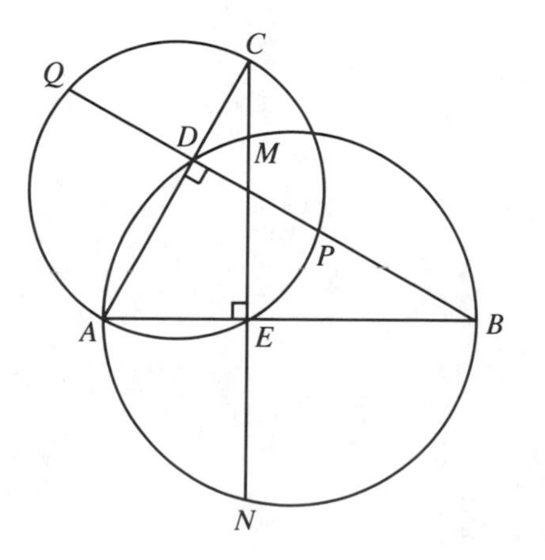

FIGURE 100

Since AB and AC are diameters, these circles actually pass through the feet D and E of the altitudes (figure 100). Now, quadrilaterals are often shown to be cyclic by proving a pair of opposite angles are supplementary. After failing in my attempts to do this, I turned to the following more basic approach.

A chord of a circle that is perpendicular to a diameter is bisected by that diameter; thus AB and AC are in fact the perpendicular bisectors of the segments MN and PQ respectively. Consequently, any circle having MN as a chord would have to have its center somewhere on the line AB, and similarly, a circle through P and Q must have its center on the line AC. Therefore, if M, N, P, and Q *are* concyclic, the center of the circle through them could only be the point A. Since the perpendicular bisectors immediately give $AM = AN$ and $AP = AQ$, it remains only to show that $AM = AP$. But this is easy.

From right-triangle AEC, we get $AE = b \cdot \cos A$. Now, diameter AB subtends a right angle at M, making ME the altitude to the hypotenuse in right-triangle AMB (figure 101). By a standard mean proportion, then, we have

$$AM^2 = AB \cdot AE = c(b \cdot \cos A).$$

Similarly, $AD = c \cdot \cos A$, and PD is the altitude to the hypotenuse in right-triangle APC. Hence

$$AP^2 = AC \cdot AD = b(c \cdot \cos A) = AM^2,$$

and the conclusion follows.

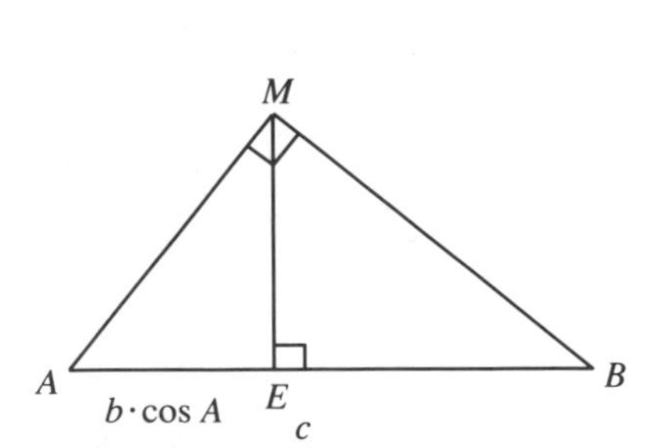

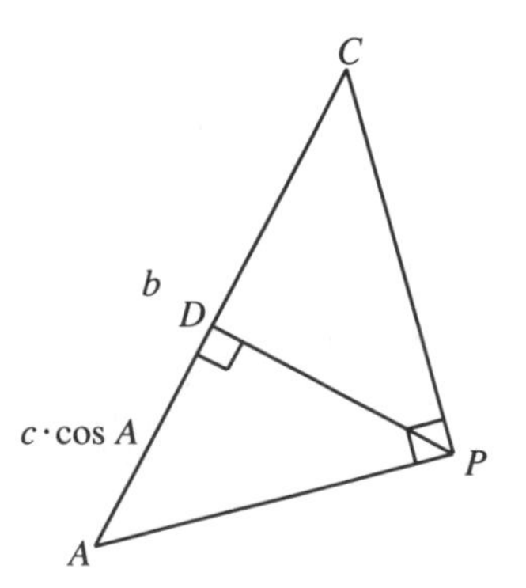

FIGURE 101

Problem 1506 from Crux Mathematicorum†

Through a fixed point P on a given circle a variable chord PQ crosses a fixed chord AB at R. Determine the locus of the circumcenter C of triangle AQR.

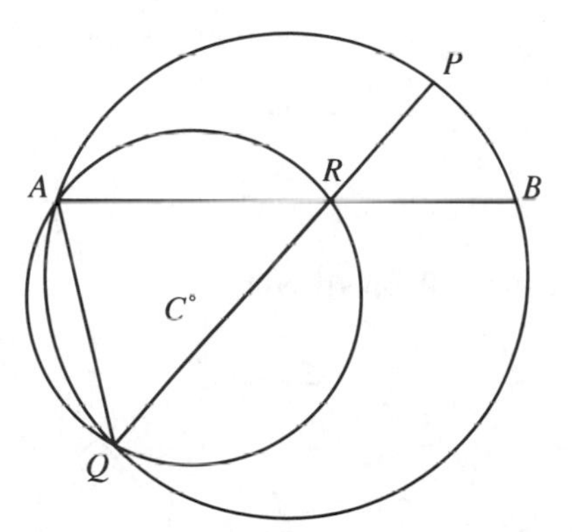

FIGURE 102

As PQ rotates about P, the angle x at Q is always just the angle in the segment of the given circle that is cut off by the chord AP (figure 103). That is to say, angle AQR is always x and therefore $\angle ACR$ at the center C is always $2x$. Since the perpendicular CK to AR bisects $\angle ACR$, $\angle ACK = x$, and the angle that AC makes with the fixed chord AB is a constant y. The locus of C, then, is simply the straight AL which is inclined to AB at the angle $y = 90° - x$.

---

† (1991, 87; proposed by Jordi Dou, Barcelona, Spain; solution by P. Penning, Delft, The Netherlands.)

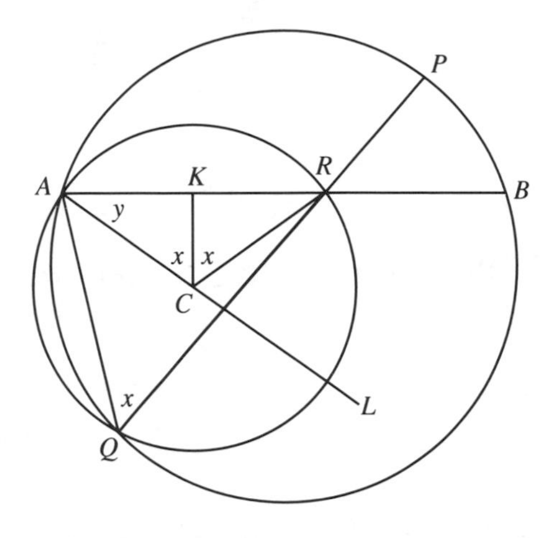

FIGURE 103

The locus is easily constructed as follows. With center B and radius AP, mark T on the circle (figure 104). This makes $BT = AP$ and consequently $\angle TAB$ equals x. Since

$$\angle TAL = x + y = x + (90^\circ - x) = 90^\circ,$$

AL is obtained by drawing the perpendicular to AT.

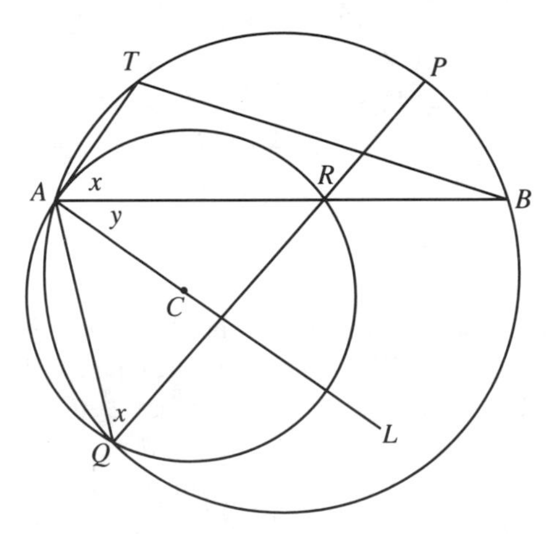

FIGURE 104

Three Unused Problems from The 1987 International Olympiad

1. Proposed by Rumania (*Crux Mathematicorum,* 1989, 41)

Show that it is possible to color the set of integers

$$M = \{1, 2, 3, \ldots, 1987\},$$

using four colors, so that no arithmetic progression with 10 terms has all its members the same color.

The following solution is due to George Evagelopoulos, Athens, Greece, and independently to Zun Shan and Edward Wang of Wilfrid Laurier University, Waterloo, Ontario.

Since each integer may be colored any of the four colors, the total number of ways of coloring the set M is 4^{1987}. If the N colorings which contain a monochromatic 10-term arithmetic progression does not claim all 4^{1987} of them, there must be at least one coloring of the desired kind. We would like to show, then, that

$$N < 4^{1987}.$$

Let A denote the number of 10-term arithmetic progressions that are contained in M. In order to guarantee that a coloring would contain a monochromatic 10-term arithmetic progression, one might proceed to

(a) choose one of the A 10-term progressions,

(b) pick one of the four colors for the selection in (a), and

(c) color the remaining 1977 integers arbitrarily.

Thus one might carelessly conclude that

$$N = A \cdot 4 \cdot 4^{1977} = A \cdot 4^{1978}.$$

However, N isn't nearly this big. This procedure gives rise to many duplications; for example, selecting $\{1, 2, \ldots, 10\}$ in step (a), coloring it red in (b), and in (c), coloring $\{11, 12, \ldots, 20\}$ blue and the rest green, to obtain

$$\underset{\text{red}}{\{1,2,\ldots,10\}}, \quad \underset{\text{blue}}{\{11,12,\ldots,20\}}, \quad \underset{\text{green}}{\{21,22,\ldots,1987\}},$$

yields the same coloring of M as the different prescription of selecting $\{11,12,\ldots,20\}$ in (a), coloring it blue in (b), and in (c), coloring $\{1,2,\ldots,10\}$ red and the rest green, to get

$$\underset{\text{blue}}{\{11,12,\ldots,20\}}, \quad \underset{\text{red}}{\{1,2,\ldots,10\}}, \quad \underset{\text{green}}{\{21,22,\ldots,1987\}}.$$

Therefore

$$N < A \cdot 4^{1978}.$$

In any case, if we were able to show that even the larger number $A \cdot 4^{1978}$ is less than 4^{1987}, or equivalently that $A < 4^9$, it would certainly follow that N is well short of 4^{1987}, establishing the desired conclusion. Let us proceed, then, to estimate A.

If $\{k, k+d, k+2d, \ldots, k+9d\}$ is a 10-term arithmetic progression in M, the final term must not exceed 1987, and therefore

$$k + 9d < 1987, \qquad \text{giving} \quad d \leq \frac{1987-k}{9}.$$

Now, $\frac{1987-k}{9}$ is not an integer for all k, and when it isn't, d is less than this fraction. In any case, there certainly cannot be more than $\frac{1987-k}{9}$ 10-term arithmetic progressions in M that begin at the integer k. Since k may be any integer from 1 to 1978 (starting later than 1978 would push the last term beyond 1987 even for d equal to its smallest possible value of 1), we have

$$\begin{aligned} A &\leq \sum_{k=1}^{1978} \frac{1987-k}{9} \\ &= \frac{1}{9}(1986 + 1985 + \cdots + 9) \\ &= \frac{1}{9}\left[\frac{1978}{2}(1986+9)\right] \\ &= \frac{1}{9} \cdot 989 \cdot 1995 \\ &< \frac{1}{8} \cdot 2^{10} \cdot 2^{11} \\ &= 2^{18} \\ &= 4^9, \end{aligned}$$

as desired.

2. (From an unspecified country: *Crux Mathematicorum,* 1987, 249)

P and Q are arbitrary points on the side BC of $\triangle ABC$, and through each of P and Q lines are drawn parallel to each of the sides AB and AC to give points P_1, P_2, Q_1, Q_2 (figure 105). Now suppose the segment PQ is slid along BC to any other position between B and C and the same construction is carried out. Prove that the sum of the areas of the two trapezoids PQQ_1P_1 and PQQ_2P_2 is always the same no matter where PQ might be on BC.

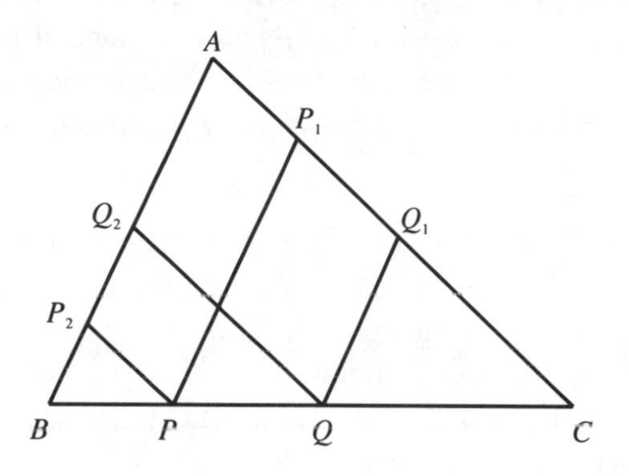

FIGURE 105

Let PP_1 and QQ_2 cross at R and let the regions have areas x, y, and z, as in figure 106. We wish to show that

$$(x+z)+(y+z)=x+y+2z$$

is a constant.

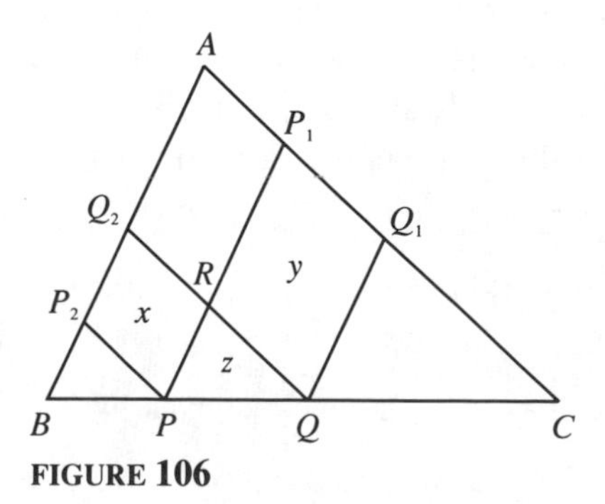

FIGURE 106

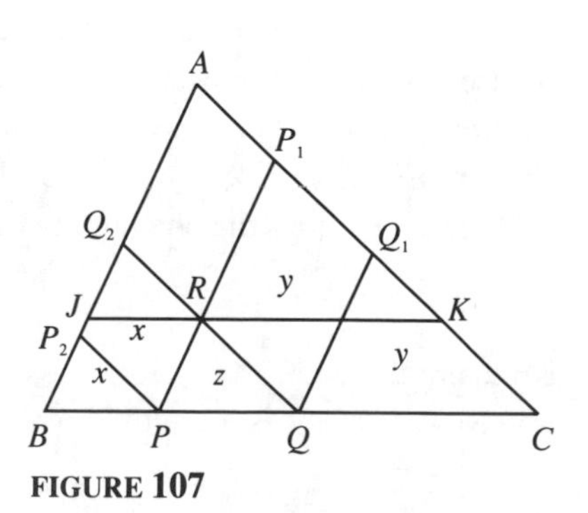

FIGURE 107

The parallel lines imply that the angles at P and Q in $\triangle PQR$ are always just the angles B and C of $\triangle ABC$, and since its base PQ is always the same length, $\triangle PQR$ never changes and simply slides along BC with PQ. Hence z is constant for all positions of PQ.

Now, parallelogram PRQ_2P_2 has base PR and lies between the parallels PR and AB. Since parallelograms on the same base and between the same parallels are equal in area, another parallelogram of area x would be $PRJB$ determined by a line JRK through R parallel to BC (figure 107). Similarly, $QRKC$ is another parallelogram of area y. Clearly, then, as PQ slides along BC, the sum of the areas x, y, and z is always given by the region $JKCB$, for the unchanging $\triangle PQR$ causes R to remain the same distance above BC and thus to run along JK. Hence $x+y+z$ is a constant, as is z itself, and therefore so is their sum $x+y+2z$. (An alternative solution is given in 1989, 69.)

3. Proposed by Yugoslavia (Crux Mathematicorum, 1989, 140)

If $r = 6+3\sqrt{3} = 11.196\ldots$, then the integer part of r is 11, which is one short of a multiple of 3. It is also true that the integer part of r^2 is one less than a multiple of 3: $[r^2] = [125.353\ldots] = 125$. Similarly, $[r^3] = [1403.479\ldots] = 1403$, is one less than a multiple of 3; in fact, the integer part of $(6+3\sqrt{3})^m$ is one less than a multiple of 3 for *all* positive integers m.

Prove that, for each positive integer $k = 2, 3, 4, \ldots$, there exists a corresponding positive *irrational* number r such that $[r^m]+1$ is a multiple of k for all positive integers m.

The following is another brilliant solution by George Evagelopoulos.

Let's begin by considering the case of $m = 1$. Since r is to be irrational, it must lie between two integers, and the number in question, $[r]+1$, would normally be determined by dropping the nonzero fractional part of r and adding 1; however, we could instead add to r the small amount s that is needed to raise it to the next integer: $[r]+1 = r+s$, where s is the appropriate number between 0 and 1. Thus the case of $m = 1$ is trivial: let s be any irrational number less than 1, like $\frac{1}{\sqrt{2}}$, and for the value of r, simply back off from any multiple kp of k by the amount s:

$$[r]+1 = r+s = (kp-s)+s = kp,$$

as required.

Now, if $r+s$ is a multiple of k, so is $(r+s)^m$, which is given by

$$(r+s)^m = r^m + mr^{m-1}s + \cdots + mrs^{m-1} + s^m,$$

that is,

$$(r+s)^m = r^m + s^m + (\text{terms divisible by } rs),$$

and we have for some integer t that

$$r^m + s^m = (r+s)^m - rst.$$

Consequently, *if* rs *were also to be a multiple of* k, it would follow that $r^m + s^m$ would be a multiple of k for *all* positive integers m. But, because s is to be less than 1, so would s^m always be less than 1, and hence $r^m + s^m$ would be the integer next above r^m and actually be the desired number $[r^m] + 1$. That is to say, for all $m = 1, 2, \ldots$,

$$[r^m] + 1 = r^m + s^m = \text{ a multiple of } k$$

for numbers r and s such that

(i) r is irrational and s is between 0 and 1,
(ii) $r + s = kp$ for some positive integer p, and
(iii) $rs = kq$ for some positive integer q.

We need to show that such r and s exist for each $k = 2, 3, \ldots$.

In view of requirements (ii) and (iii), r and s would be the roots of the equation

$$x^2 - kpx + kq = 0,$$

and therefore be given by

$$\frac{kp \pm \sqrt{k^2p^2 - 4kq}}{2.}$$

Since $kp > \sqrt{k^2p^2 - 4kq}$, the roots are both positive, and since the relation $r \leq s$ (< 1) would make $[r^m] + 1 = 1$ and not a multiple of k (since $k \geq 2$), it must be that $r > s$, and hence

$$r = \frac{kp + \sqrt{k^2p^2 - 4kq}}{2}, \qquad s = \frac{kp - \sqrt{k^2p^2 - 4kq}}{2}.$$

The existence of suitable r and s, then depends on the existence of positive integers p and q that satisfy condition (i), which requires that

(a) $k^2p^2 - 4kq$ be positive (to make $r > s$) but *not* a perfect square (to ensure the irrationality of r), and
(b) $0 < kp - \sqrt{k^2p^2 - 4kq} < 2$ (to place s between 0 and 1).

But it is easy to see that these requirements can always be met.

Condition (a) can be satisfied simply by taking $q = k$: in this case, $k^2p^2 - 4kq = k^2(p^2 - 4)$ which, to be a positive perfect square, would require $p^2 - 4$ to be a positive perfect square as well as p^2 itself, which is impossible. In order to satisfy (b), we need to choose p so that

$$kp - 2 < \sqrt{k^2p^2 - 4kq}$$

$$k^2p^2 - 4kp + 4 < k^2p^2 - 4kq$$

$$-kp + 1 < -kq$$

$$kp - 1 > kq$$

$$p > q + \frac{1}{k}, \quad \text{i.e., } p \geq q + 1.$$

Thus $q = k$ and any $p \geq k + 1$ provides a suitable r and s and the solution is complete.

In closing we observe that, for $k = 3$, the values

$$(p, q) = (k + 1, k) = (4, 3)$$

give r and s from the equation

$$x^2 - 12x + 9 = 0,$$

making

$$r = \frac{12 + \sqrt{144 - 36}}{2}$$
$$= 6 + 3\sqrt{3},$$

as noted in the statement of the problem.

Two Problems from the 1981 Leningrad High School Olympiad

1. (*Crux Mathematicorum,* 1987, 142; solution by Su Chan, J. S. Woodsworth Secondary School, Nepean, Ontario)

A square is partitioned into rectangles whose sides are parallel to the sides of the square (figure 108). For each rectangle, the ratio of its shorter side to its longer side is determined. Prove that the sum S of these ratios is always at least 1.

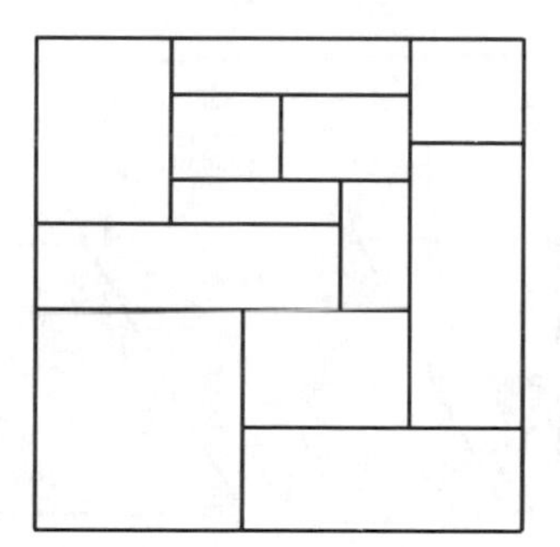

FIGURE 108

Suppose there are n rectangles and that their shorter sides are $a_1, a_2, \ldots, a_n$ and their longer sides are respectively $b_1, b_2, \ldots, b_n$. Then

$$S = \sum_{i=1}^{n} \frac{a_i}{b_i}.$$

Now, if x is the side of the square, then the areas of all the rectangles add up to x^2:

$$\sum_{i=1}^{n} a_i b_i = x^2.$$

With the observation that no rectangle can have a side b_i bigger than x itself, we have

$$
\begin{aligned}
S = \sum_{i=1}^{n} \frac{a_i}{b_i} &= \sum_{i=1}^{n} \frac{a_i b_i}{b_i^2} \geq \sum_{i=1}^{n} \frac{a_i b_i}{x^2} \\
&= \frac{1}{x^2} \sum_{i=1}^{n} a_i b_i \\
&= \frac{1}{x^2}(x^2) = 1.
\end{aligned}
$$

2. (*Crux Mathematicorum,* 1991, 37)

If P is a point inside a parallelogram $ABCD$, the sum of the four perpendiculars from P to the sides of $ABCD$ (or their extensions if necessary) is a constant, namely the sum of the two altitudes of the parallelogram (figure 109).

Prove the converse of this result: if, for every point P inside a convex quadrilateral $ABCD$, the sum of the perpendiculars to the four sides (or their extensions if necessary) is a constant, then $ABCD$ is a *parallelogram.*

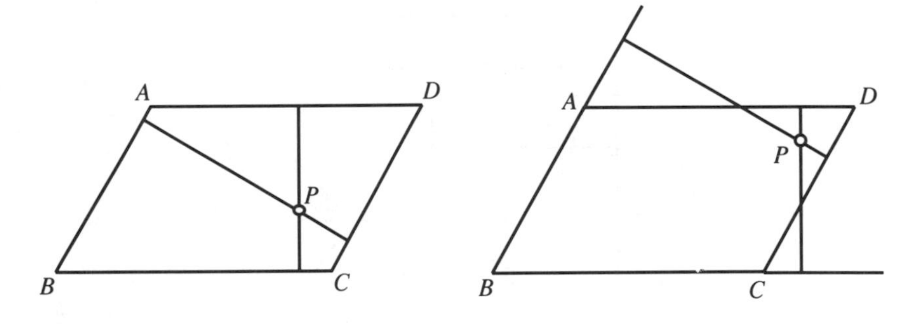

FIGURE 109

The following beautiful solution is due to Murray Klamkin, University of Alberta.

(a) The basis of Murray's solution is a brilliant way of representing the sum of the perpendiculars $(PX + PY)$ to the arms of an angle (LMN) from a point (P) inside the angle (figure 110). The key is to draw the chord QR through P which is perpendicular to the bisector MK of angle M in order to make $\triangle MQR$ isosceles with equal arms MQ and MR. Then PX and PY are altitudes in triangles PMQ and PMR, and from their areas we obtain

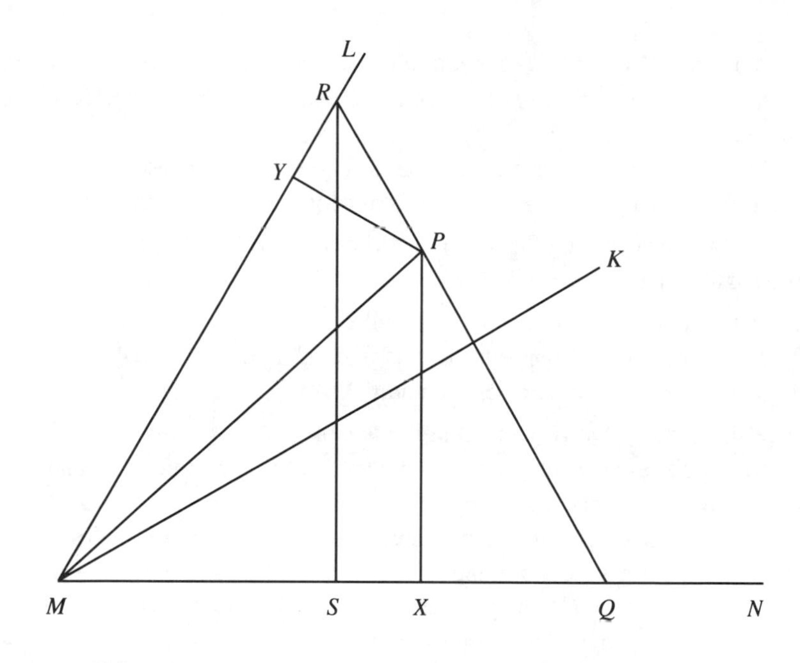

FIGURE 110

$$\triangle PMQ = \frac{1}{2} \cdot MQ \cdot PX,$$

implying

$$PX = \frac{2}{MQ} \triangle PMQ,$$

and

$$\triangle PMR = \frac{1}{2} \cdot MR \cdot PY,$$

giving

$$PY = \frac{2}{MR} \triangle PMR = \frac{2}{MQ} \triangle PMR,$$

since $MQ = MR$. Hence

$$\begin{aligned} PX + PY &= \frac{2}{MQ}(\triangle PMQ + \triangle PMR) \\ &= \frac{2}{MQ} \triangle MQR \\ &= \frac{2}{MQ}\left(\frac{1}{2} \cdot MQ \cdot RS\right) \\ &= RS. \end{aligned}$$

That is to say, $PX + PY$ *is the altitude to an arm in the isosceles triangle* MQR *determined by drawing through* P *the chord* QR *that is perpendicular to the bisector of angle* M.

Moreover, since the same isosceles triangle MQR arises for all points P on QR, the sum $PX + PY$ is a constant for all points P on QR.

(b) Now, concerning the given quadrilateral $ABCD$, there are only the three possibilities

(i) neither pair of opposite sides is parallel,
(ii) exactly one pair of opposite sides is parallel, and the desired
(iii) both pairs of opposite sides are parallel.

Let us show, then, that (i) and (ii) are untenable.

(i) Suppose that neither pair of opposite sides is parallel and that AB and CD meet at E and AD and BC meet at F (figure 111). Through any point P in $ABCD$ draw the perpendiculars to the bisectors of angles E and F to determine the isosceles triangles EQR and FTS, and through a second point M on QR draw UV perpendicular to the bisector of angle F. Then, as argued above, the sum of the perpendiculars to the sides AB and CD is the same for all points on QR, in particular from P and M. Since the sum of the perpendiculars to all four sides of $ABCD$ is given to be the same for all points in the quadrilateral, this grand total is the same for P and M, and it follows that the sum of the perpendiculars to the other pair of opposite sides, AD and BC, is also the same for P and M. Accordingly, the altitudes SL and UN in the isosceles triangles FTS and FVU must be equal. But clearly

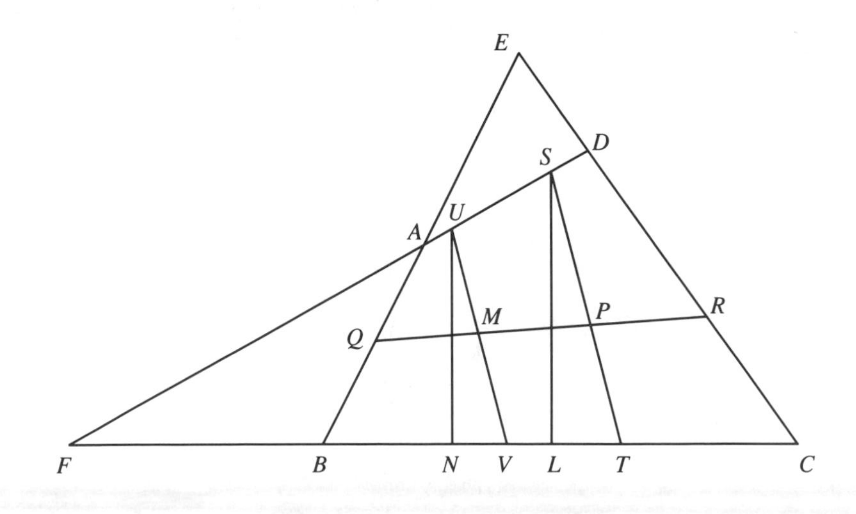

FIGURE 111

this is not so, since one of UN, SL must be closer to F than the other. Hence possibility (i) is inadmissible.

(ii) Finally, suppose that AB and CD are parallel but AD and BC meet at F (figure 112). In this case, the sum of the perpendiculars PQ and PR to the parallel sides AB and CD is simply the distance between these parallel sides for *all* points P in the quadrilateral. Consequently, since the sum of the perpendiculars to the four sides of $ABCD$ is constant, it follows that the sum of the perpendiculars to the pair of opposite sides AD and BC is also a constant for *all* points in $ABCD$. But, as we saw in the discussion of case (i), the sum of the perpendiculars to AD and BC, namely SL and UN, is *not* the same for different points P and M on QR, and our solution is complete.

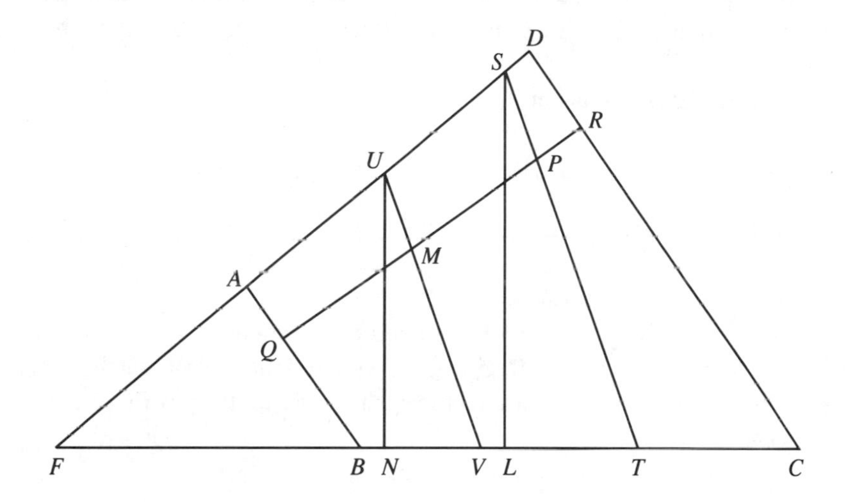

FIGURE 112

Four Problems from the Pi Mu Epsilon Journal—Fall 1992

1. Problem 763. Proposed by Russell Euler, Northwest Missouri State University, Maryland, Missouri; solution by Barbara Lehman, Brigantine, New Jersey.)

Find all real solutions of

$$(x^2 - 7x + 11)^{x^2-11x+30} = 1.$$

It comes readily to mind that the value of a^b is 1 when either

(i) $a = 1$ or

(ii) $b = 0$, provided $a \neq 0$.

Case (i) quickly gives $x^2 - 7x + 10 = 0$ and the solutions $x = 2$ and 5. Case (ii) leads to $x^2 - 11x + 30 = 0$, giving $x = 5$ and 6, neither of which makes $x^2 - 7x + 11$ equal to zero. Since this seems to cover all the possibilities, one is content to answer

$$x = 2, 5, \text{ and } 6.$$

However, there is a third case; can you think of it? It is easy to miss the possibility that a can be -1 when b is even. This yields $x^2 - 7x + 12 = 0$, giving $x = 3$ and 4, for which $x^2 - 11x + 30$ takes the even values 6 and 2 respectively. Therefore the full slate of solutions is $x = 2, 3, 4, 5, 6$. (I suppose we ought to be thankful we weren't asked to prove that the product of the roots of the equation $(x^2 - 7x + 11)^{x^2-11x+30} = 1$ is $6!$.)

The following nice way of taking in all the possibilities from a single point of view was proposed by Kandasamy Muthuvel, University of Wisconsin-Oshkosh:

> if $a^b = 1$, then so does $|a|^b = 1$, giving $b \log |a| = 0$, which is achieved by either $b = 0$ or $a = \pm 1$.

It remains only to verify that the solutions obtained from these values of a and b actually satisfy the given equation.

2. The following two problems were proposed and solved by John Wetzel, University of Illinois, Urbana, Illinois. The first is a problem in combinatorial geometry and the second was originally proposed by the renowned French geometer Emil Lemoine, who touched off the modern revival of Euclidean geometry with his contributions to a conference in Lyons in 1873.

Problem 759. A Universal Cover for Special Arcs. Let's call a plane arc "special" if its length is 1 and it lies entirely on one side of the straight line through its endpoints (figure 113).

Prove that every special arc can be covered by an isosceles right-angled triangle of hypotenuse 1.

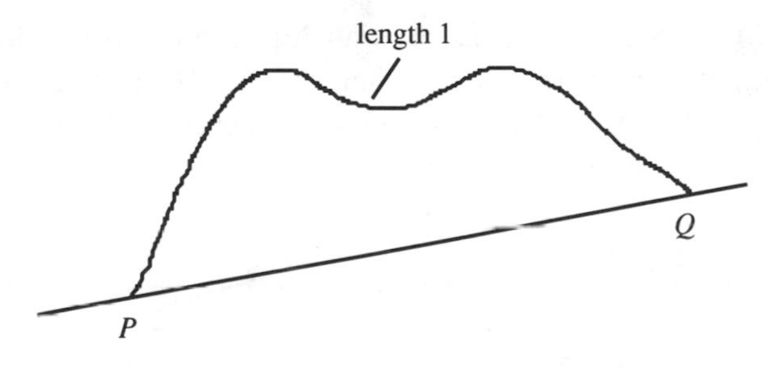

FIGURE 113

Let P and Q be the endpoints of a special arc c. Let c be enclosed in an isosceles right-angled triangle XYZ (figure 114: imagine a line ST at 45° to PQ to be slid along with S on PQ until it just touches c; YZ is similarly

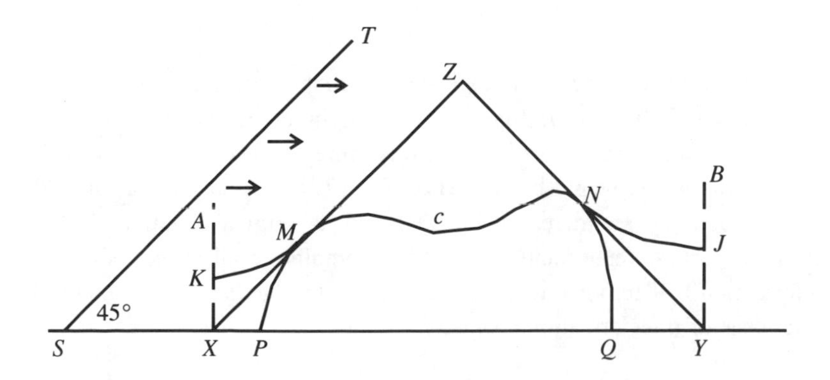

FIGURE 114

obtained on the other side of c). It could happen that either or both points of contact M and N are P and Q themselves, but that's alright.

Now, the reflection in XZ carries the base line XY into the perpendicular XA (since $\angle ZXY = 45°$) and the arc MP into an image arc MK; similarly, reflection in YZ carries arc NQ into an image arc NJ, where J lies on the perpendicular YB. Since an image has the same length as its antecedent, the length of the composite curve $KMNJ$, made up of the two image-arcs and the part of c between M and N, is the same as the length of c, namely 1. But clearly, any path from XA to YB must be at least as long as the perpendicular distance XY across the strip between them, and we have $XY \leq 1$, as desired.

3. Problem 760. Let equilateral triangles be drawn outwardly on the sides of triangle ABC to determine triangle DEF (figure 115). Starting with just $\triangle DEF$, how would you construct $\triangle ABC$?

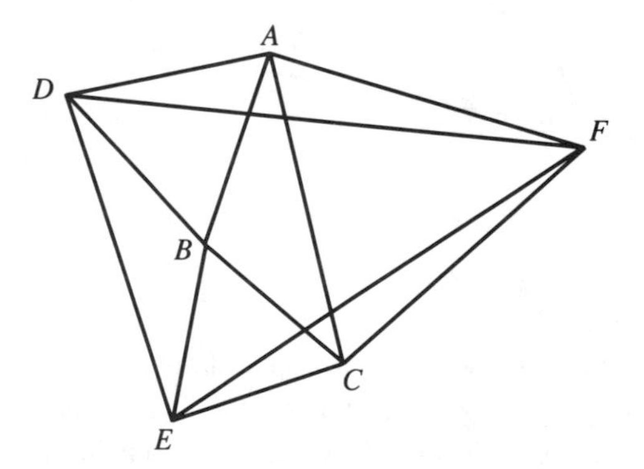

FIGURE 115

Because the constructed triangles are equilateral, it is easy to see, when the vertices are labeled as in figure 115, that a rotation through $60°$ about the center F would carry A to C, and if this were followed by a $60°$-rotation about E, the image would be carried on from C to B, and that a final $60°$-rotation about D would return the image to its original position A. Now, the effect of these three rotations can be accomplished all at once simply by performing a half-turn about A. In order to see this, let's review some of the basic ideas of transformation geometry.

a. Translations and Rotations in the Plane. It is surprising that a translation or a rotation can be accomplished by performing an appropriate pair of

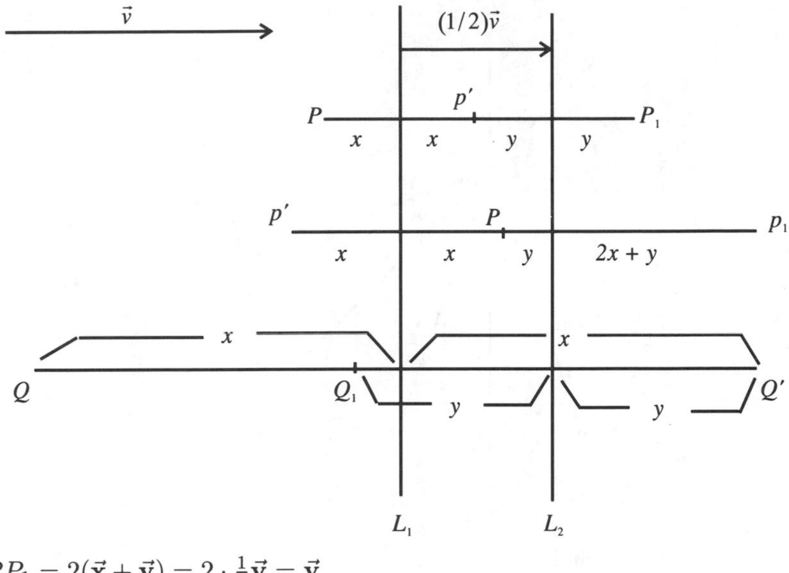

$PP_1 = 2(\vec{\mathbf{x}} + \vec{\mathbf{y}}) = 2 \cdot \frac{1}{2}\vec{\mathbf{v}} = \vec{\mathbf{v}},$
$QQ_1 = 2\vec{\mathbf{x}} - 2\vec{\mathbf{y}} = 2(\vec{\mathbf{x}} - \vec{\mathbf{y}}) = 2 \cdot \frac{1}{2}\vec{\mathbf{v}} = \vec{\mathbf{v}}.$

FIGURE 116

reflections (implying that all rigid motions in the plane can be effected using just reflections).

Suppose you want to perform the translation specified by a vector $\vec{\mathbf{v}}$, that is, through a distance $|\vec{\mathbf{v}}|$ in the direction $\vec{\mathbf{v}}$. First reflect in any line l_1 which is perpendicular to $\vec{\mathbf{v}}$ and then reflect in the line L_2 which is also perpendicular to $\vec{\mathbf{v}}$, where the directed distance *from* L_1 *to* L_2 is $\frac{1}{2}\vec{\mathbf{v}}$ (figure 116). It is easy to check that these reflections carry each point P of the plane through an intermediate image P' to a final image P_1 such that the vector $PP_1 = \vec{\mathbf{v}}$.

The great liberty afforded by this procedure is that it doesn't matter where L_1 might be placed, that is, the pair of lines L_1 and L_2 can be slid rigidly across the plane in the direction of $\vec{\mathbf{v}}$ to any convenient position.

Similarly, the rotation $O(\alpha)$ through an angle α about the center O is accomplished by reflecting in any line L_1 through O and then reflecting in the line L_2 through O such that the directed angle from L_1 to L_2 is $\frac{1}{2}\alpha$ (figure 117). Again, it doesn't matter where L_1 is taken; of course, specifying L_1 also determines L_2, and vice-versa. (These considerations shed light on the view that translations and rotations are the same kind of transformation, a translation being a rotation with center at infinity.)

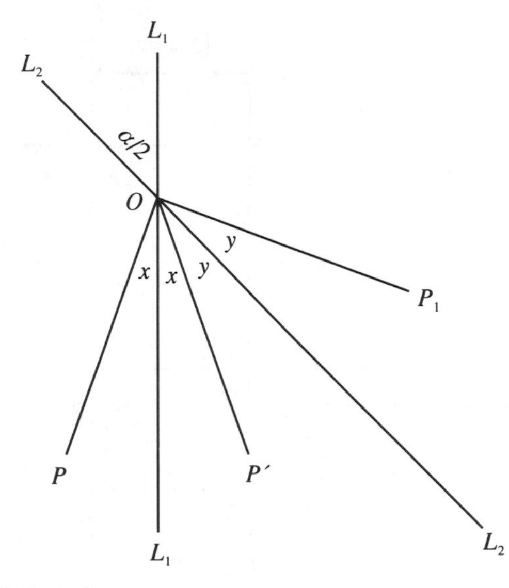

$OP = OP' = OP$, and
$\angle POP = 2(x+y) = 2\left(\frac{1}{2}\right)\alpha = \alpha$
FIGURE 117

(b) Consecutive Rotations. Now consider the compound transformation consisting of the rotation $O_1(\alpha)$ followed by a second rotation $O_2(\beta)$, which we shall denote by the product $O_1(\alpha) \cdot O_2(\beta)$. Each rotation calls for two reflections, the first in a pair of lines (L_1, L_2) through O_1 and the second in a pair of lines (M_1, M_2) through O_2. Clearly, consecutive reflections in the same line simply cancel each other. Therefore if we use the line O_1O_2 itself as the second line L_2 of the first pair and again as the first line M_1 of the second pair, the four reflections $L_1 \cdot L_2 \cdot M_1 \cdot M_2$ reduce to the two reflections $L_1 \cdot M_2$, which constitute the two components of the rotation about the point of intersection O of L_1 and M_2 through twice the directed angle θ from L_1 to M_2.

Having selected the line O_1O_2 to serve as L_2 and M_1, the proper position of L_1 is obtained by rotating about O_1 *backwards* from L_2 (i.e., O_1O_2) through an angle of $\frac{\alpha}{2}$ (i.e., rotating through the angle $-\frac{\alpha}{2}$); similarly, M_2 is found by rotating about O_2 *ahead* from O_1O_2 through the angle $+\frac{\beta}{2}$. The case of α and β both positive is illustrated in figure 118. Being an exterior angle of triangle O_1O_2O, the directed angle θ from L_1 to M_2 is simply $\frac{\alpha+\beta}{2}$.

A check of the variations in the signs and relative sizes of α and β reveals that θ is always either an exterior angle of $\triangle O_1O_2O$ or an angle in

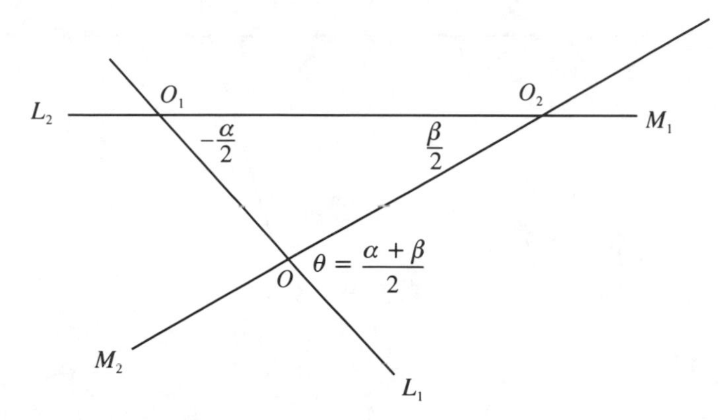

FIGURE 118

the triangle itself, implying that θ is given in all cases by the formula $\frac{\alpha+\beta}{2}$; for example, for $\alpha = -70°$, $\beta = 100°$, we have $\theta = \frac{\alpha+\beta}{2} = 15°$.

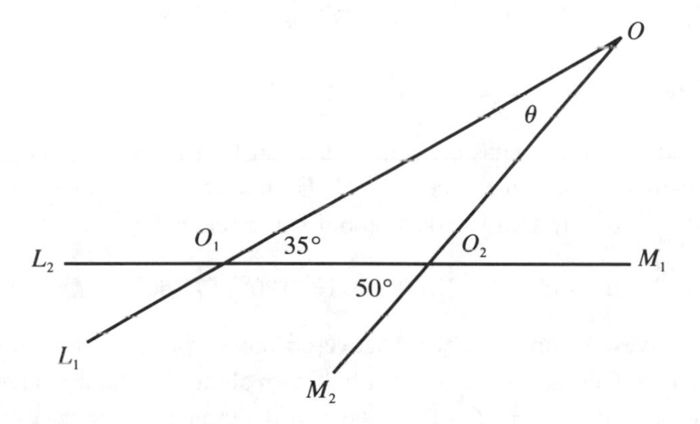

FIGURE 119

Finally, then, since reflections in two lines through O at an angle θ rotates the plane through the angle 2θ, the two rotations $O_1(\alpha)$ and $O_2(\beta)$ simply amount to a single rotation $O(2\theta)$, i.e., the rotation $O(\alpha+\beta)$.

In the event that L_1 is parallel to M_2, the center O is at infinity, in which case the resulting rotation $O(\alpha+\beta)$ degenerates to a translation (determined, of course, by reflections in the parallel lines L_1 and M_2). The complete result is

> $O_1(\alpha) \cdot O_2(\beta)$ is a rotation $O(\alpha+\beta)$ except when $\alpha+\beta = 360°$ or a multiple of 360°, in which case it is a translation (it turns out that L_1 and M_2 are parallel if and only if $\alpha+\beta = 360°$ or a multiple of 360°).

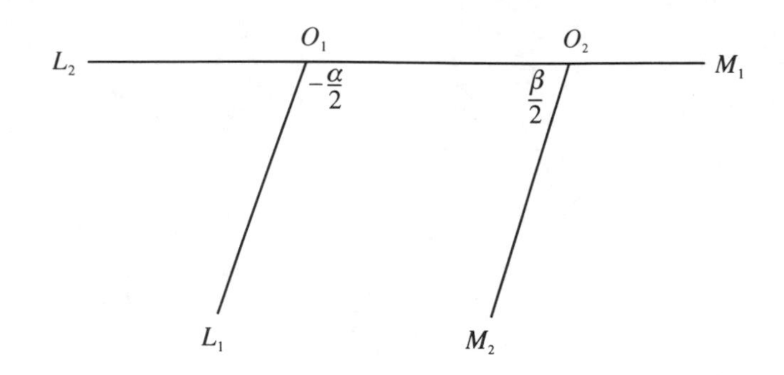

FIGURE 120

For example, for positive α and β such that $\alpha + \beta = 360°$ (figure 120), we have

$$\left|-\frac{\alpha}{2}\right| + \left|\frac{\beta}{2}\right| = \frac{\alpha}{2} + \frac{\beta}{2} = \frac{\alpha+\beta}{2} = 180°,$$

giving $L_1 \parallel M_2$.

Back to the Solution. Thus the compound transformation T consisting of the three $60°$-rotations about the centers F, E, and D, in that order, is quickly seen to be equivalent to a half-turn about some center O:

$$T = F(60°) \cdot E(60°) \cdot D(60°) = O'(120°) \cdot D(60°) = O(180°).$$

Since T leaves A unmoved (we discovered this earlier), it must be that the center O is in fact the vertex A, for a half-turn about O actually moves every point in the plane except O. Thus we can determine the effect of T upon any point P either by tracing it through the intermediate stages of the three $60°$-rotations or by locating its final image at one stroke by noting where it would be carried by a half-turn about A.

Now we ask where the point E would be carried by T (observe that E is the "middle" center of the three rotations in T). The first rotation, center F, would take E to the third vertex X of an equilateral triangle drawn outwardly on the side EF of $\triangle DEF$ (figure 121). Next, the rotation about E would take X on to F; thus the first two rotations merely take E over to F. The third rotation $D(60°)$, then, would give the final image at the third vertex Y of an equilateral triangle drawn outwardly on the side DF of $\triangle DEF$. We conclude, therefore, that a half-turn about A would carry E to Y. Since Y can be constructed from the given triangle DEF, the desired point A can be found at the midpoint of EY.

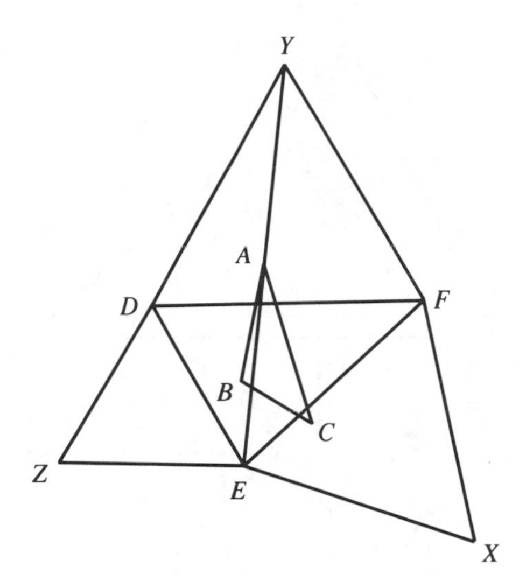

FIGURE 121

Similarly, B and C are the midpoints of FZ and DX, where Z is the third vertex of an equilateral triangle drawn outwardly on the side DE of $\triangle DEF$. (To get B, observe that $60°$-rotations about D, F, and E leave B invariant; now trace the movements of the "middle" center F in this compound transformation. Analogously for C.)

It is interesting to note that one does almost the same thing to $\triangle DEF$ that was done to $\triangle ABC$ in order to produce $\triangle DEF$ in the first place:

> the vertices A, B, and C are simply the midpoints of the segments joining the third vertices of equilateral triangles drawn outwardly on the sides of $\triangle DEF$ to the opposite vertices of $\triangle DEF$.

4. Problem 769. (Proposed by R. S. Luthar, University of Wisconsin Center, Janesville, Wisconsin; solved by George P. Evanovich, Saint Peter's College, Jersey City, New Jersey.)

If ABC is a triangle in which $c^2 = 4ab \cos A \cos B$, prove the triangle is isosceles.

Many years ago my teacher Sam Beatty used to say that you can get a lot of mileage out of the simple triangle relation (figure 122)

$$c = a \cos B + b \cos A.$$

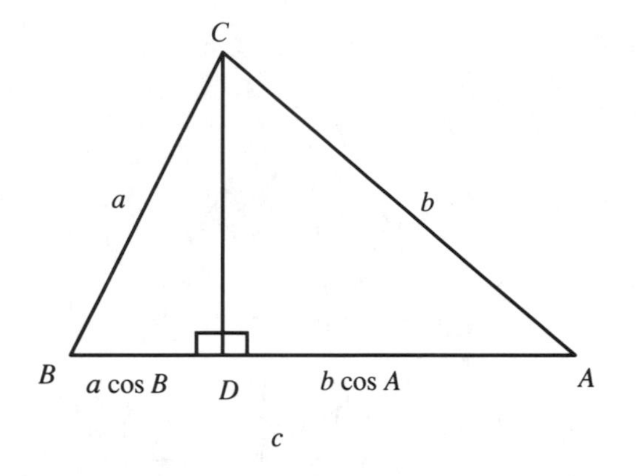

FIGURE 122

Hence

$$c^2 = a^2 \cos^2 B + 2ab \cos A \cos B + b^2 \cos^2 A,$$

and when

$$c^2 = 4ab \cos A \cos B,$$

then

$$a^2 \cos^2 B - 2ab \cos A \cos B + b^2 \cos^2 A = 0,$$

i.e.,

$$(a \cos B - b \cos A)^2 = 0.$$

It follows that

$$a \cos B = b \cos A,$$

making the point D in figure 122 the *midpoint* of AB. Thus C lies on the perpendicular bisector of AB and we have $AC = BC$.

An Elegant Solution To Morsel 26

The subject of Morsel 26 in my *More Mathematical Morsels* (p. 133) is the following problem on box-packing:

> A 3-brick is a $1 \times 1 \times 3$ block of 3 unit cubes in a row. In a $7 \times 7 \times 7$ box B there is room for $7^3 = 343$ unit cubes. Since 343 is not divisible by 3, B cannot be packed with just 3-bricks. However, 114 3-bricks can be packed into B so as to occupy all the unit cells except one; that is, there will be a single unit-cell hole H someplace. In fact, B can be packed like this in lots of ways; figure 123 shows the hole H at a corner. H could also go in the middle of an edge or at the center of a face. However, it is curious that if H is not on the outside surface, but buried somewhere in the $5 \times 5 \times 5$ inner core, then the only place it can possibly be is the very center cell itself; prove this.

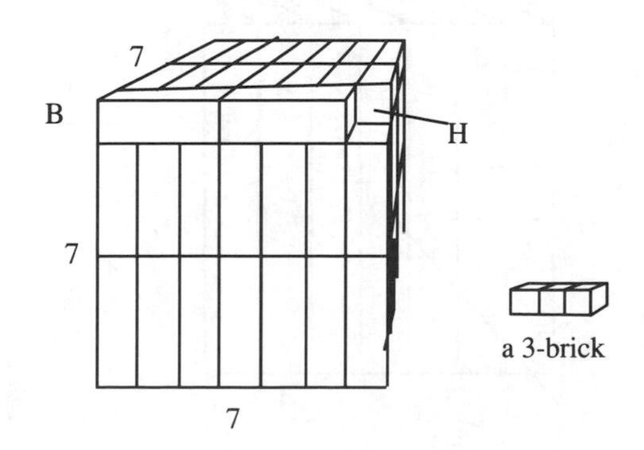

FIGURE 123

The published solution uses an appropriate coloring of the 343 cells in the box—a device that has wide application in combinatorics. In the present case, however, the following elegant solution, which exploits the obvious symmetries of a cubical box, is much simpler and prettier. It was kindly brought to my attention by Professor Will Self of Eastern Montana College, Billings, Montana.

The Solution. Suppose the box is placed in coordinate 3-space so that each coordinate plane slices it right down the middle; in this way the *centers* of its 343 unit cells occur at the lattice points $(0,0,0)$, $(1,0,0)$, $(-1,0,0)$, $\ldots$, that is, the points (x,y,z) where each of x, y, and z runs through the integers $\{-3,-2,-1,0,1,2,3\}$.

Now, let the box be sliced into 7 slabs of unit thickness by planes perpendicular to the x-axis (figure 124); thus the centers of the 49 unit cells in a slab all have the same x-coordinate—in the slab at the end they are all -3, in the next slab -2, and so on across the box. Since the box is symmetrical about the yz-coordinate plane, the *sum* of the 343 x-coordinates of all the cells in the box is zero, which we note in passing is also congruent to $0 \pmod 3$.

Clearly a 3-brick can be placed into the box in only two ways:

either it lies entirely in a slab,
or it runs through 3 consecutive slabs.

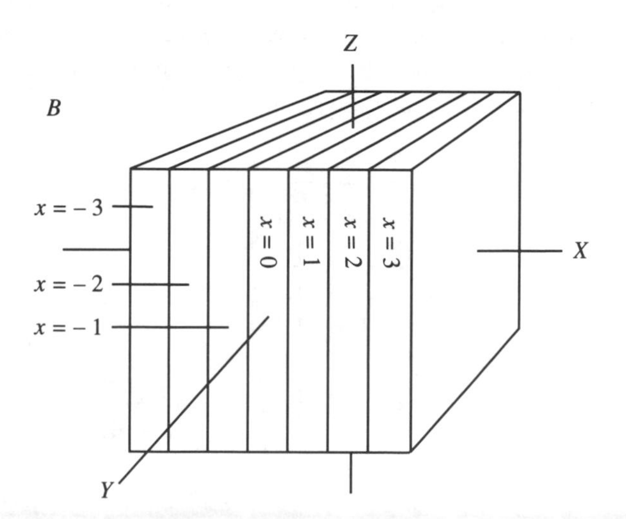

FIGURE 124

The x-coordinates of the centers of its 3 unit cells, then, are

either	all the same integer m,
or	three consecutive integers n, $n+1$, $n+2$.

In any case, the *sum* of these three coordinates is congruent to $0 \pmod 3$. Holding for each 3-brick, it follows that the sum of the x-coordinates for all 342 cells of the 114 3-bricks is also congruent to $0 \pmod 3$. But we have already seen that, for *all* 343 cells of the box, this sum is congruent to $0 \pmod 3$, and therefore the hole H must also occur at a cell whose x-coordinate is congruent to $0 \pmod 3$. This limits its value to -3, 0, or 3, and since 3 or -3 would place H on an outside face of the box, which we know is *not* the case, its x-coordinate can only be 0. Similarly the y- and z-coordinates of the center of H must also be 0, making H the very center cell at $(0, 0, 0)$.

Two Euclidean Problems from the Netherlands

These problems are not very hard, which serves to remind us that things don't have to be difficult to be enjoyable. References and credits are given at the end.

1. $ABCD$ is a rectangle with $AB < BC$. Construct points X and Y on BC, between B and C, so that $AX = XY = YD$ (figure 125).

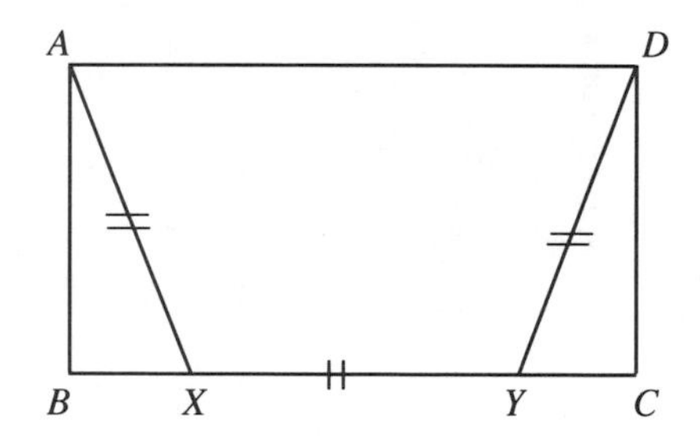

FIGURE 125

Solution 1. It is pretty obvious that the key here is to exploit the considerable symmetry of the figure. In figure 126, it is evident that the midpoint M of BC is also the midpoint of XY, revealing that the sides AX and XM in $\triangle AXM$ are in the ratio of $2 : 1$. Now, if a line were drawn through B parallel to AX, it would cross MA extended at a point P which would make triangle PBM similar to triangle AXM and thereby preserve the $2 : 1$ ratio in the fraction $\frac{PB}{BM}$. But BC is twice BM, and so PB would be equal to BC, and the solution is clear: with center B and radius BC, mark off the

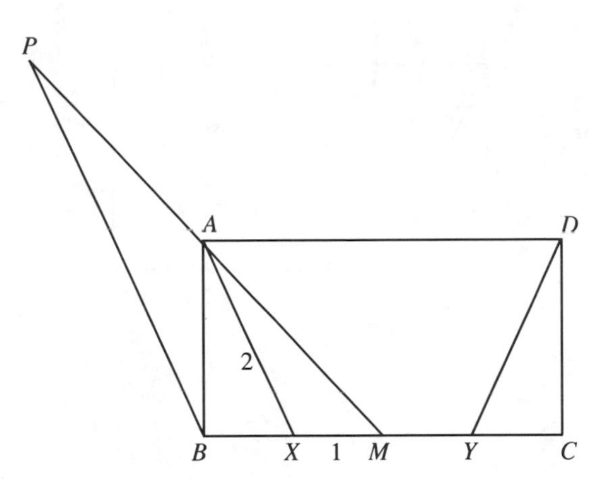

FIGURE 126

point P on MA extended, and through A draw the line parallel to PB to give X on BM. Extending XM its own length gives Y.

Solution 2. As in solution 1, after determining that $AX = 2 \cdot XM$, one might proceed at a more sophisticated level by noting that X is therefore a point of intersection of BC and the circle of Apollonius which has fixed points A and M and ratio 2 : 1 (figure 127; recall that this circle has diameter JK, where J and K divide AM internally and externally in the ratio 2 : 1).

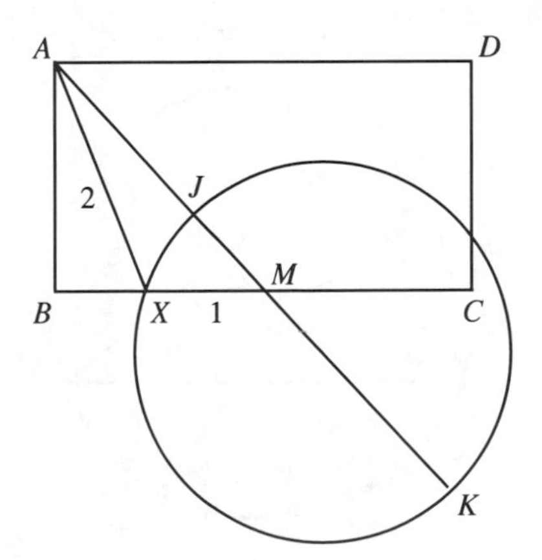

FIGURE 127

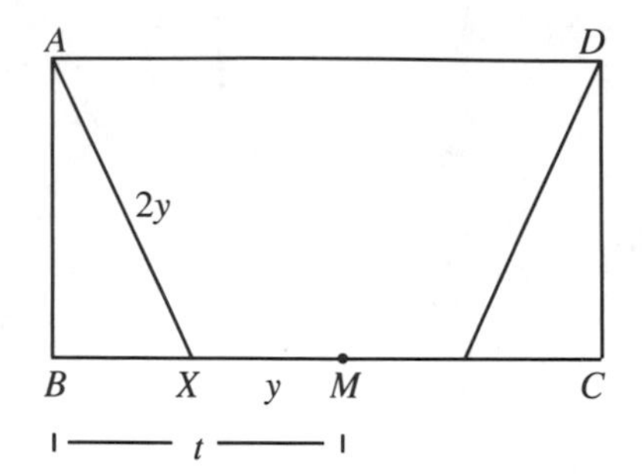

FIGURE 128

Solution 3. We include the following direct algebraic approach to show that the problem is really very elementary and can be solved without having to come up with anything ingenious or to know advanced topics.

Let $BM = t$ and $XM = y$ (figure 128). Then, since $AX = 2 \cdot XM = 2y$ as above, and $BX = t - y$, the Pythagorean theorem applied to $\triangle ABX$ gives

$$4y^2 = AB^2 + (t - y)^2,$$

which can be solved for y since AB and t are known.

2. ABC is an acute-angled triangle (equivalently, the center O of its circumcircle lies *inside* the triangle) (figure 129). L and M are the midpoints of AB and AC respectively, and N is the midpoint of the minor arc BC. The perpendicular from N meets the line AB

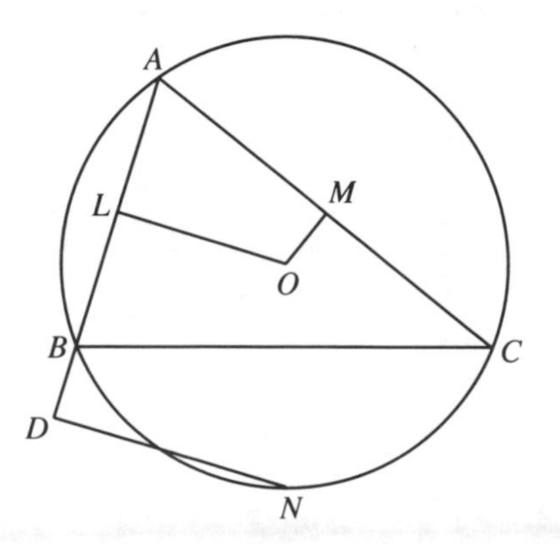

FIGURE 129

at D. Prove the following engaging properties of the quadrilateral $ALOM$:

(i) the sum of the lengths of the two sides at A is AD, i.e., $AL + AM = AD$; and

(ii) the sum of the lengths of the two sides at O is DN, i.e., $OL + OM = DN$.

Since L and M are the midpoints of chords, OL and OM are perpendicular to AB and AC. Now, observing that AL is common to AD and the sum $AL + AM$, the first property follows upon showing that $LD = AM$. Similarly, if LO is laid off along DN to complete rectangle $LDKO$, the second property requires us to show $KN = OM$ (figure 130). However, in rectangle $LDKO$, opposite sides OK and LD are equal, and therefore our problem would be solved completely if we could show that $OK = AM$ and $KN = OM$. Evidently, then, the problem reduces to showing that triangles AOM and KON are congruent.

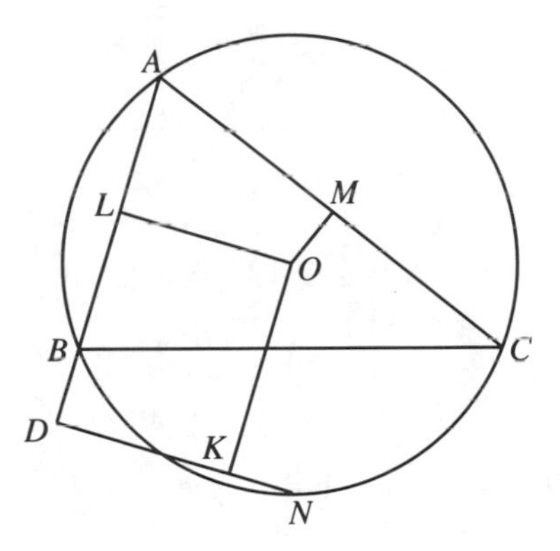

FIGURE 130

Clearly these are right-angled triangles which have equal hypotenuses OA and ON. Since we want to conclude the legs are respectively equal, we shall have to show that the triangles contain another pair of equal angles. Let us show that $\angle KON = \angle OAM$.

If the angles are labeled x, y, z, t, and s (figure 131), we wish to show that $t = z$. Noting that the base angles x in isosceles triangle ANO are equal, and that since N bisects arc BC, AN bisects $\angle A$, we have

$$s = x + z.$$

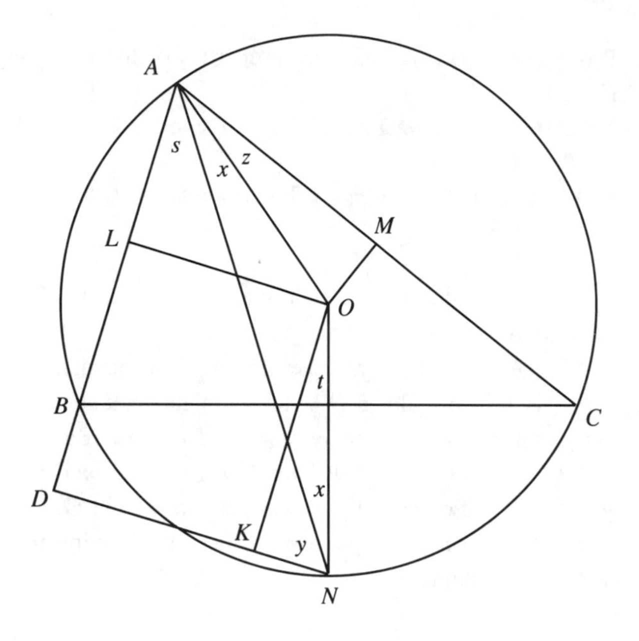

FIGURE 131

Thus the sum

$$\begin{aligned}\angle ONK + \angle OAM &= y + x + z \\ &= y + s \\ &= 90^\circ\end{aligned}$$

in right-triangle ADN. But, in right-triangle KON,

$$y + x + t = 90^\circ$$

and so

$$y + x + t = y + x + z$$

giving $t = z$, as required.

References

1. Problem 1 (*Crux Mathematicorum,* 1990, 86, problem 1407) was proposed by G. R. Veldkamp, De Bilt, The Netherlands. Solution 1 was given by each of Toshio Seimiya, Kawasaki, Japan; Leonardo Pastor, San Paulo, Brazil; and the Proposer.

Solution 2 was given by each of Marcin Kuczma, Warsaw, Poland; Hans Engelhaupt, Bamberg, Germany; Herta Freitag, Roanoke, Virginia; and Kee-Wai Lau, Hong Kong.

2. Problem 2 is a slightly recast version of problem 1411 (*Crux Mathematicorum,* 1990, 92) which was posed by D. J. Smeenk, Zaltbommel, The Netherlands. The above solution, credited to Wilson da Costa Areias, Rio de Janiero, Brazil, was also given by several other solvers who were not identified.

Two Problems from the Singapore Mathematical Society Interschool Competitions

1. From the 1988 Competition† Let $f(x)$ be a polynomial of degree n such that

$$f(k) = \frac{k}{k+1}$$

for each $k = 0, 1, 2, \ldots, n$. What is the value of $f(n+1)$?

From $f(k) = \frac{k}{k+1}$, it follows that $(k+1)f(k) - k = 0$, and hence the given values of the function assert that the $(n+1)$th-degree polynomial equation

$$(x+1)f(x) - x = 0$$

has the $n+1$ roots $0, 1, 2, \ldots, n$. In factored form, then, we have, for some constant c, the identity

$$(x+1)f(x) - x = cx(x-1)(x-2)\cdots(x-n)$$

For $x = -1$, this yields

$$1 = c(-1)(-2)\cdots\big[-(n+1)\big] = c(-1)^{n+1}(n+1)!,$$

giving

$$c = \frac{1}{(-1)^{n+1}(n+1)!} = \frac{(-1)^{n+1}}{(n+1)!}.$$

---

† (*Crux Mathematicorum,* 1991, 233; Solution by Seung-Jin Bang, Republic of Korea, and Murray Klamkin, University of Alberta)

Hence, for $x = n + 1$, we get

$$(n+2)f(n+1) - (n+1) = \frac{(-1)^{n+1}}{(n+1)!}(n+1)(n)(n-1)\cdots 2\cdot 1$$
$$= (-1)^{n+1},$$

from which

$$f(n+1) = \frac{n+1+(-1)^{n+1}}{n+2}.$$

2. From the 1989 Competition† Prove that the tangents to the circumcircle of a triangle ABC at its vertices meet the opposite sides in three collinear points L, M, and N.

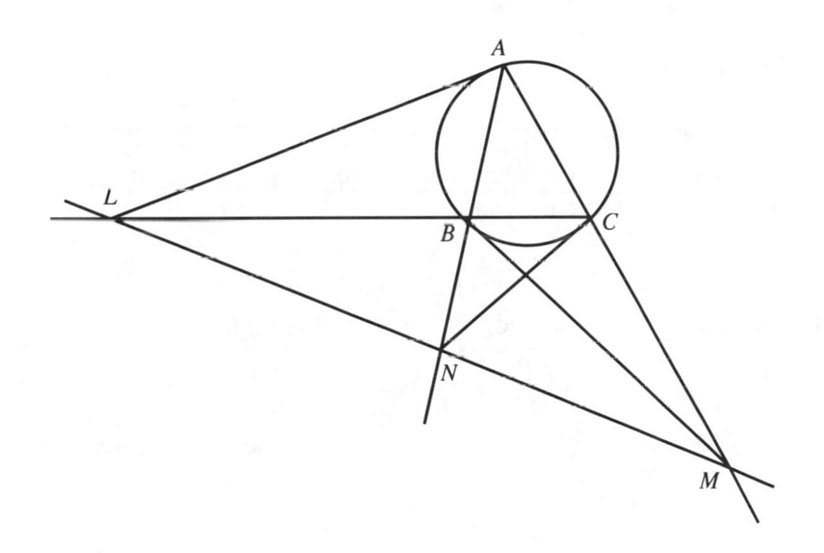

FIGURE 132

This is another opportunity to use the theorem that the angle between a tangent and a chord of a circle is equal to the angle in the segment on the opposite side of the chord. Thus, in figure 133,

$$\angle LAB = \angle ACB = (= x),$$

† (*Crux Mathematicorum,* 1991, 67)

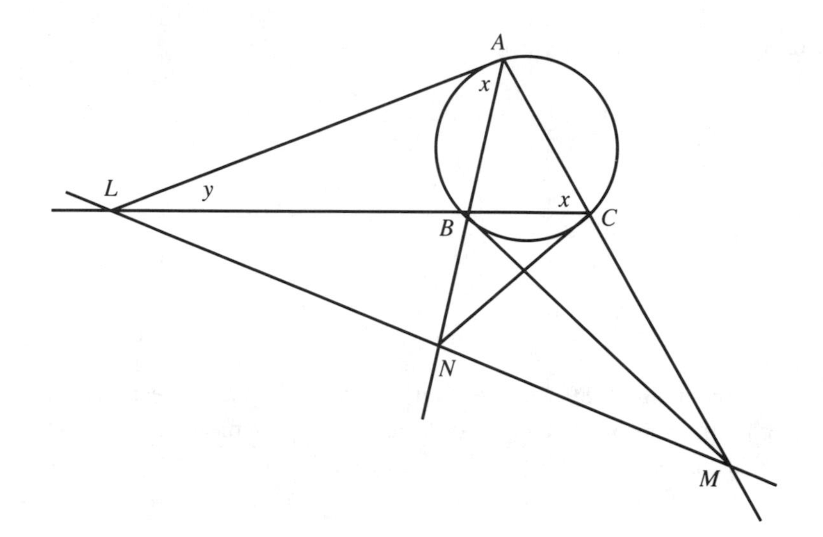

FIGURE 133

and with the angle y at L being common to triangles LAB and LAC, the triangles are similar. From the proportional sides we get

$$\frac{LB}{LA} = \frac{LA}{LC} = \frac{AB}{AC} = \frac{c}{b},$$

and therefore L divides the side BC externally in the ratio

$$-\frac{LB}{LC} = -\frac{LB}{LA} \cdot \frac{LA}{LC} = -\frac{c^2}{b^2}.$$

Similarly, M and N divide the other sides of the triangle in the ratios

$$\frac{CM}{MA} = -\frac{a^2}{c^2} \quad \text{and} \quad \frac{AN}{NB} = -\frac{b^2}{a^2}.$$

Therefore the product of the ratios into which L, M, and N divide the sides of $\triangle ABC$ is

$$\frac{AN}{NB} \cdot \frac{BL}{LC} \cdot \frac{CM}{MA} = \left(-\frac{b^2}{a^2}\right)\left(-\frac{c^2}{b^2}\right)\left(-\frac{a^2}{c^2}\right) = -1,$$

implying L, M, and N are collinear by the celebrated theorem of Menelaus. (An alternative solution is given in 1992, 170.)

Problem M1046 from Kvant (1987)†

In $\triangle ABC$, $\angle A = 60°$ and the altitudes BD and CE cross at H inside the triangle (figure 134). Prove that the line HO to the circumcenter O is the bisector of $\angle EHB$.

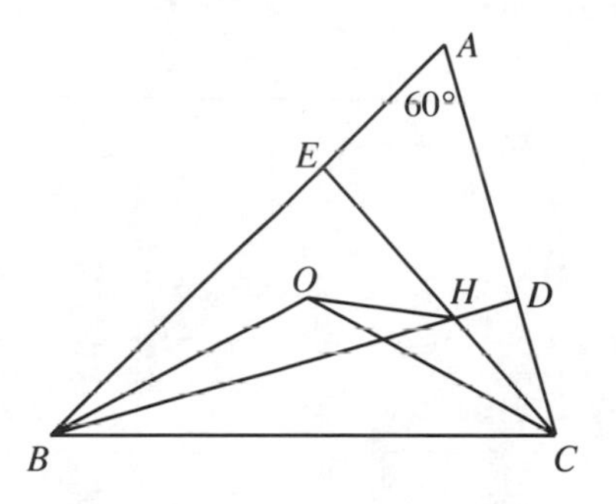

FIGURE 134

The right angles at D and E make $AEHD$ a cyclic quadrilateral, implying that the exterior angle EHB is equal to the interior and opposite angle A. Thus $\angle EHB = 60°$, and its supplement $\angle BHC = 120°$. It remains to show that $\angle OHB = 30°$.

Since O is the circumcenter of $\triangle ABC$, the chord BC subtends at O an angle that is twice the angle it subtends at A, and we have $\angle BOC = 120°$,

---

† (due to V. Progrebnyak, Vinnitsa; reported in *Crux Mathematicorum,* 1990, 103; solution by George Evagelopoulos.)

making it equal to $\angle BHC$. Thus $BOHC$ is cyclic, and in its circumcircle

$$\angle OHB = \angle OCB$$

on chord OB. But $\angle OCB$ is a base angle in isosceles triangle OBC, and we have

$$\angle OHB = \angle OCB = \frac{1}{2}(180^\circ - \angle BOC) = \frac{1}{2}(180^\circ - 120^\circ) = 30^\circ.$$

Two Theorems on Convex Figures

Somewhere in the writings of George Pólya he says that he happened to read a famous book by Ernst Mach just at the right time of life. One is indeed fortunate to come across a great book at an opportune moment in one's development and I've been lucky enough to have it happen a few times myself. One of these was my discovery of *Convex Figures* by Yaglom and Boltyanskii, alas now long out of print. It is filled with all kinds of wonderful things about convex figures and I would like to tell you about two of the results that have continued to intrigue me through the years.

(a) The Theorem of Winternitz

Let F be any plane convex figure and G its center of gravity. Then every chord through G will divide F into two parts such that the area of the smaller one is always at least four-fifths of the larger one.

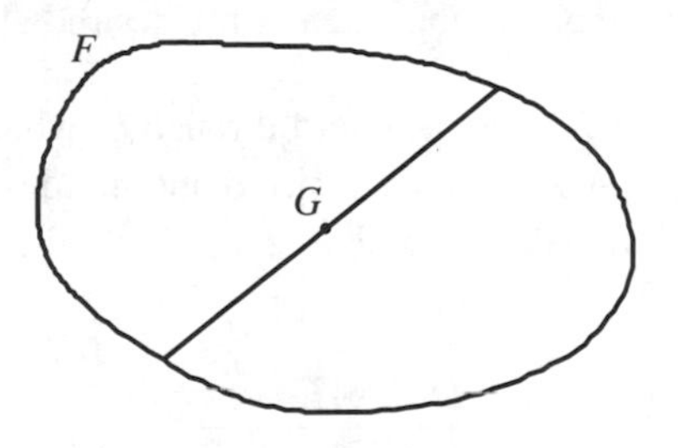

FIGURE 135

The proof is completely elementary, although somewhat detailed, and can be found in full in *Convex Figures*. It is quite easy, however, to establish the property for *triangles,* and we shall content ourselves with an elegant proof of this particular case.

Winternitz's Theorem for Triangles. (i) Suppose P is a point inside a convex figure F and L is a chord through P. We begin by asking which chord L will cut from F a region R of minimum area (figure 136)?

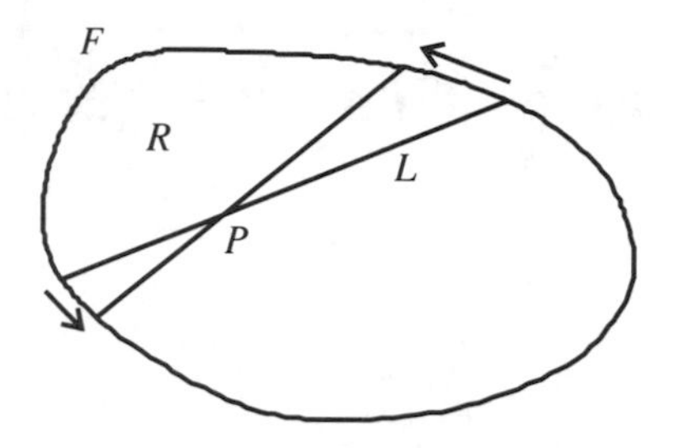

FIGURE 136

Since F is continuous and convex, unless L is *bisected* by P, a slight rotation of the chord in the appropriate direction will cause R to lose a larger wedge on one side of P than it gains on the other, showing that R does not have minimum area (figure 136). Thus we need concern ourselves only with chords that are bisected at P. The trouble is that there may be many such chords; in fact, if F is a circle and P its center, every chord through P is bisected there. In any case, since the area of R varies continuously as L rotates about P, a minimum region must be encountered in some direction. Each time P bisects the chord a local minimum is achieved and the desired absolute minimum is the smallest of these local values.

In the case of a triangle ABC, there are mercifully only three times when the center of gravity G bisects a chord through it, namely when the chord is parallel to a side of the triangle; however, as we shall see, all three local minima are the same, implying that each is the desired absolute minimum.

(ii) To this end, let DE be the chord through G which is parallel to BC (figure 137). Then, because G is the centroid and divides the median AA' in the ratio $2:1$, it follows that D and E also divide their sides in the same ratio, and we have

$$\frac{AD}{AB} = \frac{AE}{AC} = \frac{2}{3}.$$

Thus the areas of the similar triangles ADE and ABC are in the ratio $4:9$ (i.e., as the squares on corresponding sides). That is to say, the chord DGE divides $\triangle ABC$ into a triangle of area four-ninths $\triangle ABC$ and a quadrilateral of area five-ninths $\triangle ABC$. Clearly, the chords through G parallel to the other sides would yield the same result.

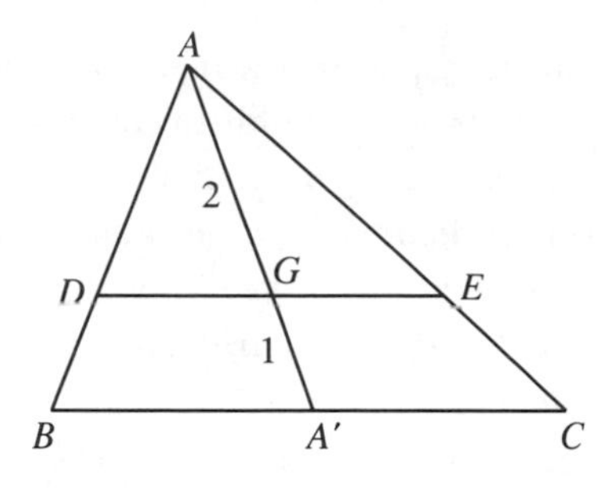

FIGURE 137

Now, when DE is parallel to BC, triangles ADG and ABA' are similar and we get $DG = \frac{2}{3}BA'$; similarly $GE = \frac{2}{3}A'C$. Since AA' is a median, then $BA' = A'C$, and hence $DG = GE$. Thus a chord through G parallel to a side is bisected by G.

On the other hand, rotating DE about G clearly increases one of the two parts DG or GE of the chord while decreasing the other, showing that only when the chord is parallel to a side is it bisected by G. Thus G bisects a chord through it *if and only if* the chord is parallel to a side. Since the three bisected chords all cut off triangles of area $\frac{4}{9} \triangle ABC$ and quadrilaterals of area $\frac{5}{9} \triangle ABC$, it follows that no chord through G can cut off a region of area smaller than $\frac{4}{9} \triangle ABC$. The larger part, then, could not exceed $(1 - \frac{4}{9} =)$ $\frac{5}{9} \triangle ABC$, and the theorem follows.

(b) The Theorem of Kovner

Kovner's theorem concerns convex figures which have *central symmetry*, that is, figures which have a "center" O in which the figure reflects into itself (figure 138; the reflection of a point P in a point O is the point P' such that O bisects PP'); clearly, a half-turn is an alternative way of looking at reflection in a point.

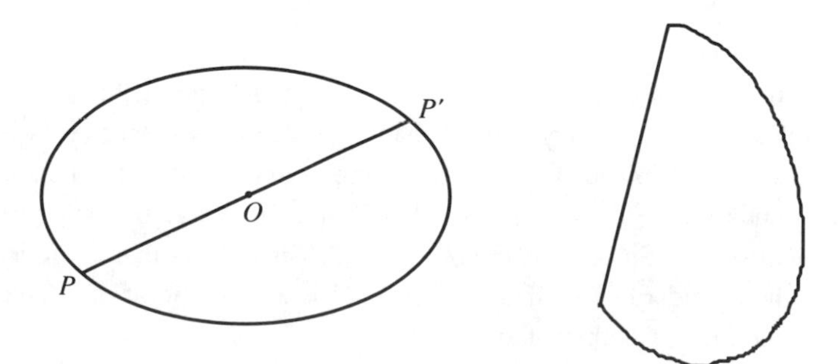

FIGURE 138

Now, there is nothing to stop a convex figure from being lopsided (figure 138). However, convexity does not allow for any bumps or dents in the boundary and so the interior of a convex figure is free of any kind of bottleneck. In fact, the interior of a convex figure enjoys such an unconstricted shape that, despite any lack of symmetry it might have, *it is always possible to inscribe in a convex figure F a* ***centrally-symmetric convex*** *figure ϕ whose area is* ***at least two-thirds*** *the area of F.*

The complete proof of this theorem of Kovner is given in Yaglom and Boltyanskii, and we shall again be content to prove it just for triangles.

Kovner's Theorem for Triangles. Before specializing to triangles, let us consider some general implications of inscribing a centrally-symmetric convex figure ϕ in a given convex figure F. Reflection in the center O of ϕ holds no surprises with regard to ϕ itself, but how about the image F' of F? Since O is inside F, and O is its own image, F and F' must overlap in some figure K (figure 139). It might come as a mild surprise to learn that K must also be a centrally-symmetric convex figure (also with center O). If F and F' cross at a point X, then they must also cross at the image X' of X:

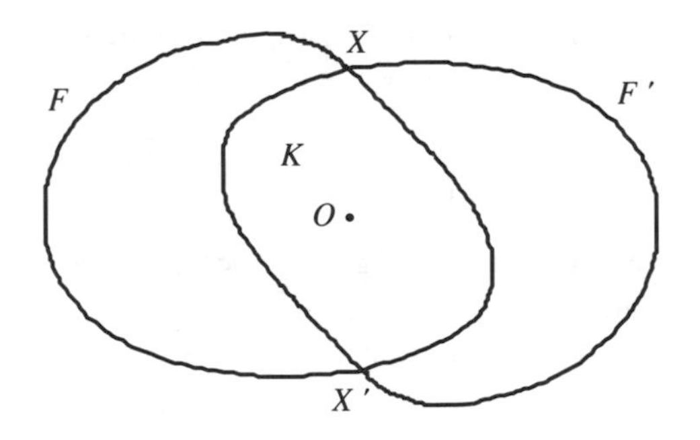

FIGURE 139

If a point A is carried into B by reflection in a point, then clearly B is also reflected into A; they simply change places. Now, since X is on the boundary of F, its image X' must lie on the boundary of F'; but X is also on the boundary of F', and therefore X is itself the image of some point on the boundary of F. Since X is the image of X', it follows that X' must be a point on the boundary of F; that is to say, X' is another point of intersection of the boundaries of both F and F'.

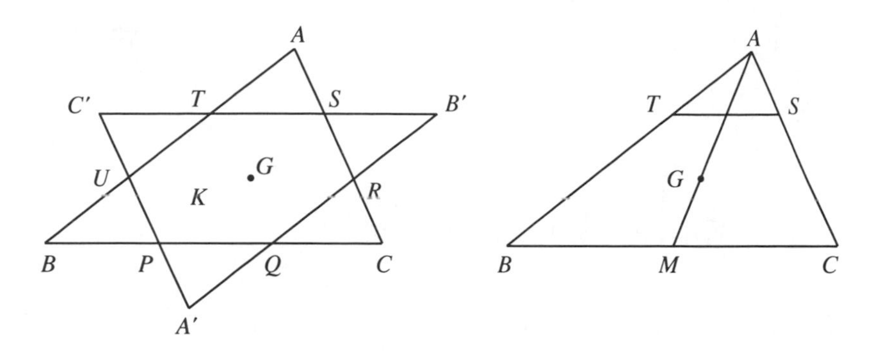

FIGURE 140

Now, the boundaries of F and F' might intersect in more than two points; for example, as in figure 140, a triangle F can intersect its image F' in six places. However, in any case, the points of intersection of the boundaries of F and F' have to go together in "opposite" object-image pairs (like X and X'), implying that each of the arcs on the boundary of K is the image of an opposite arc of K, and hence that K must be symmetric about the center O; and since two convex figures always have a *convex* intersection (an easy exercise), we conclude that K is a centrally-symmetric convex figure inscribed in F (with the same center O as ϕ).

Thus, in searching for large centrally-symmetric convex figures in F, it is K that claims our attention, not the figure ϕ we started with, and the search for large K is clearly just the problem of locating favorable centers O in which to reflect F.

Consequently, as we cast our minds over the noteworthy points of a triangle, it might occur to us to try O at the centroid G. Accordingly, let $\triangle ABC$ be reflected in its centroid G to yield the centrally-symmetric hexagon of intersection $K = PQRSTU$ (figure 140).

It is a simple calculation to show that the area of K is exactly $\frac{2}{3}\triangle ABC$, for K cuts from $\triangle ABC$ a little triangle of area $\frac{1}{9} \triangle ABC$ at each vertex. Consider the triangle ATS (figure 140). Clearly the sides TS and PQ of K are parallel segments which are equidistant from the center G, on opposite sides of it. Since G divides median AM in the ratio $2 : 1$, TS must bisect AG and therefore trisect AM. Since TS is parallel to BC, then $\triangle ATS$ is similar to $\triangle ABC$ and the ratio of corresponding sides is $\frac{1}{3}$. Thus $\triangle ATS = \frac{1}{9} \triangle ABC$, and the theorem follows.

The Infinite Checkerboard

Suppose that a nonnegative integer is put into each square of a checkerboard that extends indefinitely upward and to the right, the numbers being entered from left to right in the rows and the rows done from the bottom so that the entry in each cell is the smallest nonnegative integer that has not been used to its left in the same row or below it in the same column.

Thus the bottom left corner starts things off with 0 and the first row and first column are each completed with the positive integers $1, 2, 3, \ldots$, in order. Succeeding rows are not so easily determined, and the problem is to find the entry in the 100th row and 1000th column.

6	7	4	5	2	3	0	1	
5	4	7	6	1	0	3	8	
4	5	6	7	0	1	2	3	
3	2	1	0	7	6	5	4	
2	3	0	1	6	7	4	5	
1	0	3	2	5	4	7	6	
0	1	2	3	4	5	6	7	

FIGURE 141

Even a small sample suggests that the diagonal contains only 0's and that the array is symmetric about the diagonal. Still, the order of the numbers in the higher rows seems very haphazard and an analytic expression for the rule of formation is well obscured. Nevertheless, it is out of the question to calculate the nearly one hundred thousand prior entries in order to see what goes in position $(1000, 100)$, and so establishing a general formula for the entry in position (m, n) seems to be the only road to a solution.

Without trying to trace the mental processes that go into the making of such a discovery, it turns out that the entry in cell (m, n) is given by the "nim-sum" of the integers $m-1$ and $n-1$, which we denote by $(m-1)\oplus(n-1)$. The nim-sum $a \oplus b$ of two integers is simple to perform—just write the integers in binary notation and add them *without carrying;* the name is due to the fact that this operation is the key to the popular game of nim. In the case at hand, then, we have

$$\begin{aligned}\text{the integer in cell } (1000, 100) &= 999 \oplus 99 \\ &= 1111100111_2 \oplus 1100011_2 \quad \text{(in binary)} \\ &= 1110000100_2 \\ &= 900 \quad \text{(in decimal notation).}\end{aligned}$$

We begin our proof with the observation that the nonnegative integers form an Abelian group under the operation nim-sum; this is established by straightforward checking and is left as an exercise. We note also that the operation nim-sum is associative and that, for all nonnegative integers n,

$$n \oplus n = 0 \quad \text{and} \quad n \oplus 0 = n.$$

Proceeding by induction, suppose the cells which precede position $(A+1, B+1)$ in its row contain, in some order, the entries

$$\{P \oplus B\} = \{0 \oplus B, 1 \oplus B, \ldots, (A-1) \oplus B\},$$

and that the cells below it in its column contain, in some order, the integers

$$\{A \oplus Q\} = \{A \oplus 0, A \oplus 1, \ldots, A \oplus (B-1)\}$$

(figure 142). To show that $A \oplus B$ goes in cell $(A+1, B+1)$ we need to show that every nonnegative integer $X < A \oplus B$ is already present in at least one of these sets and that $A \oplus B$ itself does not belong to either of them.

The latter property is easily established indirectly. Suppose that $A \oplus B$ is to the left of cell $(A+1, B+1)$ in its row, that is, for some nonnegative

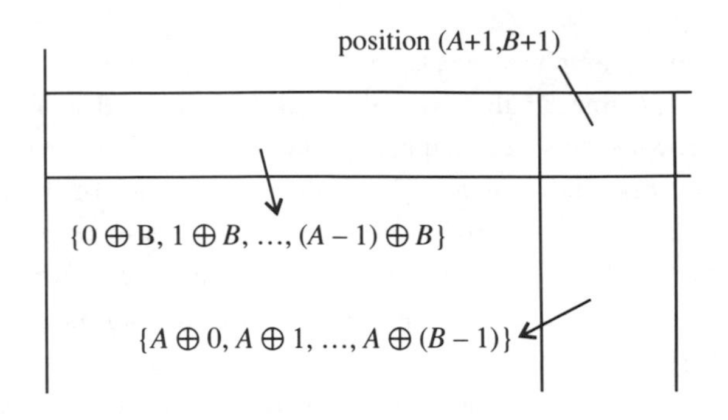

FIGURE 142

integer $P < A$, that $A \oplus B = P \oplus B$. Then

$$(A \oplus B) \oplus B = (P \oplus B) \oplus B$$

$$A \oplus (B \oplus B) = P \oplus (B \oplus B) \quad (\oplus \text{ is associative})$$

$$A \oplus 0 = P \oplus 0 \quad (\text{recall } n \oplus n = 0)$$

$$A = P, \quad \text{a contradiction.}$$

Thus $A \oplus B$ can't occur earlier in the row of cell $(A+1, B+1)$, nor similarly can it occur lower in its column.

Now, let $A \oplus B = C$, and suppose that X is any nonnegative integer less than C. It remains to show that X has already been used ahead of cell $(A+1, B+1)$ in its row or below it in its column.

To this end, let the binary representations of A, B, C, and X be stacked as shown, padding out any shorter ones with 0's on the left to make them all the same length:

$$\begin{aligned} & \qquad\qquad\qquad\quad\;\; \downarrow \\ A &= a_n\ a_{n-1}\ \dots\ a_{i+1}\ a_i\ a_{i-1}\ \dots\ a_1\ a_0 \\ B &= b_n\ b_{n-1}\ \dots\ b_{i+1}\ b_i\ b_{i-1}\ \dots\ b_1\ b_0 \\ C = A \oplus B &= c_n\ c_{n-1}\ \dots\ c_{i+1}\ c_i\ c_{i-1}\ \dots\ c_1\ c_0 \\ X &= x_n\ x_{n-1}\ \dots\ x_{i+1}\ x_i\ x_{i-1}\ \dots\ x_1\ x_0. \end{aligned}$$

Now compare the corresponding digits from the left in the numbers C and X. Their digits might be the same for a distance, but since $X \neq C$, a pair of unequal digits c_i and x_i must be encountered for a *first* time in some position i; and because $X < C$, it could only be that these binary digits are $c_i = 1$ and $x_i = 0$. But, in order to have $c_i = 1$, the corresponding digits a_i and b_i in the nim-sum $A \oplus B$ $(= C)$ must be different. Suppose, then, that $a_i = 1$ and $b_i = 0$.

Because X and C have the same make-up to the left of the ith place, the digits in these positions in the sums

$$B \oplus X \quad \text{and} \quad B \oplus C$$

will also be the same place-for-place up to the ith digit. However, in position i itself, $B \oplus X$ will have the digit

$$b_i \oplus x_i = 0 \oplus 0 = 0,$$

while $B \oplus C$ will have

$$b_i \oplus c_i = 0 \oplus 1 = 1,$$

implying that

$$B \oplus X < B \oplus C.$$

But clearly

$$\begin{aligned} B \oplus C &= B \oplus (A \oplus B) = (B \oplus A) \oplus B \\ &= (A \oplus B) \oplus B = A \oplus (B \oplus B) = A \oplus 0 = A, \end{aligned}$$

and we conclude that $B \oplus X < A$.

Finally, then, letting $B \oplus X = P$, we have $P < A$, placing it among the numbers $\{0, 1, \ldots, A - 1\}$; and, in view of the fact that

$$P \oplus B = (B \oplus X) \oplus B = X,$$

it follows that X is indeed one of the numbers $P \oplus B$ where $P < A$, making X one of the entries to the left of cell $(A + 1, B + 1)$ in its row.

In the alternative case , $a_i = 0$ and $b_i = 1$, a similar argument reveals that X is among the integers below cell $(A + 1, B + 1)$ in its column, and with the 0 in the bottom left corner of the checkerboard to seed the induction, the argument is complete.

Two Problems from the 1986 Swedish Mathematical Competition[†]

1. Consider the set S_n of all pairs (a,b), with $a < b$, that can be constructed from the integers $\{1,2,3,\ldots,n\}$, $n \geq 3$. Prove that there are always the same number of pairs in which $b < 2a$ as there are in which $b > 2a$. For example, for the integers $\{1,2,3,4,5\}$,

$$S_5 = \{(1,2),(1,3),(1,4),(1,5),(2,3),(2,4),(2,5),(3,4),(3,5),(4,5)\},$$

containing 4 pairs of each kind:

$$b < 2a: \quad (2,3),\ (3,4),\ (3,5),\ (4,5);$$

$$b > 2a: \quad (1,3),\ (1,4),\ (1,5),\ (2,5).$$

Let X_n, Y_n, and Z_n denote the subsets containing, respectively, the pairs (a,b) in which $b < 2a$, $b > 2a$, and $b = 2a$. Clearly, if (a,b) is a pair in S_n, then it also occurs in S_{n+1}; more specifically, if (a,b) belongs to X_n, it also belongs to X_{n+1}, i.e., $X_n \subseteq X_{n+1}$. Similarly, $Y_n \subseteq Y_{n+1}$.

The pairs in S_{n+1} that are not in S_n are simply the pairs containing $n+1$: $(1,n+1),(2,n+1),(3,n+1),\ldots,(n,n+1)$. Therefore, if $|X_n| = |Y_n|$ and if these new pairs split evenly over X_{n+1} and Y_{n+1}, it would follow that $|X_n + 1| = |Y_n + 1|$.

Now, n is either odd or even. If $n+1 = 2k+1$, the new pairs are

$$(1,2k+1),(2,2k+1),\ldots,(k,2k+1) \mid (k+1,2k+1),\ldots,(2k,2k+1),$$

† (*Crux Mathematicorum,* 1989, 34)

the first k of which go into Y_{n+1} and the latter k into X_{n+1}; if $n+1=2k$, the new pairs are

$$(1,2k),(2,2k),\ldots,(k-1,2k) \mid (k,2k) \mid (k+1,2k),\ldots,(2k-1,2k),$$

the first $k-1$ of which go into Y_{n+1} and the latter $k-1$ into X_{n+1}, with $(k,2k)$ going into Z_{n+1}.

Thus the new pairs *do* split evenly over X_{n+1} and Y_{n+1}, and therefore, whenever $|X_n| = |Y_n|$, we have $|X_{n+1}| = |Y_{n+1}|$. Since $X_3 = \{(2,3)\}$ and $Y_3 = \{(1,3)\}$ (with $Z_3 = \{(1,2)\}$), then $|X_3| = |Y_3|$ and it follows by induction that $|X_n| = |Y_n|$ for all $n \geq 3$. (An alternative solution is given in 1990, 295.)

2. Now for a more difficult problem.

If the unit interval $[0,1]$ is covered by a finite set of closed intervals S, prove that some subset of *pairwise disjoint* intervals in S has total length which is at least $\frac{1}{2}$.

If the points of $[0,1]$ which are covered by an interval I of S are also covered by one or more of the other intervals in S, the set $S-I$ still covers the unit interval. Accordingly, let all such inessential intervals be discarded to yield a set T in which each interval J covers some subinterval along $[0,1]$ which is not covered by any other member of T. (Such a reduction of S needs to be done one interval at a time, for an inessential interval at one stage can become essential as the multiple coverings are stripped away; in figure 143 both b and c are inessential as things stand, but discarding either of them promotes the other.)

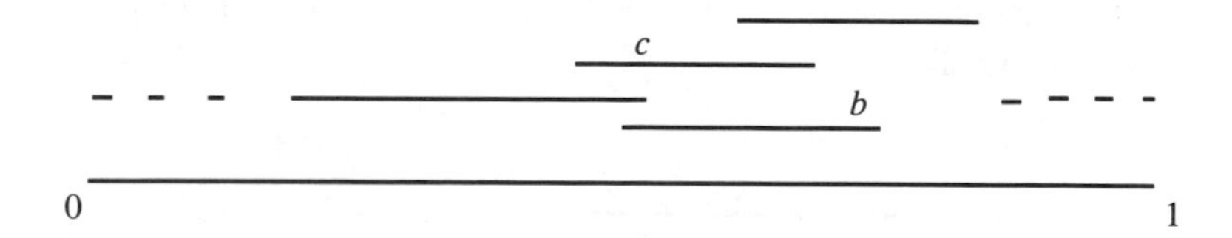

FIGURE 143

In the resulting set T, then, no two intervals can begin at the same point and no two intervals can end at the same point for, in either case, the shorter one would be inessential (figure 144).

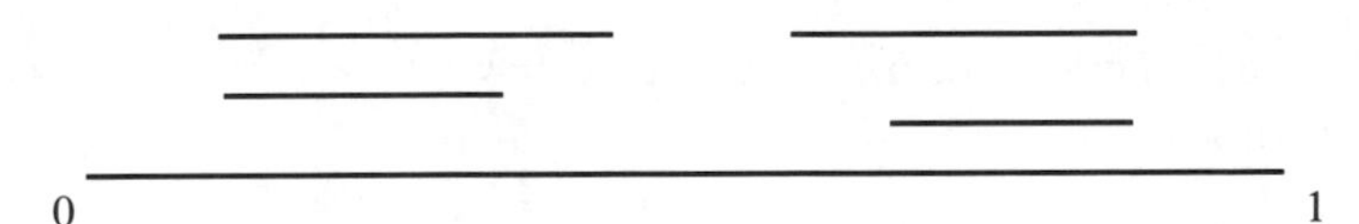

FIGURE 144

Suppose the initial points of the intervals in T are

$$a_0 = 0 < a_1 < a_2 < \cdots < a_n,$$

and that the interval that starts at a_k is called i_k (figure 145).

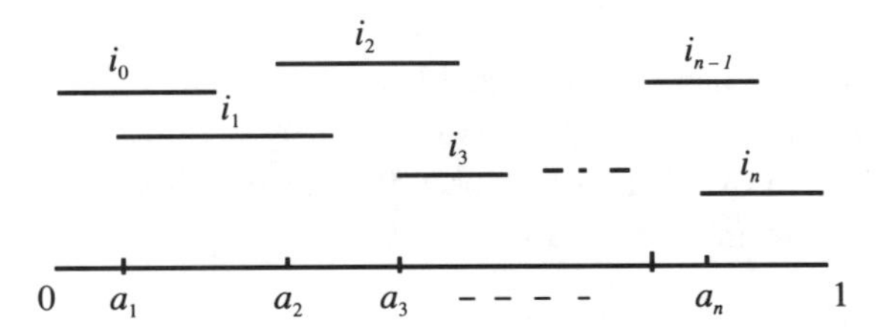

FIGURE 145

The crucial property of T is that *the intervals* i_k *and* i_{k+2} *are always disjoint.* To see this, suppose to the contrary that i_k and i_{k+2} overlap (figure 146). Now, the interval i_{k+1} must begin at a_{k+1} somewhere between a_k and a_{k+2}, and

(i) if i_{k+1} were to stop short of the terminal end of i_{k+2}, then i_{k+1} itself would be inessential, a contradiction,

(ii) if i_{k+1} goes beyond the terminal end of i_{k+2}, the i_{k+2} would be inessential, a contradiction.

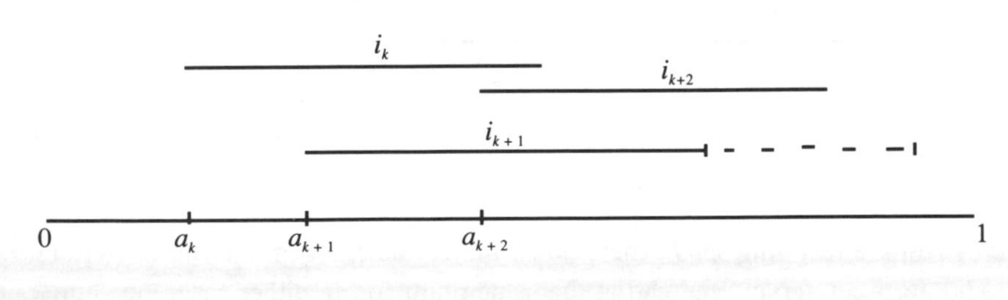

FIGURE 146

Thus i_k and i_{k+2} must be disjoint.

Clearly, if i_k can't overlap i_{k+2}, it can't extend as far as a_{k+2}, let alone to any of the intervals $i_{k+3}, i_{k+4}, \ldots$, which start even later. It follows that the subset

$$X = \{i_0, i_2, i_4, \ldots\},$$

containing all the intervals having even subscripts, is a set of pairwise disjoint intervals, and so is the complementary subset

$$Y = \{i_1, i_3, i_5, \ldots\}.$$

Since X and Y together cover the entire unit interval, we have

$$\text{(the total length of } X) + \text{(the total length of Y)} \geq 1,$$

implying that at least one of X and Y has a total length of at least $\frac{1}{2}$.

(An alternative solution is given in 1990, 296.)

A Brilliant 1-1 Correspondence

There are 15 ways of mixing two or fewer y's into a row of three x's:

$$xxx,\ yxxx,\ xyxx,\ xxyx,\ xxxy,\ yyxxx,\ xyyxx,\ xxyyx,$$

$$xxxyy,\ yxyxx,\ yxxyx,\ yxxxy,\ xyxyx,\ xyxxy,\ xxyxy.$$

In how many ways can r or fewer y's be mixed into a row of n x's?

It is easy to see that the number of ways of mixing in *exactly* k y's is $\frac{(n+k)!}{n!k!}$: of the $n+k$ letters altogether, n are alike of one kind and k are alike of a second kind.

This is just the binomial coefficient $\binom{n+k}{n}$, which is appropriately the number of ways of picking n of the $n+k$ places for the x's, and we quickly have that the required number is

$$N = \sum_{k=0}^{n} \binom{n+k}{n} = \binom{n}{n} + \binom{n+1}{n} + \binom{n+2}{n} + \cdots + \binom{n+r}{n}.$$

The problem is to evaluate this sum.

In my *More Mathematical Morsels* (pp. 126–7), it is shown how this series adds up to $\binom{n+r+1}{n+1}$:

Clearly there are $\binom{n+r+1}{n+1}$ ways of selecting a subset of $n+1$ elements from $\{1, 2, \ldots, n+r+1\}$; that is, $\binom{n+r+1}{r}$ ways, since $\binom{t}{s} = \binom{t}{t-s}$. Now, the greatest element in such a subset can be as small as $n+1$ and as big as $n+r+1$, and any value in between. The number of selections

- whose greatest element is $n+1$ is $\binom{n}{n}$ (the other n elements must come from the smaller elements $\{1, 2, \ldots, n\}$),
- whose greatest element is $n+2$ is $\binom{n+1}{n}$ (the others are from $\{1, 2, \ldots, n+1\}$),

....

- whose greatest element is $n+r+1$ is $\binom{n+r}{n}$.

This is undoubtedly a very perceptive observation, but Alexander Soifer has pointed out a brilliant 1-1 correspondence that obviates the need to add up this series [1].

Let the y's that are missing from each arrangement be tacked on the end to bring it up to the full complement of r y's. Of course, this obscures the identity of most of the original arrangements for we can't tell which of any final y's that occcur were added and which were there to begin with. In order not to lose its identity, we need to mark the end of an arrangement before adding on the y's. To use another y for this purpose is certainly out of the question; however, an x will do nicely. Thus, let each arrangement be capped with an x before attaching the missing y's. This gives a total of $n+1$ x's, and the last one and all the y's after it plainly constitute the appendage. For example, in the case of three x's and two y's,

$xxxy \to xxxyxy$ since one of the two y's is unused,

$xyyx \to xyyxx$ since no y was unused,

$xxx \to xxxxyy$ sincc both y's arc unused.

Conversely, if the tail consisting of the final x and all the y's that follow it are cut from an arrangement containing $n+1$ x's and *exactly* r y's, an arrangement of the kind under consideration results. Thus there is a 1-1 correspondence between the arrangements in question and the $\binom{n+r+1}{r}$ ways of arranging $n+1$ x's and r y's. The conclusion follows.

1. A. Soifer: *Mathematics As Problem Solving* (1988). This problem was used in the final round (an oral round) of the 1985 competition Math Counts (a national competition).

The Steiner-Lehmus Problem Revisited

At the mention of the Steiner-Lehmus problem, some people might have to suppress an involuntary "Not that old problem again!". Admittedly the problem has been solved in dozens of ways since it was posed in 1840. Nevertheless, I beg your indulgence in presenting a solution by Mowaffaq Hajja, Kuwait University, in the hope that a really nice piece of work is always welcome. (Professor Hajja informs me that he is evidently not the only one to have discovered this solution, for after finding it himself, he came across it in the Russian volume *Problems in Elementary Mathematics* by Lidsky, Ovsyannikov, Tulaikov, and Shabunin (Mir Publishers, Moscow), as problem 342 on pages 52 and 224.)

> In this famous problem it is given that the angle-bisectors BD and CE of triangle ABC have the same length inside the triangle ($x = y$ in figure 147), and one is required to prove that the triangle is isosceles ($AB = AC$).

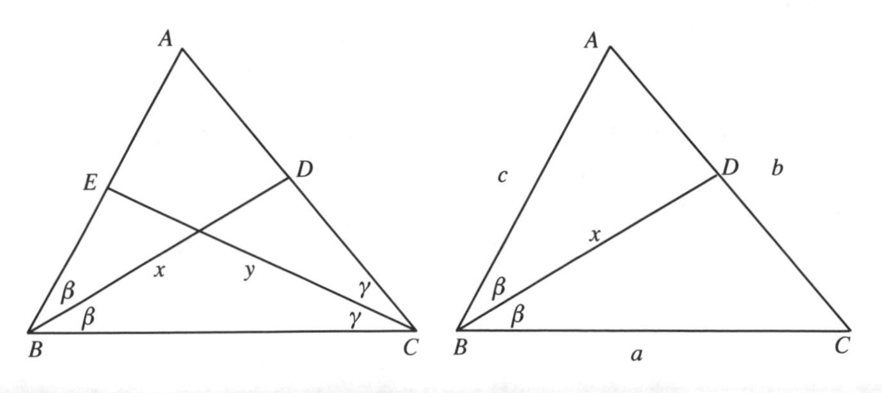

FIGURE 147

Let $\angle B = 2\beta$ and $\angle C = 2\gamma$ and let the sides of $\triangle ABC$ be a, b, c, as usual. Then, using the formula $\frac{1}{2}pq\sin R$ for the area of a triangle, the obvious relation

$$\triangle ABC = \triangle ABD + \triangle DBC$$

gives

$$\frac{1}{2}ac\sin 2\beta = \frac{1}{2}cx\sin\beta + \frac{1}{2}xa\sin\beta.$$

Since $\sin 2\beta = 2\sin\beta\cos\beta$, dividing through by $\frac{1}{2}\sin\beta$, we get

$$2ac\cdot\cos\beta = cx + xa,$$

from which

$$\cos\beta = \frac{1}{2}x\left(\frac{1}{a}+\frac{1}{c}\right).$$

Similarly, for the bisector y of angle C $(= 2\gamma)$, we have

$$\cos\gamma = \frac{1}{2}y\left(\frac{1}{a}+\frac{1}{b}\right).$$

Now suppose that the sides b and c are not equal; for definiteness suppose $b > c$. In this case, the angle B, opposite the longer side b, is greater than angle C. Thus $\beta > \gamma$, and we have

$$\cos\beta < \cos\gamma,$$

giving

$$\frac{1}{2}x\left(\frac{1}{a}+\frac{1}{c}\right) < \frac{1}{2}y\left(\frac{1}{a}+\frac{1}{b}\right).$$

But x and y are equal, and therefore this yields

$$\frac{1}{c} < \frac{1}{b},$$

making $b < c$, contradicting our supposition that $b > c$. Hence b and c cannot be unequal and triangle ABC must be isosceles.

Professor Hajja also noted the following corollary.

A Corollary. *From the relation $\cos\gamma = \frac{1}{2}y(\frac{1}{a}+\frac{1}{b})$, it follows that*

$$\frac{2}{\frac{1}{a}+\frac{1}{b}} = \frac{y}{\cos\gamma},$$

*that is to say, $\frac{y}{\cos\gamma}$ is the **harmonic mean** between a and b.*

Therefore, if γ were to be made equal to $60°$, the harmonic mean h would be given by

$$h = \frac{y}{\frac{1}{2}} = 2y.$$

Thus a neat construction for the harmonic mean h between a and b is illustrated in figure 148:

(i) lay off $AC = b$ and $CB = a$ on the arms of a $120°$-angle C,
(ii) bisect the angle to get S on AB (see figure 148), and
(iii) extend CS its own length to T.

Then

$$CT = 2y = h.$$

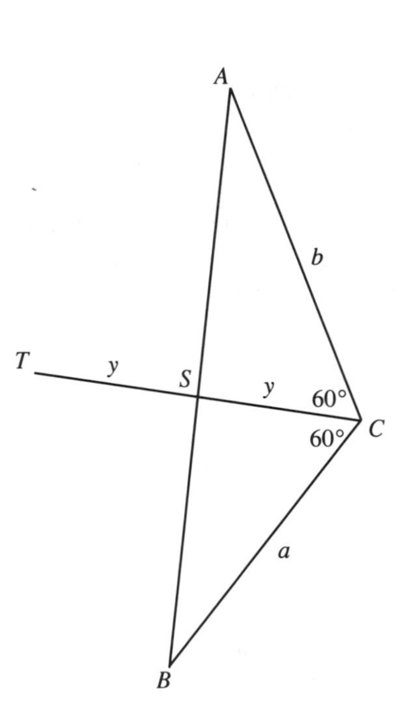

FIGURE 148

Two Problems from the 1987 Bulgarian Olympiad

1. (*Crux Mathematicorum,* 1989, 33) If one begins a sequence of positive integers $\{x_n\}$ with $x_1 = x_2 = 1$ and derives succeeding terms by the unassuming procedure of multiplying the preceding term by 14, subtracting the term ahead of that, and then subtracting 4, that is, for $n \geq 1$,

$$x_{n+2} = 14x_{n+1} - x_n - 4,$$

the next few terms turn out to be

$$x_3 = 14(1) - 1 - 4 = 9 = 3^2,$$
$$x_4 = 14(9) - 1 - 4 = 121 = 11^2,$$
$$x_5 = 14(121) - 9 - 4 = 1694 - 13 = 1681 = 41^2,$$
$$x_6 = 14(1681) - 121 - 4 = 23534 - 125 = 23409 = 153^2,$$

all of which are perfect squares.

Prove, in fact, that x_n is *always* a perfect square.

The following ingenious solution is due to my colleague David Jackson. The recurrence gives

$$x_{n+2} = 14x_{n+1} - x_n - 4$$

and

$$x_{n+1} = 14x_n - x_{n-1} - 4,$$

and subtraction yields

$$x_{n+2} - x_{n+1} = 14x_{n+1} - 15x_n + x_{n-1},$$

i.e.,

$$x_{n+2} - 15x_{n+1} + 15x_n - x_{n-1} = 0.$$

Solving the auxiliary equation

$$x^3 - 15x^2 + 15x - 1 = 0,$$

we get

$$(x-1)(x^2 - 14x + 1) = 0 \quad \text{and} \quad x = 1, 7 \pm 4\sqrt{3},$$

giving

$$x_n = (7 + 4\sqrt{3})^n \cdot a + 1^n \cdot b + (7 - 4\sqrt{3})^n \cdot c,$$

where a, b, and c are constants.

From here the straightforward procedure would be to determine a, b, and c by solving the system of equations obtained by substituting $x_n = 1$, 1, and 9 for $n = 1$, 2, and 3 respectively, and then show that the resulting formula for x_n gives a perfect square for all n. However, David avoids the need to undertake this tedious path by employing the following elegant argument.

Observe that we are not required to produce a formula for x_n, but just to prove that x_n is always a perfect square. We begin by asking what are the prospects of expressing our general solution as a perfect square:

$$\begin{aligned} x_n &= (7 + 4\sqrt{3})^n \cdot a + 1^n \cdot b + (7 - 4\sqrt{3})^n \cdot c \\ &= p^2 + 2pq + q^2 = (p+q)^2? \end{aligned}$$

Are $7 \pm 4\sqrt{3}$ the squares of any reasonably simple numbers? Having thought to ask this question is the crucial step, for once asked, it is a small step to the answer

$$7 \pm 4\sqrt{3} = (2 \pm \sqrt{3})^2.$$

Noting also that the product $(2+\sqrt{3})(2-\sqrt{3}) = 1$, which is the third root of the auxiliary equation, we have

$$x_n = \left[(2+\sqrt{3})^n\right]^2 \cdot a + (2+\sqrt{3})^n(2-\sqrt{3})^n \cdot b + \left[(2-\sqrt{3})^n\right]^2 \cdot c$$

suggesting that x_n is the square of an expression of the form

$$y_n = (2+\sqrt{3})^n \cdot u + (2-\sqrt{3})^n \cdot v,$$

where u and v are appropriate constants.

Now, any such sequence $\{y_n\}$ would be the solution of a recurrence whose auxiliary equation is

$$\begin{aligned} \left[y - (2+\sqrt{3})\right]\left[y - (2-\sqrt{3})\right] &= 0, \\ y^2 - 4y + 1 &= 0, \end{aligned}$$

that is, of the recurrence

$$y_n = 4y_{n-1} - y_{n-2}.$$

From this it follows that, if two consecutive values of y_n are integers, then all succeeding terms will also be integers. Therefore, let values be assigned to the constants u and v so that $y_1 = y_2 = 1$. This would insure that $\{y_n\}$ is a sequence of integers, would make $y_1^2 = 1 = x_1$ and $y_2^2 = 1 = x_2$, and would also give $y_3 = 4(1) - 1 = 3$, making $y_3^2 = 9 = x_3$, which is encouraging.

With such u and v, then, setting

$$z_n = y_n^2 = (7 + 4\sqrt{3})^n \cdot u^2 + 2uv \cdot 1^n + (7 - 4\sqrt{3})^n \cdot v^2,$$

gives a sequence $\{z_n\}$ of integer squares which begins $\{1, 1, 9\}$, *just as* $\{x_n\}$ *does,* and which satisfies the recurrence whose auxiliary equation has roots $x = 1,\ 7 \pm 4\sqrt{3}$, that is, the recurrence

$$x_{n+2} = 15x_{n+1} - 15x_n + x_{n-1},$$

just as $\{x_n\}$ *does.* Thus $\{z_n\}$ is locked in step with $\{x_n\}$ and $x_n = z_n = y_n^2$ for all n. (An alternative solution is given in 1990, 292.)

We note that variations on this theme provide a general approach to a whole class of problems. Here are two pretty results due to my colleague Ian McGee.

1. Prove that the sequence $\{x_n\}$ defined by $x_1 = x_2 = 1$ and, for $n \geq 1$,

$$x_{n+2} = 3x_{n+1} - x_n - 2(-1)^n,$$

thus beginning $\{1, 1, 4, 9, 25, \ldots\}$ is none other than the squares of the Fibonacci numbers $\{f_n^2\}$.

2. Prove that

$$x_{n+2} = 4x_{n+1} - x_n - 2(-1)^n,$$

with $x_1 = x_2 = 1$, generates a sequence, beginning

$$\{1, 2, 9, 32, 121, 450, 1681, 6272, \},$$

in which x_{2n+1} is always a perfect square and x_{2n} is always twice a perfect square; show also that

$$\sqrt{x_{2n+1}} = x_n + x_{n+1}.$$

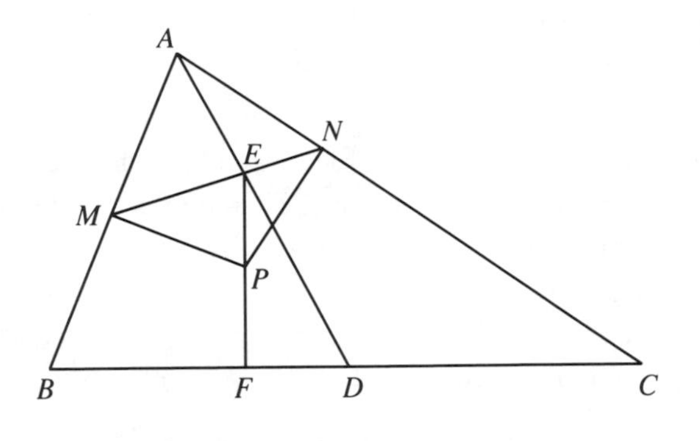

FIGURE 149

2. (*Crux Mathematicorum,* 1990, 293) From a point E on median AD of $\triangle ABC$ the perpendicular EF is dropped to BC, and a point P is chosen on EF. Then perpendiculars PM and PN are drawn to the sides AB and AC.

Now, it is most unlikely that M, E, and N will lie in a straight line, but in the event that they do, prove that AP bisects $\angle A$.

Our solution is by Toshio Seimiya of Kawasaki, Japan.

With right angles at M and N, $AMPN$ is a cyclic quadrilateral, and in its circumcircle

$\angle MAP = \angle MNP \ (= x)$ on chord MP, and

$\angle PAN = \angle PMN \ (= y)$ on chord PN (figure 150).

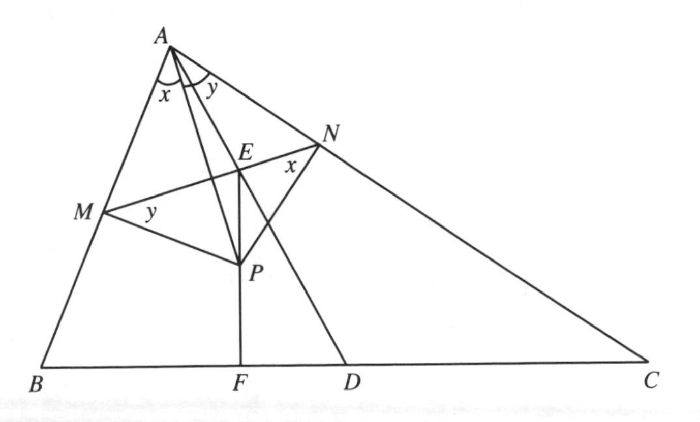

FIGURE 150

The problem of proving $x = y$, then, reduces to showing

$$(y =)\ \angle PMN = \angle PNM\ (= x).$$

Now, it is not at all evident how to use the fact that E lies on the median AD. However, Toshio Seimiya proceeds brilliantly as follows.

Because E is on the median AD, it is the midpoint of the segment KEL across the triangle which is parallel to the side BC (figure 151): similar triangles yield $\frac{KE}{BD} = \frac{EL}{DC}$, and since $BD = DC$, then $KE = EL$.

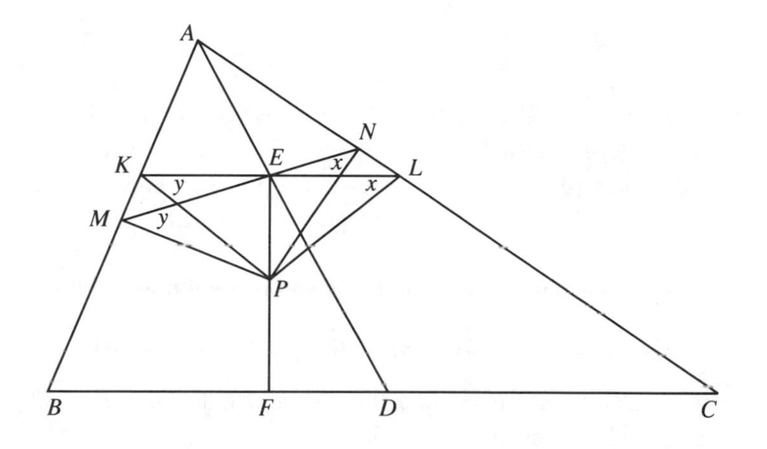

FIGURE 151

Since EF is perpendicular to BC and KL is parallel to BC, then KL is also perpendicular to EF, and therefore PL subtends a right angle at each of E and N, making $EPLN$ cyclic. Hence

$$\angle ELP = \angle ENP = x$$

on chord EP. Similarly the right angles at M and E make $KMPE$ cyclic and

$$\angle EKP = \angle EMP = y$$

(again on chord EP). But, with E the midpoint of KL, EP is the perpendicular bisector of KL, and therefore $PK = PL$, making triangle PKL isosceles and its base angles x and y equal.

Two Problems from the 1987 Hungarian National Olympiad

1. (*Crux Mathematicorum,* 1989, 100) Prove that every selection of 1325 integers from $M = \{1, 2, \dots, 1987\}$ must contain some three numbers $\{a, b, c\}$ which are relatively prime in pairs, but that this can be avoided if only 1324 integers are selected.

The set M contains 993 even numbers and 662 multiples of 3:

$$A = \{2, 4, 6, \dots, 1986\}, \text{and} \quad B = \{3, 6, 9, \dots, 1986\}.$$

Since every sixth integer is a multiple of both 2 and 3, the number of different integers in A and B altogether is only

$$993 + 662 - \frac{1986}{6} = 1655 - 331 = 1324.$$

Now, any triple of numbers taken from the union of A and B must contain two numbers from A or two from B and thus contain a pair which fails to be relatively prime. Hence the subset $A \cup B$ is an example of 1324 integers from M in which no three are pairwise relatively prime.

Suppose, however, that S is a subset of 1325 integers from M. If the prime number 1987 belongs to S, then it is easy to pick two numbers to go with it to form a pairwise relatively prime triple: since 1324 is more than half of 1986, among the other 1324 members of S there must be two that are consecutive, t and $t+1$, and therefore relatively prime; then $(t, t+1, 1987)$ is a pairwise relatively prime triple, for neither t nor $t+1$ can have a nontrivial common factor with the larger prime number 1987.

Suppose, then, that 1987 does not belong to S. In this case, S is selected from the set $N = \{1, 2, \dots, 1986\}$, which consists of 331 abutting intervals of six integers each:

$$N = \{1, 2, 3, 4, 5, 6; 7, 8, 9, 10, 11, 12; 13, \dots; 1981, 1982, 1983, 1984, 1985, 1986\}.$$

Now, 1324 is exactly two-thirds of 1986, and therefore, with its 1325 members, S contains more than two-thirds of the integers in N. Consequently, since $331 \cdot 4$ is only 1324, at least one of these abutting intervals must contribute at least five of its six integers to S. As we shall see, any interval that does this provides S with a desired pairwise relatively prime triple.

Observing that the first number in each interval is odd, such an interval may be expressed

$$I = O_1, E_1, O_2, E_2, O_3, E_3.$$

Now, among strings of consecutive integers there are more than just adjacent pairs that are relatively prime. Because any common divisor d of two integers must divide their difference, *consecutive odd* numbers are also relatively prime (d must divide 2, making it 1 or 2, and 2 does not divide an odd number). Thus a consecutive triple (O, E, O) is pairwise relatively prime, as is the set of three consecutive odd numbers $(O, -, O, -, O)$. Therefore, in the event that S were to receive all six of the integers from an interval I, S would come into a desired pairwise relatively prime triple (O, E, O) among others.

Suppose, then, that I contributes only five of its six integers to S. If the integer r that is retained in I is even, then S gets a desired triple (O, E, O) of consecutive integers. In fact, a moment's reflection confirms that S receives such a consecutive triple in all cases except when the reserved integer r is the middle odd number O_2, giving S the five integers

$$O_1, E_1, -, E_2, O_3, E_3.$$

Now, (O_1, E_1) and (O_1, O_3) are relatively prime pairs, and therefore O_3 would complete a desired triple $T = (O_1, E_1, O_3)$ when E_1 and O_3 are relatively prime. Unfortunately, E_1 and O_3 are not always relatively prime. But since their difference is 3, their greatest common divisor can fail to be 1 only when they are both multiples of 3. When this is the case, however, the neighboring integers O_1 and E_2 cannot also be multiples of 3, and are therefore relatively prime, implying $T = (O_1, E_2, O_3)$ is a desired triple. Thus one way or the other S always has a pairwise relatively prime triple T.

2. (*Crux Mathematicorum,* 1991, 69) What is the minimum value of the function

$$f(x) = \sqrt{a^2 + x^2} + \sqrt{(b - x)^2 + c^2}$$

where a, b, and c are positive real numbers?

Our ingenious solution is by Mangho Ahuja, Southeast Missouri State University, and independently by D. J. Smeenk, Zaltbommel, The Netherlands. (An alternative solution is given in 1992, 171.)

While it might be quite obvious that $\sqrt{a^2+x^2}$ is the length of the hypotenuse of a right triangle having sides a and x, and that $\sqrt{(b-x)^2+c^2}$ is the hypotenuse of a right triangle with sides $(b-x)$ and c, the idea of placing such triangles so that they abut along a common base line AB of length b is very clever indeed (figure 152). The function $f(x)$, then, is simply the length of a polygonal path from C to D that bounces off AB, and the determination of the minimum such path is an old problem that many will remember solving in their study of mirrors in high school physics.

It is easy to see that the position of P on AB which yields the minimum path CPD is the point of intersection of AB and the line joining C to the image D' obtained when D is reflected in AB (figure 153). Let Q be any

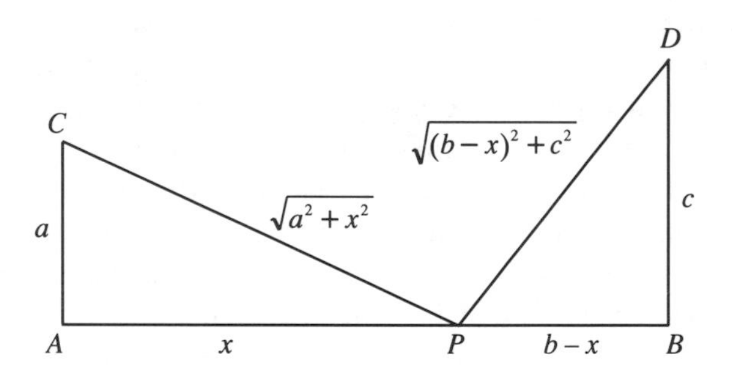

FIGURE 152

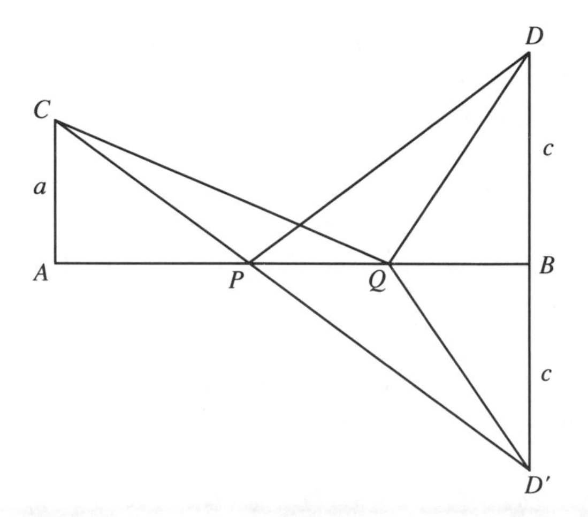

FIGURE 153

other point on AB. Now, clearly the mirror AB is the perpendicular bisector of DD', making $PD = PD'$ and $QD = QD'$. Therefore

$$\text{path } CPD = CP + PD = CP + PD' = CD',$$

while

$$\begin{aligned} \text{path } CQD &= CQ + QD \\ &= CQ + QD' \\ &> CD' \quad \text{(by the triangle inequality)} \\ &= \text{path } CPD. \end{aligned}$$

Thus, if CR is drawn perpendicular to DD' (figure 154), we have $BR = a$ and

$$\begin{aligned} \text{the minimum path } CPD &= CD' \\ &= \text{the hypotenuse of right triangle } CRD' \\ &= \sqrt{b^2 + (a + c)^2}. \end{aligned}$$

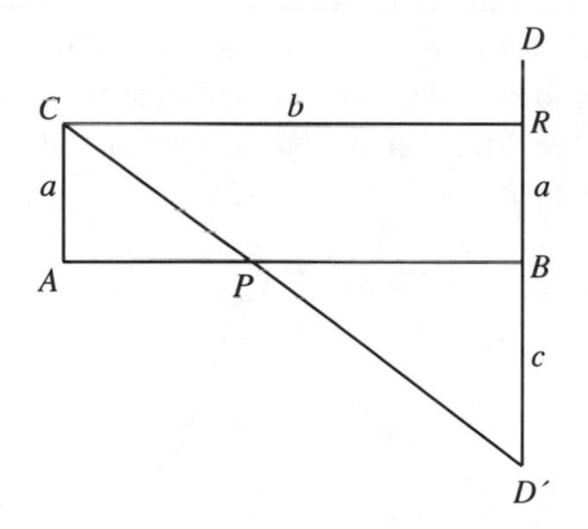

FIGURE 154

A Problem from the 1987 Canadian Olympiad†

An odd number of people, each armed with a water gun, are spread around a field so that, for each of them, each of the others is a different distance away. At a given signal, each person fires at and hits his nearest neighbor. Prove that some lucky person doesn't get wet at all.

Clearly, if there are just three people (A, B, C), one of them must remain dry (figure 155): if AB is the shortest side of $\triangle ABC$, which the different distances guarantee to be scalene, then A and B shoot each other and nobody shoots C. It is a little disappointing that this wonderful-sounding problem is so easily solved by induction.

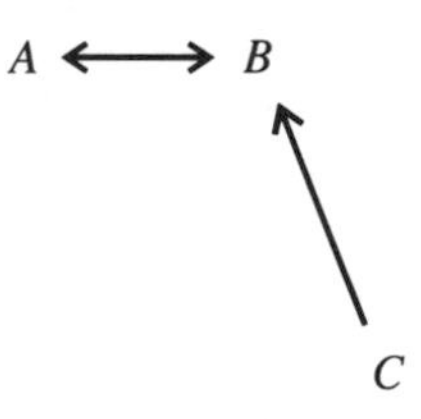

FIGURE 155

Suppose the property in question holds for any $n-2$ suitably located people, n odd, and that S is a set of n such people. Of all the distances determined by pairs of people in S, let the distance between A and B be one of those which is minimal (presumably there could be more than one realization of the minimum distance). Since there is no one closer to A than B and no one closer to B than A, they must shoot each other.

† (*Crux Mathematicorum,* 1987, 172)

Now suppose this game were played by the $n - 2$ people in the subset $R = S - A - B$. For each person in R, each of the others in R is still a different distance away, and since $n - 2$ is also odd, the induction hypothesis implies that some person X in R does not get wet in this restricted version of the game.

However, the presence of A and B, who shoot each other, would not induce anyone to shoot X; if anything, the only effect of introducing A and B into the game would be to draw off the fire of those members of R who happen to have A or B as their nearest neighbor. In any case, X remains dry.

Since the claim holds for $n = 3$, the conclusion follows by induction for all odd n.

Problem 1123 from Crux Mathematicorum†

Suppose ABC is a scalene triangle. Let BD and CE bisect angles B and C, and let BD and CE be laid off along BA and CA, respectively, to give points F and G (figure 156). Prove that, if angle A is $60°$, then FG will be parallel to BC.

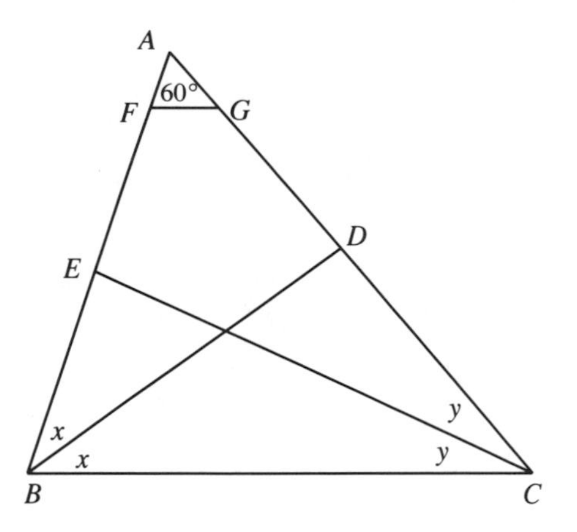

FIGURE 156

Actually this is the converse of Problem 1123, which was posed by J. T. Groenman (Arnhem, The Netherlands), and the following solution is the reverse of the published solution by D. J. Smeenk (Zaltbommel, The Netherlands).

† (*Crux Mathematicorum,* 1987, 199)

Let angles B and C be $2x$ and $2y$ respectively (figure 156). Then, since $\angle A = 60°$, we have $2x + 2y = 120°$, giving $3x + 3y = 180°$.

As usual, let a, b, and c denote the lengths of the sides BC, AC and AB. We shall show that FG is parallel to BC by establishing the proportion

$$\frac{BF}{BA} = \frac{CG}{CA}.$$

Since $BF = BD$ and $CG = CE$, this is just

$$\frac{BD}{c} = \frac{CE}{b}, \quad \text{or equivalently,} \quad \frac{BD}{CE} = \frac{c}{b}.$$

From the law of sines we have

$$\frac{c}{b} = \frac{\sin C}{\sin B} = \frac{\sin 2y}{\sin 2x},$$

and so we would like to show that

$$\frac{BD}{CE} = \frac{\sin 2y}{\sin 2x}.$$

But this is easy.

From $\triangle BDC$, the law of sines gives

$$\frac{BD}{BC} = \frac{\sin C}{\sin \angle BDC}.$$

Since $\angle BDC + (x + 2y) = 180°$ in $\triangle BDC$, we have $\sin \angle BDC = \sin(x + 2y)$, yielding

$$\frac{BD}{BC} = \frac{\sin 2y}{\sin(x + 2y)},$$

and giving

$$BD = \frac{BC \cdot \sin 2y}{\sin(x + 2y)}.$$

Similarly, from $\triangle BEC$ we get

$$\frac{CE}{BC} = \frac{\sin 2x}{\sin(2x + y)} \quad \text{and} \quad CE = \frac{BC \cdot \sin 2x}{\sin(2x + y)}.$$

Hence

$$\frac{BD}{CE} = \frac{\sin 2y \cdot \sin(2x + y)}{\sin 2x \cdot \sin(x + 2y)}.$$

But

$$(x + 2y) + (2x + y) = 3x + 3y = 180°,$$

making these angles supplementary, and therefore $\sin(x+2y)=\sin(2x+y)$. Hence

$$\frac{BD}{CE}=\frac{\sin 2y}{\sin 2x},$$

as desired.

A Problem from the 1987 AIME†

The unavoidable 1 and n are trivial divisors of a positive integer n. In keeping with this, the divisors between 1 and n are nontrivial, and if n is equal to the product P of its nontrivial divisors, let's call it a *nice* number. The smallest nice number is $6 = 2 \cdot 3$, and in this question we are asked simply to determine the sum of the first 10 nice numbers.

Obviously any positive integer which is the product of exactly two different prime numbers, $n = pq$, is nice. Thus

$$2 \cdot 3,\ 2 \cdot 5,\ 2 \cdot 7,\ 3 \cdot 5,\ 3 \cdot 7,\ 2 \cdot 11,\ 2 \cdot 13,\ 3 \cdot 11,\ 5 \cdot 7,\ \text{and } 3 \cdot 13$$

are 10 nice numbers, and their sum is

$$6 + 10 + 14 + 15 + 21 + 22 + 26 + 33 + 35 + 39 = 221.$$

Unfortunately, the correct answer is 182, and so we must have skipped over some nice numbers of a different type. In fact, we missed two of them, namely 8 and 27, making the correct sum

$$6 + 8 + 10 + 14 + 15 + 21 + 22 + 26 + 27 + 33 = 182.$$

This underscores the value of looking at a few early cases in detail. If we had tested the integers between 6 and 10, we would have caught the 8 and perhaps realized that a second form for nice numbers is the cube of a prime number, $n = p^3$, with its divisors 1, p, p^2, and p^3.

Of course we still need to consider the possibility of other forms:

† (*Crux Mathematicorum*, 1987, 109)

(i) For higher powers of a single prime, $n = p^{3+t}$, $t \geq 1$, the nontrivial divisors p^2 and p^{2+t} by themselves make the product $P \geq p^{4+t} > n$, closing off this possibility;

(ii) For n the product of two or more primes with at least one occurring more than once, $n = p^m q \ldots$, $m \geq 2$, the two nontrivial divisors p^m and p are enough to provide P with a factor p^{m+1}, which exceeds p^m, making $n = P$ impossible; and

(iii) Similarly, if $n = pqr \ldots$, the product of more than two primes, each occurring to the first degree, the divisors p and pq give P a factor of p^2, exceeding the power of p in n.

Thus nice numbers are only given by the forms pq and p^3, assuring us that the first 10 nice numbers are indeed as stated above.

A Generalization of Old Morsel 3

In my first book of morsels (*Mathematical Morsels,* Dolciani Series, MAA, 1978), problem 3 asks for the number of regions into which the interior of a circle is divided by a certain set of chords, no three of which are concurrent inside the circle. On the way to the solution, the slightly more general formula, $C + P + 1$, is derived for any set of C such chords, where P is the number of points of intersection inside the circle. Clearly, transforming the bounding circle to a closed convex curve would not alter the number of regions in the interior; also, if the sections along the chords between consecutive points of intersection were replaced by simple arcs, no change in the number of regions would result.

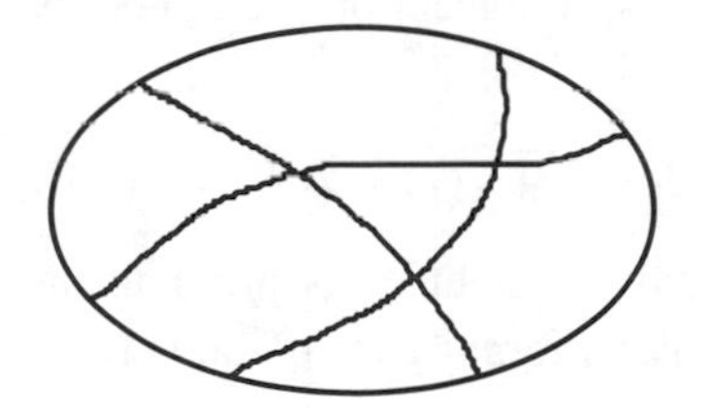

FIGURE 157

It is an entirely different matter, however, if the chords are replaced by arbitrary curves and the concurrency stipulation is dropped. It takes a little thought to appreciate the extensive complications that attend this generalization. Not only can many curves concur at a point but the same two curves can cross each other any number of times. A much greater complication, however, is due to the fact that this admits *self-intersecting* curves, which can interact with other curves and also loop around themselves to produce any number of points of intersection on their own (figure 158).

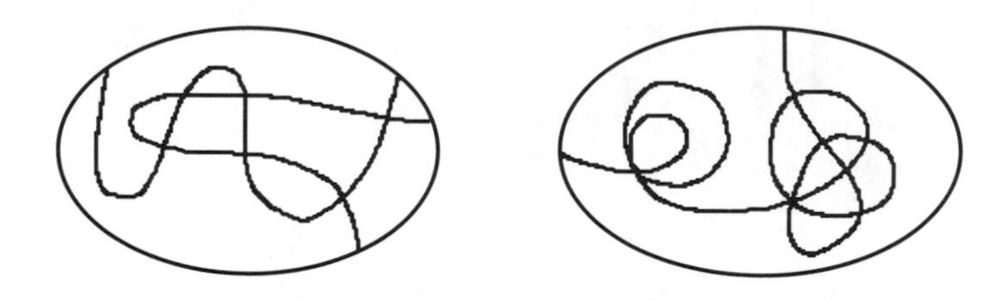

FIGURE 158

It is a great pleasure, therefore, to report that I. Hermann, of Krasnojarsk, Russia, has written to say that only a slight generalization of the formula $C+P+1$ is needed to account for all these possibilities. It suffices to extend the middle term P to the sum

$$\sum_{n\geq 2}(n-1)p_{2n},$$

where p_{2n} is the number of points of intersection of degree $2n$. Hermann's formula for the number of regions R is therefore

$$R = C + \sum_{n\geq 2}(n-1)p_{2n} + 1.$$

For example, in figure 159 there are 3 curves (the number of curves is one-half the number of points of contact on the boundary), $p_6 = 2$, $p_4 = 7$, and we have

$$R = 3 + 1(7) + 2(2) + 1 = 15.$$

Proof. Mr. Hermann did not include his proof in his letter but the formula follows nicely from Euler's formula for plane networks

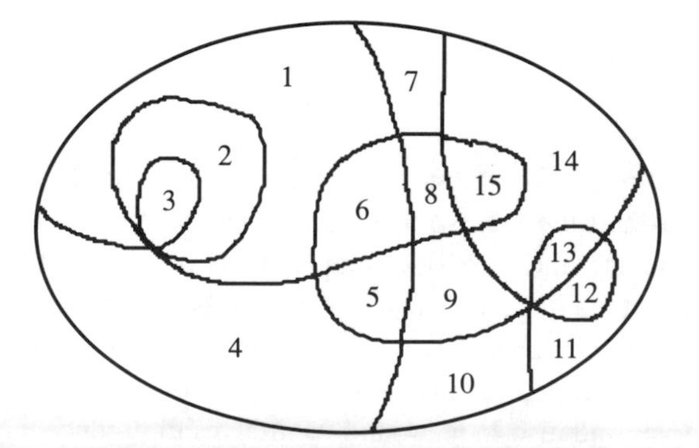

FIGURE 159

$$V - E + F = 1,$$

where F, to our liking, does *not* count the infinite outer face.

Since unwanted complications arise if we allow curves to cross on the boundary, let us insist that all points of intersection occur in the *interior* of the figure. Furthermore, let us require each curve to have *two distinct endpoints* on the boundary. (This is not really essential, for it can always be achieved without changing the number of regions in the figure by a careful splitting of the "multiple" endpoints into sufficiently nearby points on the boundary.)

Clearly, to solve Euler's formula for F, we need to calculate the values of V and E. Since a point of intersection lies in the interior of the figure, each time a curve goes through it, it must both enter and leave the point thus making its degree an *even* number; furthermore, in order to mark an intersection, the degree must be at least 4. Therefore the possible degrees of the points of intersection are $4, 6, 8, \ldots$, up to some maximum $2k$ for a given figure.

Now, the total number of points of intersection is simply

$$p_4 + p_6 + p_8 + \cdots + p_{2k} = \sum_{n \geq 2} p_{2n}.$$

However, V counts all the points in the figure, and when the $2C$ endpoints on the boundary are included, we have

$$V = 2C + \sum_{n \geq 2} p_{2n}.$$

Since each arc and each loop contributes 2 to the sum S of all the degrees in the figure, E is just $\frac{1}{2}S$. Now, at an endpoint on the boundary the degree is 3, and therefore the $2C$ endpoints contribute $6C$ to S. Obviously each of the p_{2n} points of degree $2n$ contributes $2n$ to the sum, and we have

$$E = \frac{1}{2}\left[6C + \sum_{n \geq 2} 2n \cdot p_{2n}\right] = 3C + \sum_{n \geq 2} n \cdot p_{2n}.$$

Hermann's formula, then, follows immediately:

$$\begin{aligned} F &= 1 + E - V \\ &= 1 + \left[3C + \sum^{n \geq 2} n \cdot p_{2n}\right] - \left[2C + \sum_{n \geq 2} p_{2n}\right] \\ &= C + \sum_{n \geq 2} (n-1) \cdot p_{2n} + 1. \end{aligned}$$

Two Problems from the 1991 Canadian Olympiad†

1. Prove that the equation

$$x^2 + y^5 = z^3$$

has infinitely many nonzero integral solutions (x, y, z).

The official solution (*Crux Mathematicorum,* 1991, 198) is a model of succinctness: after noting $(3, -1, 2)$ is a solution, it is simply observed that, whenever (x, y, z) is a solution, so is $(k^{15}x, k^6y, k^{10}z)$ for all integers k.

At first glance this observation strikes one as being exceedingly perceptive; however, it is based on quite a simple idea: if (x, y, z) satisfies

$$x^2 + y^5 = z^3,$$

then, multiplying through by k^m, where m is the least common multiple of the exponents, in this case 30 ($= \operatorname{lcm}(2, 5, 3)$), we get

$$k^{30}x^2 + k^{30}y^5 = k^{30}z^3,$$

that is,

$$(k^{15}x)^2 + (k^6y)^5 = (k^{10}z)^3.$$

This problem is therefore a "gift" to any contestant who covered this lcm approach in his training sessions. Its presence on the olympiad indicates that the examiners felt this would not be the case very often and that most of the contestants would have to invent the method on the spot or proceed in some other way. While there is certainly no topping this lcm approach, perhaps the following alternative line of thought might be of some interest.

---

† (*Crux Mathematicorum,* 1991, 161)

The difficulty, of course, is that x^2 and y^5 can't be added together as literal expressions. However, if x^2 and y^5 happen to be equal, their sum can be written as the single term $2x^2$ (or $2y^5$); and this is certainly possible, for it occurs when they are the same 10th power: when

$$x^2 = u^{10} = y^5 \quad \text{(i.e., } x = u^5,\ y = u^2\text{)}$$

then

$$x^2 + y^5 = 2u^{10}.$$

Therefore, let us look further into the ramifications of $x^2 = y^5$.

In this case, the equation would reduce to

$$2x^2 = z^3,$$

which suggests making x a power of 2 so that the coefficient can be absorbed into the exponent: if $x = 2^t$, then $2x^2 = 2^{2t+1}$, calling for $2t + 1$ to be divisible by 3 in order to make 2^{2t+1} a perfect cube. Clearly $t = 3k + 1$ is satisfactory for any integer k:

$$2^{2t+1} = 2^{6k+3} = (2^{2k+1})^3.$$

So far, then, we have

$$x = 2^t = 2^{3k+1} \quad \text{satisfies} \quad x^2 + y^5 = (2^{2k+1})^3,$$

provided $x^2 = y^5$, i.e., $2^{2t} = y^5$. It remains only to have t divisible by 5, that is,

$$t = 3k + 1 \equiv 0 \pmod 5.$$

Hence it suffices to take $k \equiv 3 \pmod 5$, i.e., $k = 5a + 3$, a an integer. Therefore

$$x = 2^t = 2^{3k+1} = 2^{15a+10}, \qquad y^5 = x^2 = 2^{30a+20}$$

yields

$$x^2 + y^5 = 2y^5 = 2 \cdot 2^{30a+20} = 2^{30a+21} = (2^{10a+7})^3,$$

and we have the infinite family of solutions

$$(x, y, z) = (2^{15a+10}, 2^{6a+4}, 2^{10a+7}),$$

a any integer.

2. Ten different numbers from the set $\{0, 1, 2, \ldots, 14\}$ are to be placed in the ten circles in figure 160. The positive differences between the pairs of numbers in adjacent circles (i.e., circles connected by an edge) are all to be different. Is it possible to do this? Justify your answer.

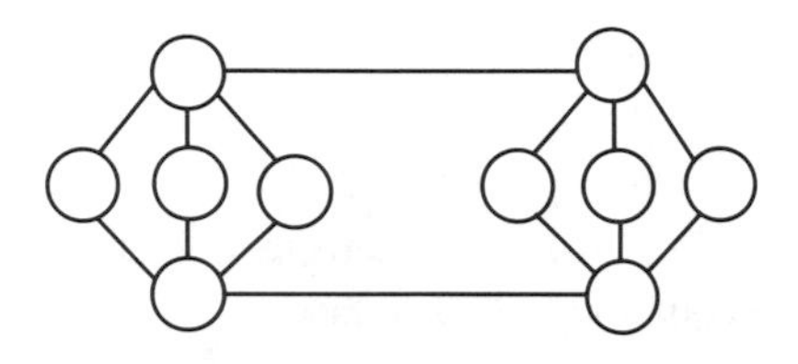

FIGURE 160

Again let us consider an alternative to the incisive official solution.

Let the numbers in the circles be labeled as in figure 161. Since there are 14 edges, indicating 14 different differences, and because the only possible differences are the 14 numbers $\{1, 2, \ldots, 14\}$, each difference must arise precisely once. Thus 7 of the differences must be odd, i.e. the difference between an odd number and an even number, and 7 must be even, i.e., the difference between two odd numbers or two even numbers.

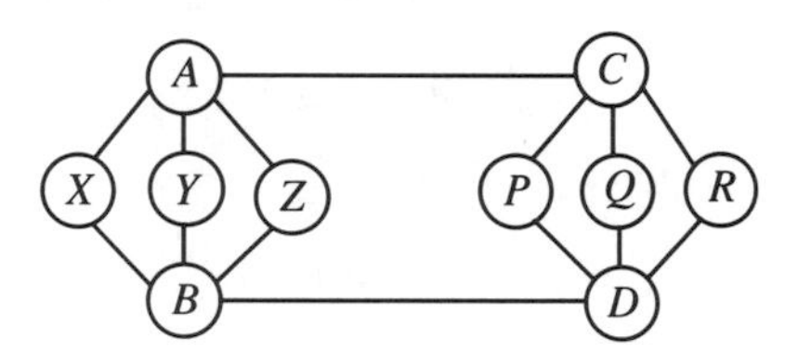

FIGURE 161

Now, *if A and B have the same parity,* then $|A - X|$ and $|B - X|$ have the same parity, and therefore entering X between A and B provides either two odd differences or two even differences; similarly for Y and Z. Similarly also for the other section containing $\{(C, D)$ and $(P, Q, R)\}$.

On the other hand, *if A and B are of opposite parity,* then X, Y, and Z each produce one odd difference and one even difference; similarly for $\{(C, D)$ and $(P, Q, R)\}$.

The only possibilities concerning the parities of A, B, C, and D are variations of the following four simple cases:

(i) (A, B) have the same parity, (C, D) have the same parity, and the parities of the two pairs are the same,
(ii) (A, B) have the same parity, (C, D) have the same parity, and the parities of the two pairs are opposite,
(iii) In one pair the numbers have the same parity and in the other the opposite parity,
(iv) The numbers in each pair have opposite parity.

Since the variations within a case are equivalent, we need examine only one example of each kind.

(i) *A,B,C,D are all even.* In this case, $|A - C|$ and $|B - D|$ are both even and the values of X, Y, Z, P, Q, R are required to provide 5 more even differences and 7 odd ones. However, each of X, Y, Z, P, Q, R produces two even or two odd differences and therefore 5 more even differences cannot be produced (nor 7 odd ones).

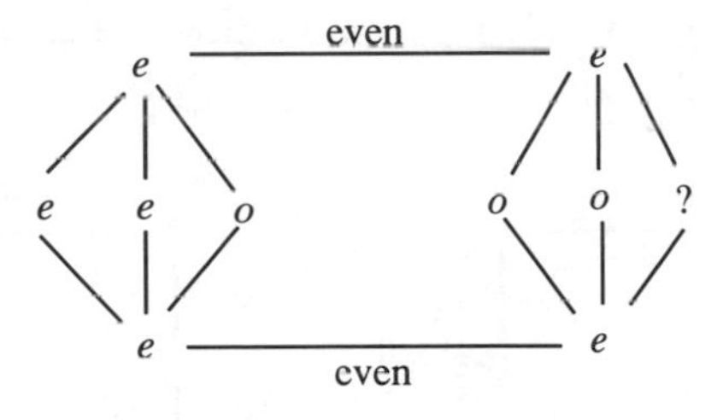

FIGURE 162. e = even entry, o = odd entry

(ii) (A, B) *even,* (C, D) *odd.* (This is essentially the same as (i).) Again X, Y, Z, P, Q, R each generates even and odd differences *two at a time* and cannot yield the additional 5 odd differences to go with $|A - C|$ and $|B - D|$.

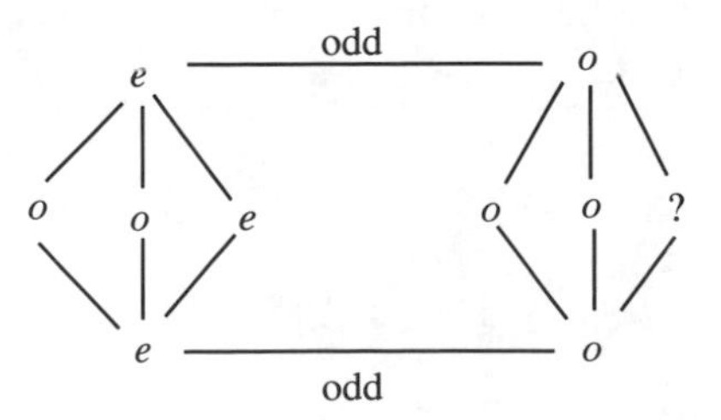

FIGURE 163

(iii) *A, B, C even, D odd.* This time $|A - C|$ is even and $|B - D|$ is odd and 6 of each kind of difference is wanting. Now, whatever the values of P, Q, R, each generates one odd difference and one even difference for a total of 3 of each kind, thus throwing the same demand on the entries X, Y, Z. However, each of X, Y, Z generates odd and even differences *two at a time* and therefore cannot supply the required 3 of each type.

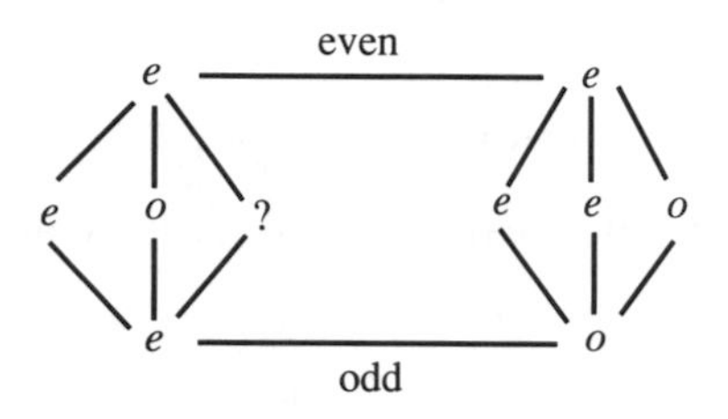

FIGURE 164

(iv) *A, C even, B, D, odd or B, C even, A, D odd.* In this case, $|A - C|$ and $|B - D|$ are either both even or both odd. Since each of the six numbers X, Y, Z, P, Q, R generates one odd and one even difference, in either case it is impossible for them to provide the required 7 remaining differences of the other kind.

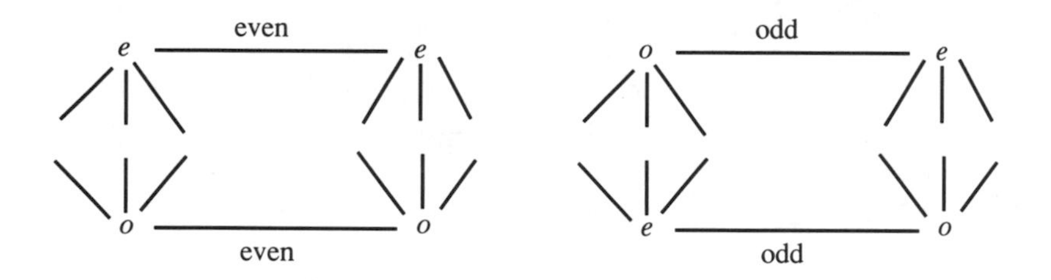

FIGURE 165

Since this covers all the possibilites, we conclude that the circles cannot be filled in as proposed.

An Old Chestnut

Before undertaking any mathematical analysis of the following situation, give your intuition a little test by taking a quick guess at the answer.

> Suppose a toy boat is hauled ashore by a boy at the edge of a pond. As the boy pulls in a yard of string, will the boat advance toward shore by more, less, or exactly one yard?

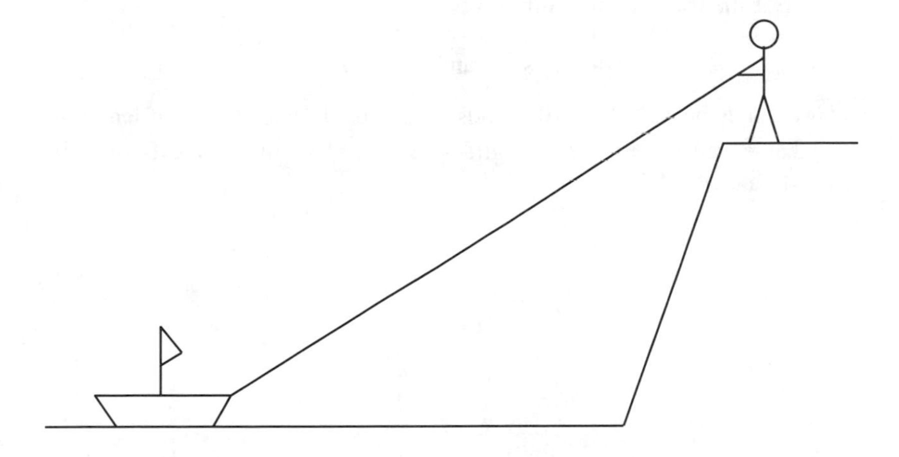

FIGURE 166

Many people are surprised by the answer, including me. Suppose, to begin with, that the boat is a distance d from shore and that the string is s units long (figure 167). When the boat is hauled in all the way, there will still

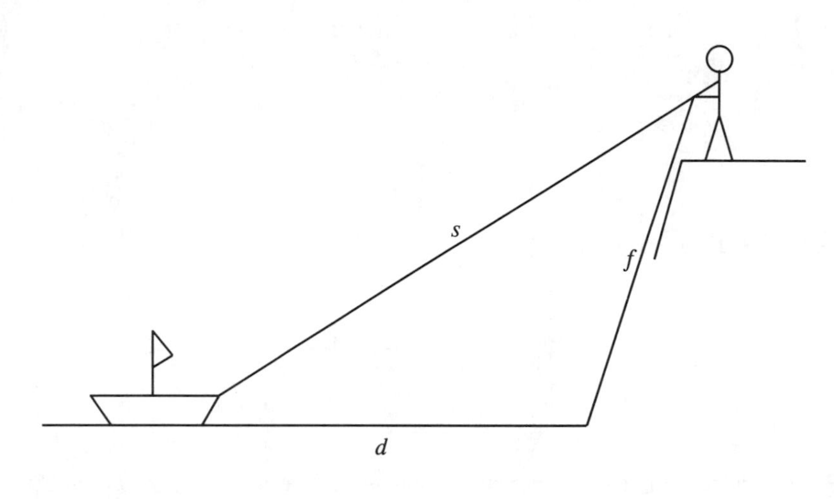

FIGURE 167

be a short length of string f between the boy and the boat because the boy, being on shore, is above the level of the water—it doesn't matter whether this piece is vertical or slanted.

Clearly, then, the total length of string pulled in is just the difference $s - f$. But the triangle inequality gives

$$d + f > s, \quad \text{and} \quad d > s - f,$$

and we conclude that the boat travels a greater distance than the length of string hauled in; indeed, the boat glides toward shore more quickly than the string shortens!!

A Combinatorial Discontinuity

1. Here is a problem of my colleague Ron Dunkley.

Consider the permutations of $\{1, 2, \ldots, n\}$ in which *no integer is less than both its immediate neighbors*. Combinatorics is full of situations in which trifling adjustments in the initial conditions give rise to monumental changes in the numbers of things. In the present instance, for example, the number of permutations of the type in question of $\{1, 1, 2, \ldots, 20\}$ is more than 130 thousand *times* as many as for the set $\{1, 2, \ldots, 20\}$. It is remarkable that simply adding an extra 1 can make such an overwhelming difference.

Determine the number of such permutations of $\{1, 1, 2, \ldots, n\}$.

It is convenient to use zig-zag figures to represent the increasing and decreasing strings in a permutation as it is scanned from left to right; for example,

1 4 5 7 9 8 6 3 2 is of the type 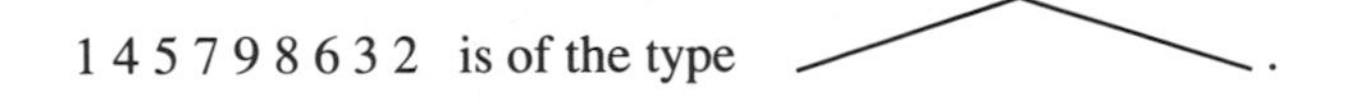 .

The forbidden configuration, then, is a decreasing-increasing pair of consecutive strings, a bottoming out, for in such a figure some number b would be bracketed by two bigger numbers a and c:

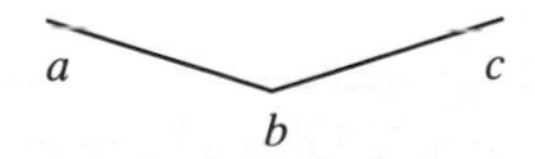

In the best mathematical tradition, let's begin by attempting the problem for the simpler set $\{1, 2, \ldots, n\}$. Each permutation can be built around the number n. Since all the other numbers are less than n, there is no avoiding a downward trend as a permutation is scanned on either side of n:

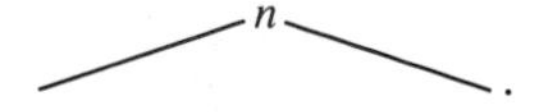

Consequently, if $n-1$ is not entered right next to n, on one side or the other, a number t smaller than $n-1$ would have to occur between n and $n-1$ and a forbidden upward change of direction would be inevitable:

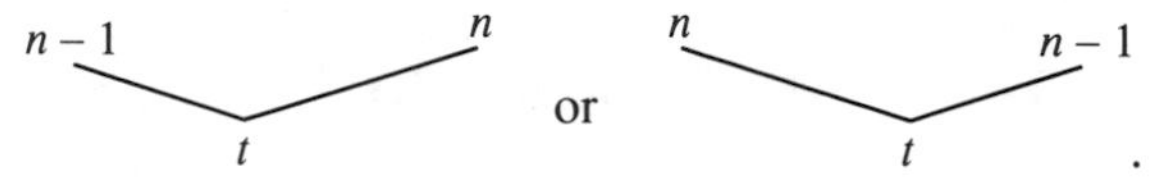

Thus $n-1$ must go immediately next to n, on one side or the other, giving 2 ways it might be entered. Similarly, any gap between $n-2$ and the block containing n and $n-1$ would spoil the permutation, implying that $n-2$ can go only at one end or the other of the block. The unavoidable downward trend on either side of the peak n must be preserved throughout the permutation in order to avert a violation. Thus, when the numbers are entered in the order $n, n-1, n-2, \ldots, 2, 1$, each number after n must be placed at one end or the other of the solid block formed by the previous entries, giving 2 ways to enter each of them, for a total of 2^{n-1} different permutations.

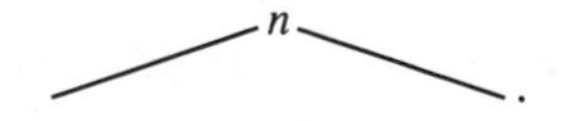

Observe that this does not neglect to count the permutations in which all the numbers are placed on the same side of n:

$$1\ 2\ 3\ \ldots\ n \quad \text{and} \quad n\ n-1\ n-2\ \ldots\ 2\ 1.$$

Now let's see how an additional 1 alters things. Obviously either the two 1's occur together or they don't. When they are separated, there is no choice but to put one at each end to yield a permutation of the type

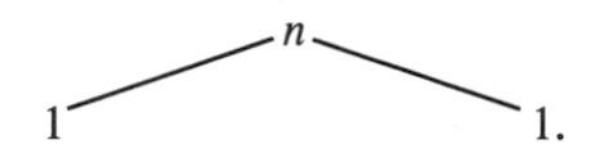

In this case, there is only one way to enter the two 1's and after inserting n between them, there are 2 choices for each of the remaining $n-2$ integers $\{2, 3, \ldots, n-1)$, giving 2^{n-2} acceptable permutations in which the 1's are split.

Now, when the 1's are together, each of them shields the other from violating the required condition, allowing them to go anywhere: for example,

$$4\ 5\ 7\ 9\ 1\ 1\ 8\ 6\ 3\ 2.$$

But, better than that, they permit fluctuations in the otherwise monotonic strings on the flanks, making legitimate all permutations conforming to the general pattern

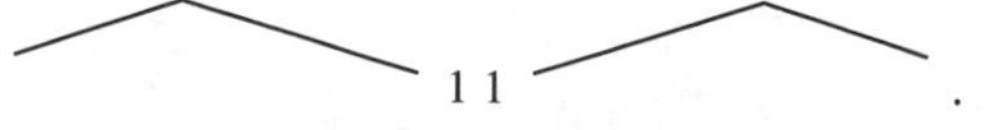

Allowing empty sections, this pattern encompasses all the new possibilities, for example,

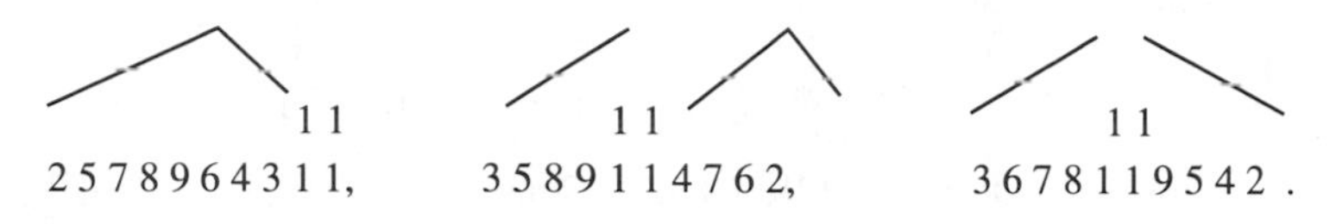

Suppose there are k integers in the section preceding the 1's and $n-1-k$ in the final section:

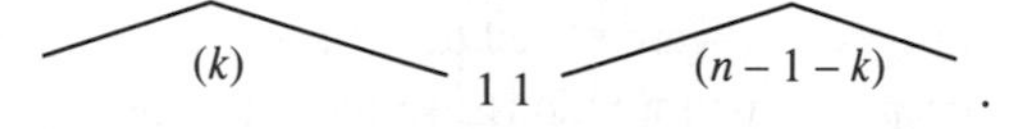

Then any of the 2^{k-1} acceptable permutations of the k numbers preceding the 1's can be paired with any of the $2^{n-1-k-1}$ acceptable arrangements of the numbers following the 1's for a total of

$$2^{k-1} \cdot 2^{n-1-k-1} = 2^{n-3}$$

acceptable permutations having *a particular k integers* in the first part.

At this point it is easy to fall prey to the mistake I made in my first attempt at the problem. For $k = 0$, the factor 2^{k-1} here gives $2^{-1} = 1/2$, which is not even an integer, and for $k = n-1$, the factor $2^{n-1-k-1}$ does the same. Hence these extreme cases must be handled separately from our general analysis. This is easily done, however, for they represent the permutations in which n and the remaining $n-2$ other numbers are all on the same side of the two 1's, and so there are clearly 2^{n-2} acceptable permutations in each case, for a total of $2 \cdot 2^{n-2} = 2^{n-1}$ permutations.

Now, for $1 \leq k \leq n-2$, there are $\binom{n-1}{k}$ ways of selecting k numbers from $\{2, 3, \ldots, n\}$ for service in the first section, with the complementary subset defaulting to the latter section, implying $\binom{n-1}{k} \cdot 2^{n-3}$ acceptable permutations in each case. Altogether, then, the total number of acceptable permutations, not forgetting those in which the 1's are split, is

$$\begin{aligned} &2^{n-2} + 2^{n-1} + \sum_{k=1}^{n-2} \binom{n-1}{k} \cdot 2^{n-3} \\ &\quad = 2^{n-2} + 2^{n-1} + 2^{n-3} \left[\sum_{k=0}^{n-1} \binom{n-1}{k} - \binom{n-1}{0} - \binom{n-1}{n-1} \right] \end{aligned}$$

(note the change of limits)

$$= 2^{n-2} + 2^{n-1} + 2^{n-3}(2^{n-1} - 2)$$
$$= 2^{n-1} + 2^{2n-4}.$$

Therefore there are

$$2^{19} + 2^{36} = 2^{19}(1 + 2^{17}) = 2^{19} \cdot 131073$$

permutations of $\{1, 1, 2, \ldots, 20\}$, 131073 times the paltry 2^{19} that are given by $\{1, 2, \ldots, 20\}$.

2. The problem is considerably more complicated when m $(n+1)$'s and m $(n+2)$'s are added to the set $\{1, 2, \ldots, n\}$, $m \geq 2$: how many acceptable permutations are there for the numbers

$$\{1, 2, \ldots, n, \underbrace{n+1, n+1, \ldots, n+1}_{\ldots\, m \text{ times}\, \ldots}, \underbrace{n+2, n+2, \ldots, n+2}_{\ldots\, m \text{ times}\, \ldots}\}\ ?$$

Again it is the big numbers $n+1$ and $n+2$ which divide the permutations into beginnings and endings. As above, the numbers $1, 2, \ldots, n$ must determine an increasing string at the beginning and a decreasing string at the end (allowing empty strings), and a moment's reflection reveals that a violation of the required property would be inevitable if the $(n+1)$'s and $(n+2)$'s were to occur anywhere except in a solid block of $2m$ places somewhere in the permutation. With the block defining the peak of the permutation, there are 2^n ways of entering the integers $\{1, 2, \ldots, n\}$ around it, and the real problem is to determine the number of ways of arranging the $(n+1)$'s and $(n+2)$'s among themselves (clearly no smaller number could go between any two of these large numbers).

To this end, let's begin to construct the block by entering all m of the $(n+2)$'s together in a row. The number in question, then, is the number of ways of mixing in the $(n+1)$'s in the $m+1$ spaces which border the $(n+2)$'s:

$$\begin{array}{c} m+1 \text{ spaces} \\ \downarrow \quad \downarrow \quad \downarrow \quad \downarrow \ldots \downarrow \quad \downarrow \quad \downarrow \\ \_\_\_\ \ X \quad X \quad X \quad \ldots \quad X \quad X \ \ \_\_\_. \\ |\ \ldots\ m\ (n+2)\text{'s}\ \ldots\ | \end{array}$$

In each of the two outer spaces any number of the $(n+1)$'s may be entered, and in each of the $m-1$ inner spaces between consecutive $(n+2)$'s any number of them can be placed *except exactly one,* which would obviously be in violation of the required condition. Now, all these options are represented precisely by the generating function

$$f(x) = (1+x+x^2+\cdots)\underbrace{(1+x^2+x^3+\cdots)\cdots(1+x^2+x^3+\cdots)}_{\cdots\ m-1 \text{ times}\ \cdots}(1+x+x^2+\cdots),$$

where the exponent of the term chosen from the kth bracket, when this product is multiplied out, signifies the number of $(n+1)$'s to be inserted in the kth of these spaces (note the term x^1 is missing from the $m-1$ interior factors). Thus the term $x^2 \cdot 1 \cdot x^4 \cdots 1 \cdot x^{m-6}$ would call for two $(n+1)$'s in the first outer space, none in any of the inner spaces except the second one, which is to get 4 of them, and the remaining $m-6$ of the $(n+1)$'s in the final outer space. Each term in which the exponents add up to m contains a prescription for allocating the $(n+1)$'s in an acceptable manner. Therefore the number of ways of entering them is the coefficient of x^m in this function $f(x)$, which we shall denote by $[x^m]f(x)$.

Now,

$$\begin{aligned} f(x) &= (1+x+x^2+\cdots)^2(1+x^2+x^3+\cdots)^{m-1} \\ &= (1-x)^{-2}\left[1+x^2(1-x)^{-1}\right]^{m-1} \\ &= (1-x)^{-2}\sum_{i=0}^{m-1}\binom{m-1}{i}x^{2i}(1-x)^{-i} \\ &= \sum_{i=0}^{m-1}\binom{m-1}{i}x^{2i}(1-x)^{-(i+2)}, \end{aligned}$$

which, recalling that

$$(1-x)^{-k} = \sum_{j\geq 0}\binom{k+j-1}{j}x^j,$$

gives

$$\begin{aligned} f(x) &= \sum_{i=0}^{m-1}\binom{m-1}{i}x^{2i}\sum_{j\geq 0}\binom{i+2+j-1}{j}x^j \\ &= \sum_{i=0}^{m-1}\binom{m-1}{i}\sum_{j\geq 0}\binom{i+j+1}{j}x^{j+2i}. \end{aligned}$$

Hence

$$[x^m]f(x) = \sum_{j+2i=m}\binom{m-1}{i}\binom{i+j+1}{j}.$$

Now, since $j + 2i = m$, then $i = \frac{m}{2} - \frac{j}{2}$, implying $i \leq \left[\frac{m}{2}\right]$ since $j \geq 0$, and therefore

$$[x^m]f(x) = \sum_{i=0}^{[m/2]} \binom{m-1}{i}\binom{m-i+1}{m-2i}$$
$$= \sum_{i=0}^{[m/2]} \binom{m-1}{i}\binom{m-i+1}{i+1}$$

(since $\binom{n}{r} = \binom{n}{n-r}$). Thus the required number of permutations is 2^n times this number, which is an expression that emphasizes the dependence of present-day combinatorics on computers.

A Surprising Theorem of Kummer

In 1852 the great German mathematician Ernst Kummer (1810–1893) determined a remarkable specification of the number of times a prime p divides the binomial coefficient $\binom{n}{r}$. He found the unlikely result that

$$\begin{aligned}&\text{the exponent of the greatest power of } p \text{ that divides } \binom{n}{r}\\&= \text{the number of } \textit{carryovers} \text{ that occur when}\\&\text{the numbers } r \text{ and } n-r \text{ are added in base } p.\end{aligned}$$

The following proof is given in Paulo Ribenboim's wonderful volume *The Book of Prime Number Records* (Springer-Verlag, 1989, 22–24).

Kummer's result follows nicely from an earlier discovery of Legendre (1808) that

$$\begin{aligned}&\text{the exponent of the greatest power of a prime } p \text{ that divides } q!\\&= \frac{1}{p-1}[q - (\text{the sum of the digits of } q \text{ in base } p)],\\&\text{which we shall denote simply by } \frac{1}{p-1}(q - s_q).\end{aligned}$$

While neither of these results is difficult to prove, they are not obvious either, and I hope you will enjoy the following arguments to establish them; although they appear somewhat formidable on the page, the logic is easy to follow.

(a) *Proof of Legendre's Result.* Scanning the factors of $q!$,

$$1 \cdot 2 \cdot 3 \cdots p \cdots 2p \cdots 3p \cdots jp \cdots q,$$

it is clear that every pth one contains a factor p. The number of these multiples of p is just $[\frac{q}{p}]$, the integer part of $\frac{q}{p}$. In addition to these, every p^2th integer contains a second factor p, and there are $[\frac{q}{p^2}]$ of them. Similarly, there are $[\frac{q}{p^3}]$ integers containing a third factor p, and so on, for a grand total of

$$T = \left[\frac{q}{p}\right] + \left[\frac{q}{p^2}\right] + \left[\frac{q}{p^3}\right] + \cdots + \left[\frac{q}{p^i}\right] + \cdots.$$

This is always just a finite series, for once p^i exceeds q all the later terms vanish. Now, we can determine that

$$T = \frac{1}{p-1}(q - s_q)$$

as follows.

Suppose in base p that

$$q = a_k p^k + a_{k-1}p^{k-1} + \cdots + a_1 p + a_0.$$

Then

$$\frac{q}{p} = a_k p^{k-1} + a_{k-1}p^{k-2} + \cdots + a_1 + \frac{a_0}{p}.$$

Being in base p, each integer $a_i \leq p-1$, and so $\frac{a_0}{p} < 1$, giving

$$\left[\frac{q}{p}\right] = a_k p^{k-1} + a_{k-1}p^{k-2} + \cdots + a_1.$$

Similarly,

$$\frac{q}{p^2} = a_k p^{k-2} + a_{k-1}p^{k-3} + \cdots + a_2 + \frac{a_1 p + a_0}{p^2},$$

where

$$a_1 p + a_0 \leq (p-1)p + p - 1 = (p-1)(p+1) = p^2 - 1 < p^2;$$

hence

$$\left[\frac{q}{p^2}\right] = a_k p^{k-2} + a_{k-1}p^{k-3} + \cdots + a_2.$$

In general,

$$\frac{q}{p^i} = a_k p^{k-i} + a_{k-1}p^{k-i-1} + \cdots + a_i + \frac{a_{i-1}p^{i-1} + a_{i-2}p^{i-2} + \cdots + a_0}{p^i},$$

where

$$\begin{aligned} a_{i-1}p^{i-1} + a_{i-2}p^{i-2} + \cdots + a_0 &\leq (p-1)(p^{i-1} + p^{i-2} + \cdots + 1) \\ &= p^i - 1 < p^i, \end{aligned}$$

implying

$$\left[\frac{q}{p^i}\right] = a_k p^{k-i} + a_{k-1} p^{k-i-1} + \cdots + a_i.$$

Altogether, then, we have

$$\begin{aligned}
\left[\frac{q}{p}\right] &= a_k p^{k-1} + a_{k-1} p^{k-2} + \cdots + a_3 p^2 + a_2 p + a_1, \\
\left[\frac{q}{p^2}\right] &= a_k p^{k-2} + a_{k-1} p^{k-3} + \cdots + a_3 p + a_2, \\
\left[\frac{q}{p^3}\right] &= a_k p^{k-3} + a_{k-1} p^{k-4} + \cdots + a_3, \\
&\cdots \\
\left[\frac{q}{p^{k-1}}\right] &= a_k p \qquad + a_{k-1} \\
\left[\frac{q}{p^k}\right] &= a_k.
\end{aligned}$$

Adding up we have

$$\begin{aligned}
T = \left[\frac{q}{p}\right] &+ \left[\frac{q}{p^2}\right] + \left[\frac{q}{p^3}\right] + \cdots + \left[\frac{q}{p^i}\right] + \cdots \\
= \quad & a_k \left(p^{k-1} + p^{k-2} + \cdots + 1\right) \\
& + a_{k-1}(p^{k-2} + p^{k-3} + \cdots + 1) \\
& + a_{k-2}(p^{k-3} + p^{k-4} + \cdots + 1) \\
& \cdots \\
& + a_2(p+1) \\
& + a_1(1) \\
&= a_k \cdot \frac{p^k - 1}{p-1} + a_{k-1} \cdot \frac{p^{k-1} - 1}{p-1} + \cdots + a_1 \cdot \frac{p-1}{p-1} \\
&= \frac{1}{p-1}\left[(a_k p^k + a_{k-1} p^{k-1} + \cdots + a_1 p) - (a_k + a_{k-1} + \cdots + a_1)\right] \\
&= \frac{1}{p-1}\left[(q - a_0) - (s_q - a_0)\right] \\
&= \frac{1}{p-1}(q - s_q).
\end{aligned}$$

(b) *Proof of Kummer's Result.* Suppose the base p representation of the addition of r and $n-r$ is

$$\begin{array}{rl} r = & a_t\, a_{t-1} \cdots a_i \cdots a_2\, a_1\, a_0 \\ n-r = & b_t\, b_{t-1} \cdots b_i \cdots b_2\, b_1\, b_0 \\ \hline n = c_{t+1} & c_t\, c_{t-1} \cdots c_i \cdots c_2\, c_1\, c_0. \end{array}$$

If necessary, let the shorter expression be padded out with 0's on the left in order to make r and $n-r$ the same length; thus at least one of a_t, b_t is nonzero; also, $c_{t+1} = 1$ if there is a carryover in the last step and 0 if not (when two numbers are added, the only possible carryover is 1).

Now consider the step in which a_i and b_i are added. If there is no carryover from the previous step, we simply add a_i and b_i; but if there *is* a carryover, we add $1 + a_i + b_i$. Thus, in either case, the sum involved may be expressed as

$$e_{i-1} + a_i + b_i,$$

where $e_{i-1} = 0$ if there is no carryover and 1 if there is. Now, the resulting digit c_i in the sum at this stage is obtained either from

$$e_{i-1} + a_i + b_i = c_i \quad \text{or} \quad e_{i-1} + a_i + b_i = p + c_i,$$

the latter implying a carryover to the next step. But, continuing with this notation, a carryover to the next step would be registered by the value of e_i, which is 1 or 0 as there is or is not a carryover. Therefore, at each step, all the possibilities concerning carryovers are neatly embodied in the single statement

$$e_{i-1} + a_i + b_i = e_i p + c_i.$$

Hence the total number of carryovers is

$$C = e_0 + e_1 + e_2 + \cdots + e_t,$$

and the stages in the addition are given by

$$\begin{aligned} a_0 + b_0 &= e_0 p + c_0, \\ e_0 \quad + a_1 + b_1 &= e_1 p + c_1, \\ e_1 \quad + a_2 + b_2 &= e_2 p + c_2, \\ \cdots \qquad & \\ e_{t-1} + a_t + b_t &= e_t p + c_t, \\ e_t \qquad\qquad &= \qquad c_{t+1}. \end{aligned}$$

Adding the columns in these equations gives

$$C + s_r + s_{n-r} = pC + s_n,$$

yielding

$$C = \frac{1}{p-1}(s_r + s_{n-r} - s_n).$$

On the other hand,

$$\binom{n}{r} = \frac{n!}{r!(n-r)!},$$

and so

the exponent of the greatest power of p to divide $\binom{n}{r}$

$=$ the exponent of the greatest power of p to divide $n!$

$-$ the exponent of the greatest power of p to divide $r!$

$-$ the exponent of the greatest power of p to divide $(n-r)!$,

which by Legendre's result is given by

$$\begin{aligned}
&\frac{1}{p-1}(n - s_n) - \frac{1}{p-1}(r - s_r) - \frac{1}{p-1}[(n-r) - s_{n-r}] \\
&\qquad = \frac{1}{p-1}[n - s_n - r + s_r - (n-r) + s_{n-r}] \\
&\qquad = \frac{1}{p-1}(s_r + s_{n-r} - s_n) \\
&\qquad = C.
\end{aligned}$$

Thus, for example, we could tell without much labor that $\binom{33}{14}$ contains only two factors 3: in base 3, for $r = 14$ and $n - r = 19$, we have

$$\begin{aligned}
r &= 112 \\
n - r &= 201 \\
\hline
n &= 1020,
\end{aligned}$$

requiring two carryovers.

A Combinatorial Problem in Solid Geometry

A rectangular solid S of integral dimensions $p \times q \times r$, where p, q, and r are relatively prime in pairs, is cut into pqr unit cubes by planes parallel to its faces. How many of these unit cubes would be cut if a corner X of S were cut off by a plane going through three neighboring vertices A, B, and C (figure 168)?

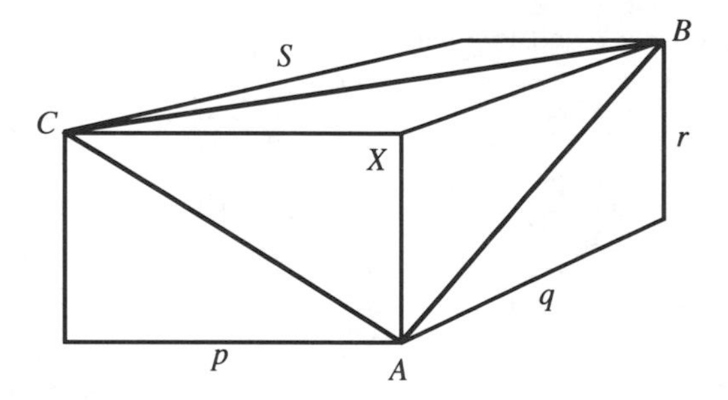

FIGURE 168

(a) First let us show that, except for A, B, and C themselves, no unit cube has a corner that lies on the cutting triangle ABC. To this end, let rectangular axes be assigned as in figure 169, making $A(p,0,0)$, $B(p,q,r)$, and $C(0,0,r)$.

If the equation of the plane of ABC is denoted by

$$dx + ey + fz = g,$$

then, since A, B, and C lie on the plane, we have

$$dp = g,$$

$$dp + eq + fr = g,$$

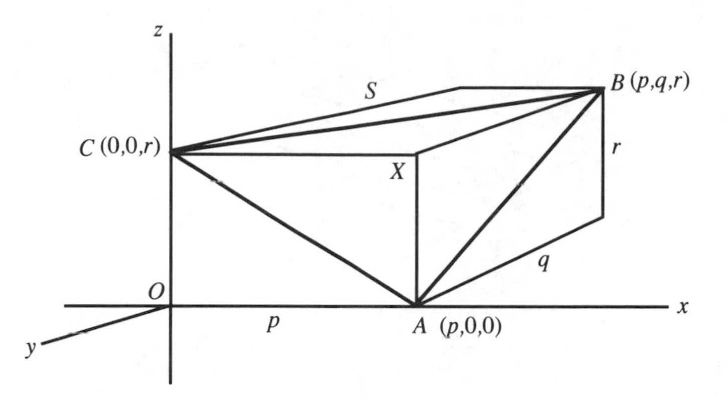

FIGURE 169

and

$$fr = g.$$

Solving for d, e, and f, we immediately have

$$d = \frac{g}{p} \qquad \text{and} \qquad f = \frac{g}{r},$$

and from

$$dp + eq + fr = g$$

we get

$$g + eq + g = g,$$

giving

$$e = -\frac{g}{q}.$$

Hence the equation is

$$\frac{g}{p}x - \frac{g}{q}y + \frac{g}{r}z = g,$$

that is,

$$qrx - rpy + pqz = pqr.$$

Now, the coordinates of the corners of the unit cubes are lattice points. Other than A, B, and C, a lattice point of $\triangle ABC$, lying in the interior of the triangle or the interior of an edge, would have to satisfy *at least two* of the conditions (see figure 169)

$$\begin{aligned} 1 \le x \le p-1 \\ 1 \le y \le q-1 \\ 1 \le z \le r-1 \end{aligned} \tag{1}$$

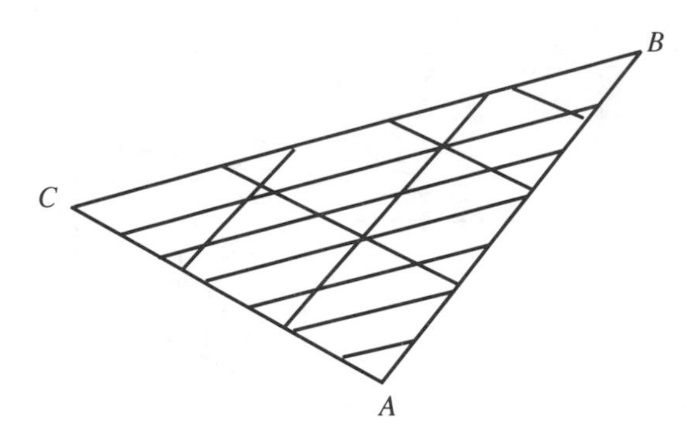

FIGURE 170

But no lattice point of $\triangle ABC$ satisfies any of these conditions. When x, y, and z are integers, the equation of the plane shows that p divides qrx, and since p, q, and r are relatively prime in pairs, this implies that p must divide x. Clearly this is impossible for $1 \le x \le p-1$, and so the x coordinate of any lattice point belonging to $\triangle ABC$ must violate the first condition in (1). Similarly, its y and z coordinates also must violate the other conditions in (1), and it follows that the only lattice points in $\triangle ABC$ are the vertices A, B, and C.

The upshot of this is that $\triangle ABC$ passes through the *interior* of any unit cube it cuts, thus intersecting the cube in a nonempty *region.* Conversely, we can imagine that each unit cube that is cut *prints* its region of intersection on the cutting triangle ABC. Thus the number of unit cubes that are cut is the number of regions in the printout on $\triangle ABC$ (figure 170).

(b) Now, the bounding edges of these regions lie in the *faces* of unit cubes, and the faces of these cubes are determined by the initial *cutting planes* which are parallel to the faces of S. Each of these planes cuts a straight line across $\triangle ABC$ and the regions we wish to count are determined by the $\left[(p-1)+(q-1)+(r-1)\right]$ planes which cut across the *interior* of the triangle, namely the planes

$$x = a, \quad y = b, \quad \text{and} \quad z = c,$$

where $1 \le a \le p-1$, $1 \le b \le q-1$, and $1 \le c \le r-1$ (figure 170).

Since the cutting planes $x = a$ are spaced at unit intervals, the $p-1$ lines across $\triangle ABC$ which are parallel to AB are also equally spaced across the triangle. Altogether the regions in question are determined by

$$p-1 \text{ equally-spaced lines parallel to } AB,$$

$q-1$ equally-spaced lines parallel to AC, and

$r-1$ equally-spaced lines parallel to BC.

(c) These regions can be counted using Euler's formula $V-E+F=1$, but let us use the following less sophisticated approach.

Let each of the lines of intersection be *peeled off* $\triangle ABC$. Each segment along a line separates the regions on opposite sides of it and when it is peeled back these regions come together to form a single larger region. Thus the number of regions *lost* in peeling off one of these lines is equal to the number of segments along it, and that is one more than the number of points of intersection along it:

the number of regions lost by peeling off a line

$= 1 +$ (the number of points of intersection on it).

It is easy to see, then, that adding up over all $\left[(p-1)+(q-1)+(r-1)\right]$ lines gives

the total number of regions lost

$= \left[(p-1)+(q-1)+(r-1)\right]$

$+$ (the total number of points of intersection).

This is so because each point of intersection would occur precisely once in this sum: peeling off one of the two lines through it would cause it to disappear from the other line as well, and so it would not be encountered again when the second line through it is peeled off (no point of intersection lies on *three* lines across the triangle, for that would imply a lattice point other than A, B, and C on the triangle).

After all the lines have been peeled off, there is still one region left, namely $\triangle ABC$ itself. Therefore the desired number of regions in $\triangle ABC$ in the first place must have been

$1+\left[(p-1)+(q-1)+(r-1)\right]+$ (the total number of points of intersection).

(d) It remains, then, to calculate the number of points of intersection of the three systems of equally-spaced parallel lines. Since no point of intersection lies on a line from each of the three systems, we can count the points by considering the systems in pairs.

Accordingly, consider the $p-1$ lines parallel to AB and the $q-1$ lines parallel to AC (figure 171). Giving the figure a half-turn about the midpoint of BC, these systems would determine $(p-1)(q-1)$ points of intersection in the resulting parallelogram. Since none of them lies on BC itself (for that

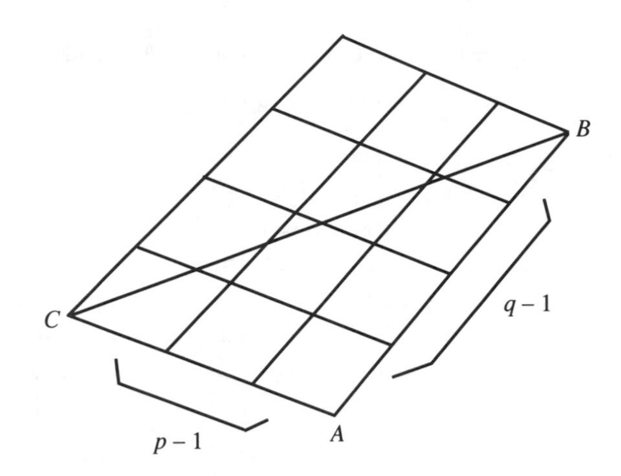

FIGURE 171

would imply a lattice point in $\triangle ABC$ in the interior of BC), half must lie on each side of BC, implying $\frac{1}{2}(p-1)(q-1)$ of them occur in $\triangle ABC$.

Similarly, the third system of $r-1$ lines gives $\frac{1}{2}(q-1)(r-1)$ additional points of intersection with one of the systems and $\frac{1}{2}(r-1)(p-1)$ with the other, for a grand total of

$$\frac{1}{2}\left[(p-1)(q-1)+(q-1)(r-1)+(r-1)(p-1)\right] \text{ points of intersection.}$$

Therefore

$$\begin{aligned}
&\text{the required number of cubes}\\
&\quad= \text{the total number of regions}\\
&\quad= 1+\left[(p-1)+(q-1)+(r-1)\right]\\
&\qquad+\frac{1}{2}\left[(p-1)(q-1)+(q-1)(r-1)+(r-1)(p-1)\right]\\
&\quad= p+q+r-2\\
&\qquad+\frac{1}{2}\left[pq-p-q+1+qr-q-r+1+rp-r-p+1\right]\\
&\quad= \frac{1}{2}(pq+qr+rp-1).
\end{aligned}$$

Two Problems from the 1989 Indian Olympiad†

1. Determine all non-square positive integers n such that

$$\left[\sqrt{n}\right]^3 | n^2,$$

where $[x]$ denotes the integer part of x.

If $\left[\sqrt{n}\right] = 1$, the condition is certainly satisfied. Thus all the integers between 1^2 and 2^2 are admissible, namely 2 and 3 (they have $\sqrt{n}$ between 1 and 2, making $\left[\sqrt{n}\right] = 1$).

In general, since n is not a square, it must lie between two consecutive squares:

$$m^2 < n < (m+1)^2.$$

All the integers in this interval have $\left[\sqrt{n}\right] = m$, and we seek those such that

$$m^3 | n^2.$$

This certainly requires $m|n$, making n a multiple of m. But there are only two multiples of m between m^2 and $(m+1)^2$, i.e., between m^2 and m^2+2m+1, namely m^2+m and m^2+2m. These values of n give

$$n^2 = (m^2+m)^2 = m^2(m+1)^2 \quad \text{or} \quad (m^2+2m)^2 = m^2(m+2)^2,$$

and to be divisible by m^3, we must have either

$$m \mid (m+1)^2 = m^2+2m+1 \quad \text{or} \quad m \mid (m+2)^2 = m^2+4m+4,$$

---

† (*Crux Mathematicorum,* 1990, 133)

implying

$$m \mid 1 \quad \text{or} \quad m \mid 4.$$

Thus the only possibilities are $m = 1$, 2, and 4. Having already considered $m = 1$, it remains only to check $m = 2$ and 4.

$m = 2$: This makes n a multiple of 2 between 2^2 and 3^2 (recall n is a multiple of m), giving $n = 6$ or 8, of which 8 is acceptable ($2^3|8^2$), but not 6 ($2^3 \nmid 6^2$);

$m = 4$: Similarly, n must be a multiple of 4 between 4^2 and 5^2, giving $n = 20$ or 24, of which 24 is acceptable ($4^3|24^2$), but not 20 ($4^3 \nmid 20^2$). Thus $n = 2$, 3, 8, and 24 are the only admissible values of n.

2. (Recall that the pedal triangle, with respect to a triangle ABC, of a point P in the plane of $\triangle ABC$, is the triangle whose vertices are the feet of the perpendiculars from P to the sides of $\triangle ABC$ or their extensions.)

Determine all points P inside acute-angled triangle ABC whose pedal triangle is *isosceles* (figure 172). For what position of P is the pedal triangle equilateral?

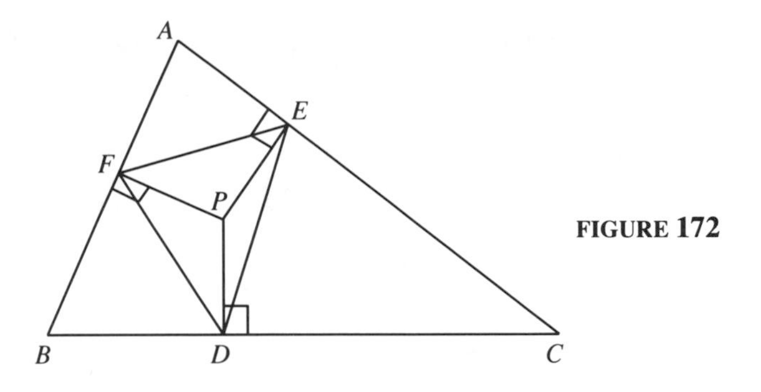

FIGURE 172

I expect this problem would be very challenging for a contestant who is not acquainted with pedal triangles. If the topic was covered in the training sessions, however, the candidate would likely have seen the following derivation of a simple formula for the length of a side of the pedal triangle.

The right angles at E and F make $AFPE$ cyclic, implying $\angle APF = \angle AEF$, denoted by α in figure 173. If the lengths of AP, BP, and CP are x, y, and z, respectively, then

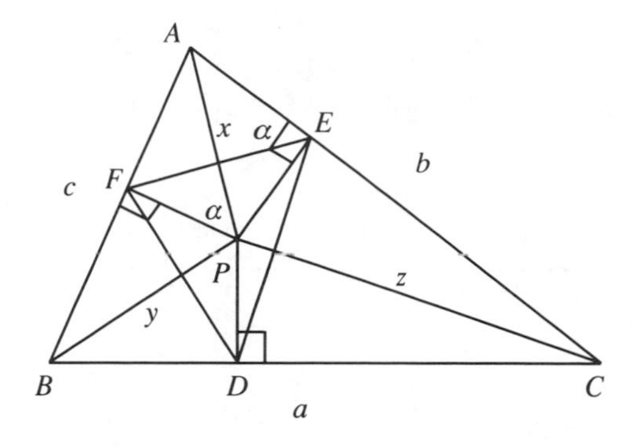

FIGURE 173

(i) in $\triangle AFE$, the law of sines gives

$$\frac{EF}{\sin A} = \frac{AF}{\sin \alpha};$$

(ii) in right triangle AFP, $\sin \alpha = \frac{AF}{x}$, giving $x = \frac{AF}{\sin \alpha}$. Hence

$$\frac{EF}{\sin A} = x, \quad \text{and} \quad EF = x \sin A.$$

Similarly, $DF = y \sin B$ and $DE = z \sin C$.

Thus $\triangle DEF$ would be isosceles with $DE = DF$ if and only if

$$y \sin B = z \sin C,$$

that is,

$$\frac{y}{z} = \frac{\sin C}{\sin B} = \frac{c}{b} \quad \text{(by the law of sines);}$$

Thus $DE = DF$ if and only if P lies on the locus determined by a point which moves so that its distances y and z to fixed points B and C are constantly in the ratio $c : b$. But such a locus is a circle of Apollonius whose diameter, in this case, is the segment UV joining the points which divide BC internally and externally in the ratio $c : b$ (figure 174). Since $AB : AC = c : b$, this circle goes through vertex A.

Therefore the required set of points consists of three circular arcs through the vertices, one for each pair of sides of the pedal triangle. It is not difficult to see that these three arcs are concurrent:

let Q be the point of intersection of two of them, say the arcs determined by the ratios

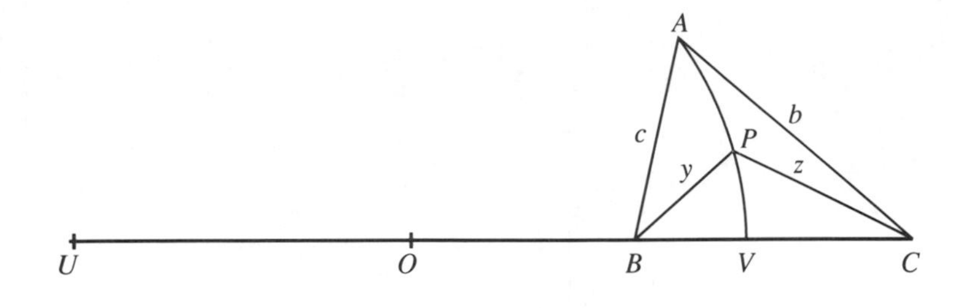

FIGURE 174

$$\frac{y}{z} = \frac{c}{b} \quad \text{and} \quad \frac{x}{y} = \frac{b}{a};$$

then multiplication gives

$$\frac{x}{z} = \frac{c}{a},$$

placing Q on the third arc.

Obviously, then, the pedal triangle is equilateral when P is at Q. We observe that, when $\triangle ABC$ itself is isosceles, say $c = b$, the circle of Apollonius through A, given by the ratio $1 : 1$, degenerates into a straight line, for the external point of divison (U) is at infinity; equivalently, this makes $y = z$, placing P on the perpendicular bisector of BC. For an scalene triangle, however, I have not been able to identify the point Q beyond its specification as the common point of the three circles of Apollonius.

A Gem from Combinatorics†

1. Suppose all the pairs of positive integers from a finite collection

$$A = \{a_1, a_2, \ldots, a_n\}$$

are added together to form a new collection

$$A_2 = \{a_i + a_j | a_i, a_j \in A,\ i \neq j\}.$$

For example, $A = \{2, 3, 4, 7\}$ would yield $A_2 = \{5, 6, 9, 7, 10, 11\}$ and $B = \{1, 4, 5, 6\}$ would give $B_2 = \{5, 6, 7, 9, 10, 11\}$. These examples show that it is possible for different collections A and B to generate the same collections A_2 and B_2. The gem promised in the title is a consequence of the identity $A_2 \equiv B_2$.

Clearly, if there are n integers in A, there will be $\binom{n}{2}$ elements in A_2, and since the numbers $\binom{n}{2}$ are strictly increasing with n,

$$\left\{\binom{n}{2}\right\} = \{1, 3, 6, 10, 15, \ldots\},$$

the identity of A_2 and B_2 could only occur when A and B are the same size to begin with.

We have deliberately said "collections" rather than "sets," for repetitions are permitted among the elements of A and B. For example,

$$A = \{2, 2, 3, 5\} \quad \text{gives} \quad A_2 = \{4, 5, 7, 5, 7, 8\},$$

† This result is due to the well-known mathematicians Paul Erdős and John Selfridge, and my acquaintance with it is due to a visit by Noga Alon of Israel. As we shall see, their proof is brief, completely elementary and straightforward—a thing of beauty and ingenuity! This essay appeared previously in the *Bulletin of the Institute of Combinatorics and Its Applications,* Vol. 1, January, 1991.

and

$$B = \{1, 3, 4, 4\} \quad \text{gives} \quad B_2 = \{4, 5, 5, 7, 7, 8\},$$

again yielding $A_2 \equiv B_2$.

Obviously, if A and B are the same collection, $A_2 \equiv B_2$ is automatic and the size of A and B could be anything. Remarkably, however,

if $A_2 \equiv B_2$ for *different* collections A and B, then the common size n of A and B *must always be a power of* 2: $n = 2^r$.

2. Suppose, then, that $A_2 \equiv B_2$ for the different collections

$$A = \{a_1, a_2, \ldots, a_n\} \quad \text{and} \quad B = \{b_1, b_2, \ldots, b_n\}.$$

The proof begins with the definitions of the generating functions

$$f(x) = x^{a_1} + x^{a_2} + \cdots + x^{a_n} = \sum x^{a_i},$$
$$g(x) = x^{b_1} + x^{b_2} + \cdots + x^{b_n} = \sum x^{b_i}.$$

Squaring these functions we get

$$\begin{aligned}[f(x)]^2 &= (x^{a_1})^2 + (x^{a_2})^2 + \cdots + (x^{a_n})^2 \\ &\quad + 2(x^{a_1+a_2} + x^{a_1+a_3} + \cdots + x^{a_{n-1}+a_n}), \\ &= (x^2)^{a_1} + (x^2)^{a_2} + \cdots + (x^2)^{a_n} + 2\sum x^{a_i+a_j}, \ (i \neq j),\end{aligned}$$

that is,

$$[f(x)]^2 = \sum (x^2)^{a_i} + 2\sum x^{a_i+a_j},$$

and similarly,

$$[g(x)]^2 = \sum (x^2)^{b_i} + 2\sum x^{b_i+b_j}.$$

Since $A_2 \equiv B_2$ implies $\sum x^{a_i+a_j} = \sum x^{b_i+b_j}$, subtraction gives

$$[f(x)]^2 - [g(x)]^2 = \sum (x^2)^{a_i} - \sum (x^2)^{b_i}.$$

Recalling that $f(x) = \sum x^{a_i}$, we have that $\sum (x^2)^{a_i}$ is simply $f(x^2)$; similarly $\sum (x^2)^{b_i} = g(x^2)$. Hence

$$[f(x)]^2 - [g(x)]^2 = f(x^2) - g(x^2),$$

and, factoring the left side, we have

$$[f(x) + g(x)][f(x) - g(x)] = f(x^2) - g(x^2).$$

Setting $f(y) - g(y) = h(y)$, this result may be written as

$$\big[f(x) + g(x)\big] \cdot h(x) = h(x^2). \tag{1}$$

Now, since A and B are not identical collections, the function $h(x)$ is not identically zero, and therefore we can divide it by $1 - x$ as often as it permits to put it into the form

$$h(x) = (1 - x)^k p(x),$$

where k is some nonnegative integer and $p(x)$, which is not identically zero, is not further divisible by $1 - x$, i.e., $p(1) \neq 0$. In these terms, we have

$$h(x^2) = (1 - x^2)^k p(x^2),$$

and relation (1) becomes

$$\big[f(x) + g(x)\big](1 - x)^k p(x) = (1 - x^2)^k p(x^2),$$

yielding

$$\big[f(x) + g(x)\big]p(x) = (1 + x)^k p(x^2). \tag{2}$$

Recalling that $f(x) = \sum x^{a_i}$ we have $f(1) = |A| = n$, and similarly $g(1) = n$. Finally, then, for $x = 1$, equation (2) yields

$$(2n)p(1) = (2^k)p(1),$$

where $p(1) \neq 0$, from which

$$n = 2^{k-1}.$$

3. We conclude by observing that if $A_2 \equiv B_2$ for different collections

$$A = \{a_1, a_2, \ldots, a_n\} \quad \text{and} \quad B = \{b_1, b_2, \ldots, b_n\},$$

then, for any positive integer m, the collections

$$A' = \{a_1, a_2, \ldots, a_n, b_1 + m, b_2 + m, \ldots, b_n + m\}$$

and

$$B' = \{b_1, b_2, \ldots, b_n, a_1 + m, a_2 + m, \ldots, a_n + m\},$$

each containing twice as many integers, also possesses the critical property $A'_2 \equiv B'_2$. Clearly, the members of A'_2 are of the three types

$$a_i + a_j, \; a_i + (b_j + m), \; \text{ and } \; b_i + b_j + 2m,$$

and, since $\{a_i + a_j\} \equiv \{b_i + b_j\}$, they are reproduced exactly by the members of B_2', which consist of the three types

$$b_i + b_j, \; b_j + (a_i + m), \quad \text{and} \quad a_i + a_j + 2m.$$

Thus, since $A = \{2, 3\}$, $B = \{1, 4\}$ constitute suitable collections of size 2, there must exist many such collections of every size $n = 2^k$, $k = 2, 3, \ldots$.

Two Problems from the 1989 Asian Pacific Olympiad†

1. If S is the sum of the positive real numbers $x_1, x_2, \ldots, x_n$, prove that

$$(1+x_1)(1+x_2)\cdots(1+x_n) \leq 1+S+\frac{S^2}{2!}+\frac{S^3}{3!}+\cdots+\frac{S^n}{n!}.$$

By the arithmetic mean - geometric mean inequality we have

$$\left[(1+x_1)(1+x_2)\cdots(1+x_n)\right]^{1/n} \leq \frac{(1+x_1)+(1+x_2)+\cdots+(1+x_n)}{n},$$

giving

$$\begin{aligned}(1+x_1)(1+x_2)\cdots(1+x_n) &\leq \left(\frac{n+S}{n}\right)^n = \left(1+\frac{S}{n}\right)^n \\ &= 1+S+\frac{n(n-1)}{2!}\cdot\frac{S^2}{n^2}+\frac{n(n-1)(n-2)}{3!}\cdot\frac{S^3}{n^3} \\ &\quad +\cdots+\frac{n(n-1)\cdots(n-n+1)}{n!}\cdot\frac{S^n}{n^n} \\ &= 1+S+1\left(1-\frac{1}{n}\right)\cdot\frac{S^2}{2!}+1\left(1-\frac{1}{n}\right)\left(1-\frac{2}{n}\right)\cdot\frac{S^3}{3!} \\ &\quad +\cdots+1\left(1-\frac{1}{n}\right)\cdots\left(1-\frac{n-1}{n}\right)\cdot\frac{S^n}{n!} \\ &\leq 1+S+\frac{S^2}{2!}+\frac{S^3}{3!}+\cdots+\frac{S^n}{n!}\end{aligned}$$

with equality only for $n=1$. (An alternative solution is given in 1991, 109.)

† (*Crux Mathematicorum,* 1989, 131)

2. Consider the set of all pairs (a, b), with $a < b$, that can be taken from $\{1, 2, 3, \ldots, n\}$. Let S be a subset consisting of any m of these pairs. Prove that there are at least

$$4m \cdot \frac{m - \frac{n^2}{4}}{3n}$$

triples (a, b, c), $a < b < c$, such that its three pairs (a, b), (a, c), and (b, c) *all belong to* S.

Our solution to this difficult problem is yet another gem by the ingenious George Evagelopoulos.

Consider a graph G having n vertices numbered $1, 2, \ldots, n$. The possible edges in G are therefore given by the pairs (a, b) from $\{1, 2, \ldots, n\}$. Let the pairs (a, b) in the subset S be used as a prescription for the edges of G. Thus each triple (a, b, c), whose three pairs all belong to S, determines a *triangle* in G, and therefore our problem is to show that, no matter which m edges might be in G, they form at least

$$4m \cdot \frac{m - \frac{n^2}{4}}{3n}$$

triangles.

Suppose (x, y) is an edge in G. George looks at the number of triangles there might be that have (x, y) as an edge. There would be a triangle for each vertex Z that is joined to *both* x and y (figure 175).

Recalling that the number of edges at a vertex v is called the degree of v and is denoted by $d(v)$, there are $d(x)$ edges emanating from x and $d(y)$ from y. Of course, one of the edges at x goes to y itself and so there are only $d(x) - 1$ edges at x that might help close a triangle at a third vertex Z. We

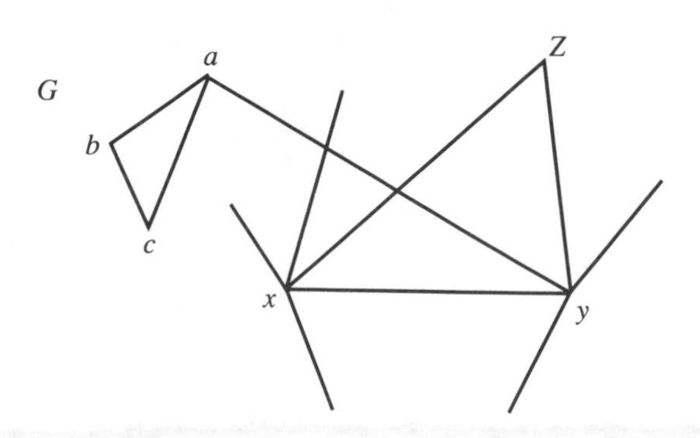

FIGURE 175

are interested, then, in considering the number of times the $d(x)-1$ edges from x and the corresponding $d(y)-1$ edges from y might come together at a common vertex Z.

It is possible that each of the $n-2$ other vertices in G might receive a single edge or no edge at all from x and y and therefore not close a triangle. However, if the combined $d(x)+d(y)-2$ edges from x and y were to number *more* than $n-2$, say $n-2+p$, then it would be impossible to avoid forming at least p triangles (putting $n-2+p$ objects into $n-2$ compartments, with not more than 2 to a compartment, must result in at least p of the compartments holding 2 of the objects). The edges at x and y, then, must complete at least

$$p=\left[d(x)+d(y)-2\right]-(n-2)=d(x)+d(y)-n$$

triangles; this formula is valid for *all* edges (x,y), for even in the event that $d(x)+d(y)-2$ is not greater than $n-2$, this number is negative or zero and would still be a correct lower bound on the number of triangles. Adding up at all m edges gives a total of at least

$$\sum_{(x,y)\in S}\left[d(x)+d(y)-n\right]$$

triangles. But, since this counts each triangle once for each of its edges, we can only assert that G must contain at least

$$N=\frac{1}{3}\sum_{(x,y)\in S}\left[d(x)+d(y)-n\right]$$

different triangles. Since this sum is taken over the m edges in G, we have

$$N=\frac{1}{3}\sum_{(x,y)\in S}\left[d(x)+d(y)\right]-\frac{1}{3}mn.$$

Now, there are $d(x)$ edges emanating from the vertex x, and since the number $d(x)$ goes into the value of N for each of them, the total contribution to N by the edges at x is $d(x)^2$. Similarly at each vertex, and we have

$$N=\frac{1}{3}\sum_{x=1}^{n}d(x)^2-\frac{1}{3}mn,$$

where the sum is taken over the *vertices* instead of the edges.

We are still faced with the nontrivial problem of relating N to the proposed number

$$4m\cdot\frac{m-\frac{n^2}{4}}{3n}.$$

To do this, George invokes the Chebyshev inequality

$$x_1^2 + x_2^2 + \cdots + x_n^2 \geq \frac{1}{n}(x_1 + x_2 + \cdots + x_n)^2.$$

This useful result follows immediately from the Power-Mean Inequality, which was proved in my earlier volume *From Erdős to Kiev*. Recall that the Power-Mean Inequality asserts, for arbitrary positive real numbers p_i and a_i that, when $s > t$,

$$\left[\frac{p_1a_1^s + p_2a_2^s + \cdots + p_na_n^s}{p_1 + p_2 + \cdots + p_n}\right]^{1/s} \geq \left[\frac{p_1a_1^t + p_2a_2^t + \cdots + p_na_n^t}{p_1 + p_2 + \cdots + p_n}\right]^{1/t}$$

Thus, for $p_i = 1$, $s = 2$, $t = 1$, and $a_i = d(i)$, the degree of vertex i, we obtain

$$\left[\frac{d(1)^2 + d(2)^2 + \cdots + d(n)^2}{n}\right]^{1/2} \geq \left[\frac{d(1) + d(2) + \cdots + d(n)}{n}\right]^{1/1}$$

and

$$\sum_{x=1}^{n} d(x)^2 \geq \frac{1}{n}\left[\sum_{x=1}^{n} d(x)\right]^2.$$

Now, the sum of the degrees in a graph is simply the total number of endpoints of the edges, and is therefore equal to twice the number of edges since each has two endpoints. Hence

$$\sum_{x=1}^{n} d(x) = 2m,$$

and therefore

$$\sum_{x=1}^{n} d(x)^2 \geq \frac{1}{n} \cdot 4m^2.$$

Accordingly, the number of triangles in G is at least

$$\begin{aligned} N &= \frac{1}{3}\sum_{x=1}^{n} d(x)^2 - \frac{1}{3}mn \\ &\geq \frac{1}{3} \cdot \frac{1}{n}4m^2 - \frac{1}{3}mn \\ &= 4m \cdot \frac{m - \frac{n^2}{4}}{3n}, \end{aligned}$$

as desired.

A Selection of Joseph Liouville's Amazing Identities Concerning the Arithmetic Functions $\sigma(n)$, $\tau(n)$, $\phi(n)$, $\theta(n)$, $\mu(n)$, $\lambda(n)$

1. Introduction

Starting on page 284 of volume one of his monumental history of the theory of numbers, Leonard Eugene Dickson lists about 50 identities concerning arithmetic functions that were published by Joseph Liouville (1809–1882) in the 1850's and 60's. Many of these are delightful surprises and I'm sure a good story could be made out of each of them, but we will content ourselves with illustrating about a third of them, proving half of these and leaving the others for your enjoyment as exercises.

The functions $\sigma(n)$, $\tau(n)$, and $\phi(n)$ are often among the attractions of a first course in number theory and I expect many readers will recall that, for positive integers n,

$$\sigma(n) = \text{ the } \textit{sum} \text{ of all the positive divisors of } n \text{ (including 1 and } n),$$

$$\tau(n) = \text{ the } \textit{number} \text{ of positive divisors of } n,$$

$$\phi(n) = \text{ Euler's } \phi\text{-function } = \text{ the number of positive integers } k \leq n$$

$$\text{which are relatively prime to } n, \text{ i.e., for which } (k, n) = 1.$$

Thus, for example,

$$\sigma(6) = 1 + 2 + 3 + 6 = 12,$$

$$\tau(6) = 4,$$

$$\phi(6) = 2$$

(among 1, 2, 3, 4, 5, 6, only 1 and 5 are relatively prime to 6).

It is difficult to imagine how functions which focus on such distinctive properties might be related in interesting ways. One surprising instance is the fact that n satisfies the condition

$$\sigma(n) + \phi(n) = n \cdot \tau(n)$$

if and only if n is a *prime* number (this is established in my *Mathematical Morsels,* Dolciani Series, 1978, page 32).

If the prime decomposition of n is $p_1^{a_1} p_2^{a_2} \cdots p_k^{a_k}$, formulas for σ, τ, and ϕ are given by

$$\sigma(n) = \prod_{i=1}^{k} \frac{p_i^{a_i+1} - 1}{p_i - 1}$$

(this is just $(1+p_1+\cdots+p_1^{a_1})(1+p_2+\cdots+p_2^{a_2})\cdots(1+p_k+\cdots+p_k^{a_k})$),

$$\tau(n) = \prod_{i=1}^{k}(a_i + 1) = (a_1 + 1)(a_2 + 1)\cdots(a_k + 1)$$

(it doesn't matter what the primes are; only the exponents count), and

$$\phi(n) = \prod_{i=1}^{k}(p_i^{a_i} - p_i^{a_i-1}).$$

Since these formulas are established in many books on elementary number theory, let us not stop to prove them.

The three other functions we shall encounter are the Möbius function $\mu(n)$ and the not so well known, but easily understood, $\theta(n)$ and $\lambda(n)$, which are defined as follows.

(a) $\lambda(n)$:

$$\lambda(n) = (-1)^{a_1+a_2+\cdots+a_k},$$

which is simply ± 1, depending on the parity of the total number of primes, counting repetitions, in the prime decomposition in n; e.g., $\lambda(1) = 1$, $\lambda(24) = \lambda(2^3 \cdot 3) = (-1)^4 = 1$, and $\lambda(12) = \lambda(2^2 \cdot 3) = (-1)^3 = -1$.

(b) $\mu(n)$: first, $\mu(1) = 1$ and, for $n \geq 2$, $\mu(n) = 0$ if any exponent $a_i > 1$, and when all $a_i = 1$, $\mu(p_1 p_2 \cdots p_k) = (-1)^k$; thus $\mu(n)$ is always 0, ± 1, and for $n > 1$, $\mu(n) = 0$ unless n is "square-free"; e.g., for a prime p, $\mu(p) = (-1)^1 = -1$, $\mu(12) = \mu(2^2 \cdot 3) = 0$, $\mu(30) = \mu(2 \cdot 3 \cdot 5) = (-1)^3 = -1$, and $\mu(51) = \mu(3 \cdot 17) = (-1)^2 = 1$.

(c) $\theta(n)$:

$\theta(n) =$ the number of ways of resolving n into an *ordered pair* of *relatively prime* factors, that is, $\theta(n)$ = the number of ordered pairs (a,b) such that $(a,b)=1$ and $ab=n$; hence $\theta(1)=1$ (for the pair $(1,1)$), $\theta(6)=4$ (for the pairs $(1,6)$, $(2,3)$, $(3,2)$, and $(6,1)$), and $\theta(12)=4$ (for the pairs $(1,12)$, $(3,4)$, $(4,3)$, and $(12,1)$, the pairs $(2,6)$ and $(6,2)$ not being relatively prime).

2. Some Examples

For easy reference, a few values of four of these functions are tabulated:

n	1	2	3	4	5	6	7	8	9	10	11	12	13	14	15	16	17	18	19	20
σ	1	3	4	7	6	12	8	15	13	18	12	28	14	24	24	31	18	39	20	42
τ	1	2	2	3	2	4	2	4	3	4	2	6	2	4	4	5	2	6	2	6
ϕ	1	1	2	2	4	2	6	4	6	4	10	4	12	6	8	8	16	6	18	8
θ	1	2	2	2	2	4	2	2	2	4	2	4	2	4	4	2	2	4	2	4

(a) Now, $n=6$ has four divisors $d=1,2,3,6$, whose complementary divisors are, respectively, $\frac{n}{d}=6,3,2,1$. If we multiply each of the four values $\phi(d)$ by its corresponding value $\tau(\frac{n}{d})$ and add up, we obtain

$$\begin{aligned} & \phi(1)\cdot\tau(6)+\phi(2)\cdot\tau(3)+\phi(3)\cdot\tau(2)+\phi(6)\cdot\tau(1) \\ = \;& 1\cdot 4 \quad + \quad 1\cdot 2 \quad + \quad 2\cdot 2 \quad + \quad 2\cdot 1 \\ = \;& 4 \quad + \quad 2 \quad + \quad 4 \quad + \quad 2 \\ = \;& 12. \end{aligned}$$

I suspect this is enough to let the cat out of thc bag and raisc your suspicion that it might always be true that

$$\sum_{d|n}\phi(d)\cdot\tau\left(\frac{n}{d}\right)=\sigma(n).$$

As unlikely as it might seem, this is indeed always the case. This is the first identity we shall prove, but at present let us take special note of the way it combines the functions ϕ and τ.

Adding up the products of the functions over all *ordered pairs* of complementary divisors of n is known as the *convolution* of the functions and is denoted by the symbol $*$; hence the above identity might be expressed as $\phi*\tau=\sigma$. Thus, in general, if $f(n)$ and $g(n)$ are functions of a positive integer n, their convolution is defined by

$$(f*g)(n)=\sum_{d|n}f(d)\cdot g\left(\frac{n}{d}\right).$$

(b) Consider the convolution of θ and τ:

$$\sum_{d|n} \theta(d)\tau\left(\frac{n}{d}\right).$$

At $n = 20$ for example, we have

$$\begin{aligned}(\theta * \tau)(20) &= \theta(1)\tau(20) + \theta(2)\tau(10) + \theta(4)\tau(5) + \theta(5)\tau(4) \\ &\quad + \theta(10)\tau(2) + \theta(20)\tau(1) \\ &= 1\cdot 6 + 2\cdot 4 + 2\cdot 2 + 2\cdot 3 + 4\cdot 2 + 4\cdot 1 \\ &= 6 + 8 + 4 + 6 + 8 + 4 \\ &= 36 \\ &= 6^2 \\ &= [\tau(20)]^2,\end{aligned}$$

an instance of the identity $\theta * \tau = \tau^2$.

(c) Calculating $\lambda * \theta$ at $n = 12$, we get

$$\begin{aligned}(\lambda * \theta)(12) &= \lambda(1)\theta(12) + \lambda(2)\theta(6) + \lambda(3)\theta(4) + \lambda(4)\theta(3) \\ &\quad + \lambda(6)\theta(2) + \lambda(12)\theta(1) \\ &= 1\cdot 4 + (-1)\cdot 4 + (-1)\cdot 2 + 1\cdot 2 + 1\cdot 2 + (-1)\cdot 1 \\ &= 4 - 4 - 2 + 2 + 2 - 1 \\ &= 1;\end{aligned}$$

in fact, $\lambda * \theta$ is always equal to 1.

(d) An identity of a different type is the remarkable

$$\sum_{d|n} d\cdot\sigma\left(\frac{n}{d}\right) = \sum_{d|n} d\cdot\tau(d).$$

Checking at $n = 10$, we have

$$\begin{aligned}\sum_{d|10} d\sigma\left(\frac{10}{d}\right) &= 1\cdot\sigma(10) + 2\cdot\sigma(5) + 5\cdot\sigma(2) + 10\cdot\sigma(1) \\ &= 1\cdot 18 + 2\cdot 6 + 5\cdot 3 + 10\cdot 1 \\ &= 18 + 12 + 15 + 10 \\ &= 55;\end{aligned}$$

and

$$\begin{aligned}\sum_{d|10} d\tau(d) &= 1 \cdot \tau(1) + 2 \cdot \tau(2) + 5 \cdot \tau(5) + 10 \cdot \tau(10)\\ &= 1 \cdot 1 + 2 \cdot 2 + 5 \cdot 2 + 10 \cdot 4\\ &= 1 + 4 + 10 + 40\\ &= 55.\end{aligned}$$

(e) In the following five cases, various exponents occur.

(i) At $n = 12$, the value of

$$\sum_{d|n} \tau(d^2)\mu\left(\frac{n}{d}\right)$$

is

$$\begin{aligned}\tau(1^2)\mu(12) + \tau(2^2)\mu(6) + \tau(3^2)\mu(4) + \tau(4^2)\mu(3) + \tau(6^2)\mu(2) + \tau(12^2)\mu(1)\\ = 1 \cdot 0 + 3 \cdot 1 + 3 \cdot 0 + 5(-1) + 9(-1) + 15 \cdot 1\\ = 0 + 3 + 0 - 5 - 9 + 15\\ = 4;\end{aligned}$$

Can you guess the other side of the identity? As a glance at the above table of values suggests (check $n = 12$), the answer is simply

$$\sum_{d|n} \tau(d^2)\mu\left(\frac{n}{d}\right) = \theta(n).$$

(ii) For any positive integer k, it is always true that

$$\sum_{d|n} \tau(d^{2k}) = \tau(n)\tau(n^k).$$

Checking at $n = 6$, with $k = 3$, we have, noting $\tau(6^6) = \tau(2^6 \cdot 3^6) = 7 \cdot 7 = 49$, that

$$\begin{aligned}\sum_{d|6} \tau(d^6)) &= \tau(1^6) + \tau(2^6) + \tau(3^6) + \tau(6^6)\\ &= 1 + 7 + 7 + 49\\ &= 64;\end{aligned}$$

and

$$\begin{aligned}\tau(6)\tau(6^3) &= 4 \cdot \tau(2^3 \cdot 3^3)\\ &= 4(4 \cdot 4)\\ &= 64.\end{aligned}$$

(iii) Also, for all positive integers k and n,

$$\sum_{d|n} \left[\theta(d)\right]^k = \tau\left(n^{2^k}\right).$$

For $k = 4$, we get, at $n = 6$, that

$$\begin{aligned}\sum_{d|6} \left[\theta(d)\right]^4 &= \theta(1)^4 + \theta(2)^4 + \theta(3)^4 + \theta(6)^6 \\ &= 1^4 + 2^4 + 2^4 + 4^4 \\ &= 1 + 16 + 16 + 256 \\ &= 289,\end{aligned}$$

and

$$\begin{aligned}\tau\left(6^{2^4}\right) &= \tau(6^{16}) = \tau(2^{16} \cdot 3^{16}) = 17 \cdot 17 \\ &= 289.\end{aligned}$$

(iv) Can you identify

$$\sum_{d^2|n} \theta\left(\frac{n}{d^2}\right),$$

where the sum is to be restricted to the divisors of n whose *squares* also divide n?

At $n = 12$, the only squares which divide n are $d^2 = 1^2$ and 2^2, giving

$$\begin{aligned}\sum_{d^2|12} \theta\left(\frac{12}{d^2}\right) &= \theta\left(\frac{12}{1}\right) + \theta\left(\frac{12}{4}\right) \\ &= \theta(12) + \theta(3) \\ &= 4 + 2 \\ &= 6.\end{aligned}$$

The identity is

$$\sum_{d^2|n} \theta\left(\frac{n}{d^2}\right) = \tau(n).$$

(v) Finally, a result known as Liouville's Generalization is the remarkable

$$\sum_{d|n} \left[\tau(d)\right]^3 = \left[\sum_{d|n} \tau(d)\right]^2.$$

For each positive integer, this yields a set of positive integers with the engaging property that

$$\text{the sum of their cubes} = \text{the square of their sum.}$$

This property is well known for the first n positive integers, and this result also generates a set for each positive integer n.

For $n = 6$, we have

$$\begin{aligned} & \tau(1)^3 + \tau(2)^3 + \tau(3)^3 + \tau(6)^3 \\ = & \ 1^3 + 2^3 + 2^3 + 4^3 \\ = & \ 1 + 8 + 8 + 64 \\ = & \ 81, \end{aligned}$$

while

$$\begin{aligned} & [\tau(1) + \tau(2) + \tau(3) + \tau(6)]^2 \\ = & \ [1 + 2 + 2 + 4]^2 \\ = & \ 9^2 \\ = & \ 81. \end{aligned}$$

3. The Common Characteristic

Although our identities are of various kinds, they can all be established by the same approach. The reason for this is that each of the functions f in these identities is *multiplicative,* that is, if m and n are relatively prime, then

$$f(mn) = f(m) \cdot f(n).$$

Once this is established for the functions in question, the proofs reduce to simple verifications.

The multiplicative nature of the functions σ, τ, and ϕ is quite clear from their definitions:

> for relatively prime m and n, no prime divisor of m occurs also in the prime decomposition of n, and vice-versa, and a glance at the formulas for these functions shows that the factors in each of them can be grouped into those arising from the primes in m and those from the primes in n to yield

$$\sigma(mn) = \sigma(m)\sigma(n),$$

$$\tau(mn) = \tau(m)\tau(n),$$

$$\phi(mn) = \phi(m)\phi(n).$$

Similarly, the multiplicative character of λ follows from its defining formula

$$\lambda(n) = (-1)^{a_1+a_2+\cdots+a_k};$$

by appropriate grouping we easily get $\lambda(mn) = \lambda(m)\lambda(n)$.

The multiplicative nature of the Möbius function μ is also quickly deduced from its definition:

suppose $m = p_1^{a_1} p_2^{a_2} \cdots p_k^{a_k}$ and $n = q_1^{b_1} q_2^{b_2} \cdots q_t^{b_t}$, where no prime p is the same as any prime q.

(i) Now, if some exponent a_i or b_i exceeds 1, then

$$\mu(mn) = \mu(p_1^{a_1} p_2^{a_2} \cdots p_k^{a_k} q_1^{b_1} q_2^{b_2} \cdots q_t^{b_t}) = 0;$$

but in this case, at least one of $\mu(m)$, $\mu(n)$ is also equal to 0, and we have

$$\mu(mn) = \mu(m)\mu(n).$$

(ii) Otherwise all exponents a_i and b_i are equal to 1 and

$$\mu(mn) = (-1)^{k+t} = (-1)^k(-1)^t = \mu(m)\mu(n).$$

Finally, consider the function $\theta(n)$, which we recall is the number of relatively prime ordered pairs (a, b) such that $ab = n$. If

$$n = p_1^{a_1} p_2^{a_2} \cdots p_k^{a_k},$$

then, in dividing up the prime divisors into factors a and b, one can't split up the primes $p_1^{a_1}$, putting some into a and the rest into b, for a and b are to be relatively prime. As far as $p_1^{a_1}$ is concerned, it goes completely into the first factor a or completely into the second one b. Hence there are 2 options concerning $p_1^{a_1}$, and similarly for each $p_i^{a_i}$, and we have the formula

$$\theta(n) = 2^k$$

(clearly the prescriptions for constructing ab and ba call for different options concerning the allocation of the primes of n and therefore this formula counts *ordered* pairs (a, b), thus conforming to the nature of $\theta(n)$). From this formula, then, we easily get

$$\theta(mn) = 2^{k+t} = 2^k \cdot 2^t = \theta(m)\theta(n).$$

4. Eight Proofs

The great merit of a multiplicative function f is that it is completely determined by its values at the prime powers p^r; in fact, peeling off the prime powers in turn, we get

$$\begin{aligned} f(p_1^{a_1}p_2^{a_2}\cdots p_k^{a_k}) &= f\left[(p_1^{a_1})\cdot(p_2^{a_2}p_3^{a_3}\cdots p_k^{a_k})\right] \\ &= f(p_1^{a_1})\cdot f(p_2^{a_2}p_3^{a_3}\cdots p_k^{a_k}) \\ &= f(p_1^{a_1})\cdot\left[f(p_2^{a_2})\cdot f(p_3^{a_3}\cdots p_k^{a_k})\right] \\ &\vdots \\ &= \prod_{i=1}^{k} f(p_i^{a_i}). \end{aligned}$$

Thus, if two multiplicative functions f and g take the same values at every prime power, $f(p^r) = g(p^r)$, it follows that $f(n)$ and $g(n)$ are identically equal. Consequently, after substantiating the multiplicative character of the functions in question, it remains only to test their values at the general prime power p^r.

For example, the right side of the proposed $\phi * \tau = \sigma$ is clearly multiplicative, but we don't know about the convolution on the left. Since it is easy to show in general that the convolution of two multiplicative functions is again multiplicative, let's digress briefly to establish this result.

Let m and n be relatively prime. Clearly a divisor d of mn is constructed by combining a factor d_1, composed of primes belinging to m, and a factor d_2 that is put together from primes in n: if

$$mn = (p_1^{a_1}\cdots p_k^{a_k})(q_1^{b_1}\cdots q_t^{b_t}),$$

then the divisors of mn are given by

$$\begin{aligned} d &= d_1\cdot d_2 \\ &= (p_1^{u_1}\cdots p_k^{u_k})(q_1^{v_1}\cdots q_t^{v_t}), \end{aligned}$$

where $0 \le u_i \le a_i$ and $0 \le v_i \le b_i$; that is, $d = d_1 d_2$ where $d_1|m$ and $d_2|n$ and d_1 and d_2 are relatively prime (if all $u_i = 0$, then $d_1 = 1$, and when all $u_i = v_j = 0$, then $d_1 = d_2 = 1$, corresponding to the divisor $d = 1$ of mn).

Conversely, any choices of divisors d_1 of m and d_2 of n give a divisor $d = d_1 d_2$ of mn in which d_1 and d_2 are relatively prime.

Therefore

$$\begin{aligned} (f*g)(mn) &= \sum_{d|mn} f(d)g\left(\frac{mn}{d}\right) = \sum_{d_1|m,\ d_2|n} f(d_1d_2)g\left(\frac{mn}{d_1d_2}\right) \\ &= \sum_{d_1|m,\ d_2|n} f(d_1)f(d_2)g\left(\frac{m}{d_1}\right)g\left(\frac{n}{d_2}\right) \end{aligned}$$

(by the multiplicative character of f and g), where the sum is taken over all choices of d_1 and d_2. But precisely these same terms are given by the product

$$\left[\prod_{d_1|m} f(d_1)g\left(\frac{m}{d_1}\right)\right] \cdot \left[\sum_{d_2|n} f(d_2)g\left(\frac{n}{d_2}\right)\right],$$

which is just

$$(f * g)(m) \cdot (f * g)(n),$$

and we conclude that $f * g$ is indeed multiplicative.

(a) $\phi * \tau = \sigma$: Thus $\phi * \tau$ is multiplicative, and to prove the identity $\phi * \tau = \sigma$, we need only show that

$$(\phi * \tau)(p^r) = \sigma(p^r).$$

But this is straightforward, since the divisors of p^r are simply $\{1, p, p^2, \ldots, p^r\}$:

$$\begin{aligned}(\phi * \tau)(p^r) &= \sum_{d|p^r} \phi(d)\tau\left(\frac{p^r}{d}\right) \\ &= \sum_{i=0}^{r} \phi(p^i)\tau(p^{r-i}) \\ &= \phi(1)\tau(p^r) + \sum_{i=1}^{r}(p^i - p^{i-1})(r - i + 1)\end{aligned}$$

(taking the case of $i = 0$ separately and using the formulas for ϕ and τ),

$$\begin{aligned}&= 1 \cdot (r+1) + \big[(p-1)r + (p^2 - p)(r-1) + (p^3 - p^2)(r-2) \\ &\quad + \cdots + (p^{r-1} - p^{r-2})(2) + (p^r - p^{r-1})(1)\big],\end{aligned}$$

which simplifies easily to

$$1 + p + p^2 + \cdots + p^r = \sigma(p^r).$$

After observing that $\tau(mn)^2 = [\tau(m) \cdot \tau(n)]^2 = \tau(m)^2 \cdot \tau(n)^2$, thus establishing the multiplicative nature of the function τ^2, the identity $\theta * \tau = \tau^2$ may be handled similarly.

(b) $\sum_{d|n} \theta(d)^k = \tau(n^{2^k})$: Since it is a simple matter of checking $f(mn)$ to verify that the functions we shall encounter are multiplicative, let us assume that this has been done throughout (this is not always a trivial matter, but still nothing more than routine). Accordingly, from now on we shall only check the functions at p^r. In the present case, then, we would like to show, for all positive integers k and prime powers p^r, that

$$\sum_{d|p^r} \theta(d)^k = \tau(p^{r \cdot 2^k}).$$

Observing that $\theta(p^0) = \theta(1) = 1$, and that for all $i \geq 1$, $\theta(p^i) = 2$ (for the ordered pairs $(1, p^i)$ and $(p^i, 1)$), we easily have

$$\sum_{d|p^r} \theta(d)^k = \sum_{i=0}^{r} \theta(p^i)^k = \theta(1)^k + \sum_{i=1}^{r} 2^k = 1 + r \cdot 2^k,$$

and

$$\tau(p^{r \cdot 2^k}) = r \cdot 2^k + 1$$

(by the formula for τ).

(c) $\sum_{d|n} \tau(d^{2k}) = \tau(n)\tau(n^k)$: For $n = p^r$, we have

$$\begin{aligned}
\sum_{d|p^r} \tau(d^{2k}) &= \sum_{i=0}^{r} \tau(p^{i \cdot 2k}) = \sum_{i=0}^{r} (i \cdot 2k + 1) \\
&= 2k \sum_{i=0}^{r} i + \sum_{i=0}^{r} 1 \\
&= kr(r+1) + (r+1) \\
&= (r+1)(kr+1),
\end{aligned}$$

and

$$\tau(p^r)\tau(p^{rk}) = (r+1)(kr+1).$$

(d) $\lambda * \theta = 1$: Clearly,

$$\begin{aligned}
(\lambda * \theta)(p^r) &= \sum_{d|p^r} \lambda(d)\theta\left(\frac{p^r}{d}\right) \\
&= \sum_{i-0}^{r} \lambda(p^i)\theta(p^{r-i}) \\
&= \lambda(1)\theta(p^r) + \sum_{i=1}^{r-1} ((-1)^i \cdot 2) + \lambda(p^r)\theta(1)
\end{aligned}$$

(taking the cases of $i = 0$ and $i = r$ separately; recall that $\theta(p^i) = 2$ for $i \geq 1$, and that r is always ≥ 1)

$$= 1 \cdot 2 + [-2 + 2 - 2 + - \cdots (-1)^{r-1} \cdot 2] + (-1)^r \cdot 1.$$

For r even, this gives $2+[-2]+1=1$, and for r odd, this is $2+[0]-1=1$, and the identity follows.

(e) $\sum_{d^2|n}\theta\left(\frac{n}{d^2}\right)=\tau(n)$: Clearly $\tau(p^r)=r+1$.

Now, for $n=p^r$, a divisor $d=p^i$ gives $d^2=p^{2i}$, and we have, since $r-2i$ must remain greater than or equal to 0, that

$$\sum_{d^2|p^r}\theta\left(\frac{p^r}{d^2}\right)=\sum_{i=0}^{[\frac{r}{2}]}\theta(p^{r-2i}),$$

where $[\frac{r}{2}]$ denotes the integer part of $\frac{r}{2}$; recalling again that $\theta(p^0)=\theta(1)=1$ and that $\theta(p^i)=2$ for all $i\geq 1$, we have for $r=2t$, that this equals

$$\sum_{i=0}^{t}\theta(p^{2t-2i})=\theta(1)+\sum_{i=0}^{t-1}2=1+2t=1+r;$$

and for $r=2t+1$, we get

$$\sum_{i=0}^{t}\theta(p^{2t+1-2i})=\sum_{i=0}^{t}2=2(t+1)=(2t+1)+1=r+1,$$

establishing the identity.

(f) $\sum_{d|n}\tau(d^2)\mu\left(\frac{n}{d}\right)=\theta(n)$: Since r is always at least 1, $\theta(p^r)=2$ in all cases. Also,

$$\sum_{d|p^r}\tau(d^2)\mu\left(\frac{p^r}{d}\right)=\sum_{i=0}^{r}\tau(p^{2i})\mu(p^{r-i}).$$

For $r=1$, this is just the two terms

$$\tau(1)\mu(p)+\tau(p^2)\mu(1)=1(-1)+3\cdot 1=2,$$

and for all $r>1$, we get

$$\sum_{i=0}^{r}\tau(p^{2i})\mu(p^{r-i})=\sum_{i=0}^{r}(2i+1)\mu(p^{r-i}),$$

where $\mu(p^{r-i})=0$ so long as $r-i$ remains greater than 1; hence this reduces to just the two terms given by $i=r-1$ and $r=i$, giving

$$\begin{aligned}[2(r-1)+1]\mu(p)+(2r+1)\mu(1)&=(2r-1)(-1)+(2r+1)\\&=2,\end{aligned}$$

completing the proof.

(g) $\sum_{d|n} \tau(d)^3 = \left[\sum_{d|n} \tau(d)\right]^2$: (Liouville's Generalization)
For $n = p^r$, we have

$$\sum_{d|n} \tau(d)^3 = \sum_{i=0}^{r} \tau(p^i)^3 = \sum_{i=0}^{r} (i+1)^3$$
$$= 1^3 + 2^3 + \cdots + (r+1)^3 = \left[\tfrac{1}{2}(r+1)(r+2)\right]^2,$$

and

$$\left[\sum_{d|n} \tau(n)\right]^2 = \left[\sum_{i=0}^{r} \tau(p^i)\right]^2 = \left[\sum_{i=0}^{r} (i+1)\right]^2$$
$$= \left[1 + 2 + \cdots + (r+1)\right]^2 = \left[\tfrac{1}{2}(r+1)(r+2)\right]^2,$$

as desired.

(h) $\sum_{d|n} d\sigma\left(\frac{n}{d}\right) = \sum_{d|n} d\tau(d)$: For $n = p^r$,

$$\sum_{d|n} d\tau(d) = \sum_{i=0}^{r} p^i \tau(p^i) = \sum_{i=0}^{r} p^i (i+1)$$
$$= 1 + 2p + 3p^2 + \cdots + (r+1)p^r;$$

also,

$$\sum_{d|n} d\sigma\left(\frac{n}{d}\right) = \sum_{i=0}^{r} p^i \sigma(p^{r-i}) = \sum_{i=0}^{r} p^i (1 + p + \cdots p^{r-i})$$
$$= 1(1 + p + p^2 + \cdots + p^r) + p(1 + p + \cdots p^{r-1})$$
$$+ p^2(1 + p + \cdots + p^{r-2}) + \cdots + p^r(1)$$
$$= 1 + 2p + 3p^2 + \cdots + (r+1)p^r,$$

completing the proof.

5. Eight Illustrated Exercises

(a) $\sum_{d|n} \theta(d) = \tau(n^2)$: For $n = 24$, we have

$$\sum_{d|24} \theta(d) = \theta(1) + \theta(2) + \theta(3) + \theta(4) + \theta(6) + \theta(8) + \theta(12) + \theta(24)$$
$$= 1 + 2 + 2 + 2 + 4 + 2 + 4 + 4 = 21,$$

and $\tau(24^2) = \tau(2^6 \cdot 3^2) = 7 \cdot 3 = 21$.

(b) $\sum_{d|n} \lambda(d)\theta(d) = \lambda(n)$: For $n = 12$,

$$\begin{aligned}\sum_{d|12} \lambda(d)\theta(d) &= \lambda(1)\theta(1) + \lambda(2)\theta(2) + \lambda(3)\theta(3) \\ &\quad + \lambda(4)\theta(4) + \lambda(6)\theta(6) + \lambda(12)\theta(12) \\ &= 1 \cdot 1 + (-1) \cdot 2 + (-1) \cdot 2 + 1 \cdot 2 + 1 \cdot 4 + (-1) \cdot 4 \\ &= 1 - 2 - 2 + 2 + 4 - 4 \\ &= -1;\end{aligned}$$

and $\lambda(12) = \lambda(2^2 \cdot 3) = (-1)^3 = -1$.

(c) $\sum_{d|n} \tau(d^2)\phi\left(\frac{n}{d}\right) = \sum_{d|n} \frac{n}{d}\theta(d)$: For $n = 12$,

$$\begin{aligned}\sum_{d|12} \tau(d^2)\phi\left(\frac{12}{d}\right) &= \tau(1)\phi(12) + \tau(4)\phi(6) + \tau(9)\phi(4) \\ &\quad + \tau(16)\phi(3) + \tau(36)\phi(2) + \tau(144)\phi(1) \\ &= 1 \cdot 4 + 3 \cdot 2 + 3 \cdot 2 + 5 \cdot 2 + 9 \cdot 1 + 15 \cdot 1 \\ &= 4 + 6 + 6 + 10 + 9 + 15 \\ &= 50,\end{aligned}$$

and

$$\begin{aligned}\sum_{d|12} \frac{12}{d}\theta(d) &= 12 \cdot \theta(1) + 6 \cdot \theta(2) + 4 \cdot \theta(3) \\ &\quad + 3 \cdot \theta(4) + 2 \cdot \theta(6) + 1 \cdot \theta(12) \\ &= 12 \cdot 1 + 6 \cdot 2 + 4 \cdot 2 + 3 \cdot 2 + 2 \cdot 4 + 1 \cdot 4 \\ &= 12 + 12 + 8 + 6 + 8 + 4 \\ &= 50.\end{aligned}$$

(d) $\sum_{d|n} \theta(d)\sigma(\frac{n}{d}) = \sum_{d|n} \frac{n}{d}\tau(d^2)$: At $n = 20$, we have

$$\begin{aligned}\sum_{d|n} \theta(d)\sigma\left(\frac{20}{d}\right) &= \theta(1)\sigma(20) + \theta(2)\sigma(10) + \theta(4)\sigma(5) + \theta(5)\sigma(4) \\ &\quad + \theta(10)\sigma(2) + \theta(20)\sigma(1) \\ &= 1 \cdot 42 + 2 \cdot 18 + 2 \cdot 6 + 2 \cdot 7 + 4 \cdot 3 + 4 \cdot 1 \\ &= 42 + 36 + 12 + 14 + 12 + 4 \\ &= 120,\end{aligned}$$

and

$$\sum_{d|20} \frac{20}{d}\tau(d^2) = 20\tau(1) + 10\tau(4) + 5\tau(16) + 4\tau(25) + 2\tau(100) + 1\tau(400)$$
$$= 20 + 30 + 25 + 12 + 18 + 15$$
$$= 120.$$

(e) $(\phi * \sigma)(n) = n\tau(n)$: For $n = 15$, we get

$$(\phi * \sigma)(15) = \phi(1)\sigma(15) + \phi(3)\sigma(5) + \phi(5)\sigma(3) + \phi(15)\sigma(1)$$
$$= 1 \cdot 24 + 2 \cdot 6 + 4 \cdot 4 + 8 \cdot 1$$
$$= 24 + 12 + 16 + 8$$
$$= 60,$$

and $15\tau(15) = 15 \cdot 4 = 60$.

(f) $\sum_{d|n} d\sigma(d) = \sum d|nd^2\sigma(\frac{n}{d})$: At $n = 12$,

$$\sum_{d|12} d\sigma(d) = 1\sigma(1) + 2\sigma(2) + 3\sigma(3) + 4\sigma(4) + 6\sigma(6) + 12\sigma(12)$$
$$= 1 \cdot 1 + 2 \cdot 3 + 3 \cdot 4 + 4 \cdot 7 + 6 \cdot 12 + 12 \cdot 28$$
$$= 1 + 6 + 12 + 28 + 72 + 336$$
$$= 455,$$

and

$$\sum_{d|12} d^2\sigma\left(\frac{12}{d}\right) = 1^2\sigma(12) + 2^2\sigma(6) + 3^2\sigma(4)$$
$$+ 4^2\sigma(3) + 6^2\sigma(2) + 12^2\sigma(1)$$
$$= 1 \cdot 28 + 4 \cdot 12 + 9 \cdot 7 + 16 \cdot 4 + 36 \cdot 3 + 144 \cdot 1$$
$$= 28 + 48 + 63 + 64 + 108 + 144$$
$$= 455.$$

(g) $\sum_{d|n} \sigma(d) = \sum_{d|n} d\tau(\frac{n}{d})$: For $n = 12$,

$$\sum_{d|12} \sigma(d) = \sigma(1) + \sigma(2) + \sigma(3) + \sigma(4) + \sigma(6) + \sigma(12)$$
$$= 1 + 3 + 4 + 7 + 12 + 28$$
$$= 55,$$

and

$$\sum_{d|12} d\tau\left(\frac{12}{d}\right) = 1\tau(12) + 2\tau(6) + 3\tau(4) + 4\tau(3) + 6\tau(2) + 12\tau(1)$$

$$= 1 \cdot 6 + 2 \cdot 4 + 3 \cdot 3 + 4 \cdot 2 + 6 \cdot 2 + 12 \cdot 1$$

$$= 6 + 8 + 9 + 8 + 12 + 12$$

$$= 55.$$

(h) $\sum_{d|n} \sigma(d)\sigma(\frac{n}{d}) = \sum_{d|n} d\tau(d)\tau(\frac{n}{d})$: For $n = 18$,

$$\sum_{d|18} \sigma(d)\sigma\left(\frac{18}{d}\right) = \sigma(1)\sigma(18) + \sigma(2)\sigma(9) + \sigma(3)\sigma(6)$$

$$+ \sigma(6)\sigma(3) + \sigma(9)\sigma(2) + \sigma(18)\sigma(1)$$

$$= 2(1 \cdot 39 + 3 \cdot 13 + 4 \cdot 12)$$

$$= 2(39 + 39 + 48)$$

$$= 2(126) = 252,$$

and

$$\sum_{d|18} d\tau(d)\tau\left(\frac{18}{d}\right) = 1\tau(1)\tau(18) + 2\tau(2)\tau(9) + 3\tau(3)\tau(6)$$

$$+ 6\tau(6)\tau(3) + 9\tau(9)\tau(2) + 18\tau(18)\tau(1)$$

$$= 1 \cdot 1 \cdot 6 + 2 \cdot 2 \cdot 3 + 3 \cdot 2 \cdot 4 + 6 \cdot 4 \cdot 2 + 9 \cdot 3 \cdot 2$$

$$+ 18 \cdot 6 \cdot 1$$

$$= 6 + 12 + 24 + 48 + 54 + 108$$

$$= 252.$$

6. References

For a very profitable abstract algebra approach to the identities involving convolutions, see the well-written and instructive paper "Number-Theoretic Functions via Convolution Rings" by S. K. Berberian, University of Texas, which appeared as the lead article in the April issue of *Mathematics Magazine,* 1992.

The text *Introduction to Arithmetical Functions* by Paul McCarthy, University of Kansas (Springer-Verlag, 1986), a reference given by Dr. Berberian, contains a wealth of material in this field.

A Problem from the 1988 Austrian-Polish Mathematics Competition†

The two integer sequences $\{a_k\}$ and $\{b_k\}$, $k \geq 0$, satisfy the relations

$$b_k = a_k + 9 \text{ and } a_{k+1} = 8b_k + 8 \text{ for all } k \geq 0.$$

Given that the number 1988 belongs to $\{a_k\}$ or to $\{b_k\}$, prove that no term of $\{a_k\}$ is a *perfect square.*

From the given relations we easily get

$$a_{k+1} = 8b_k + 8 = 8(a_k + 9) + 8 = 8a_k + 80,$$

revealing that $\{a_k\}$ is governed by

$$a_{k+1} - 8a_k = 80.$$

Thus

$$a_{k+2} - 8a_{k+1} = 80,$$

and by subtraction we have

$$a_{k+2} - 9a_{k+1} + 8a_k = 0 \quad \text{for all } k \geq 0.$$

Solving this recursion by a standard procedure, we obtain the auxiliary equation $x^2 - 9x + 8 = 0$, whose roots are 1 and 8, giving $a_k = p \cdot 8^k + q \cdot 1^k$ for some constants p and q. For $k = 0$, $a_0 = p+q$, and for $k = 1$, $a_1 = 8p+q$. Hence $a_1 - a_0 = 7p$. But $a_1 = 8a_0 + 80$, and so $7p = a_1 - a_0 = 7a_0 + 80$, making $p = a_0 + \frac{80}{7}$, and then $q = a_0 - p = -\frac{80}{7}$. Therefore we have the

---

† (*Crux Mathematicorum,* 1989, 290)

formula

$$a_k = \left(a_0 + \frac{80}{7}\right) \cdot 8^k - \frac{80}{7},$$

i.e.,

$$a_k = a_0 \cdot 8^k + 80 \cdot \frac{8^k - 1}{7}. \tag{1}$$

From a_k, b_k is obtained by adding 9 (recall $b_k = a_k + 9$).

Now, if some member of $\{b_k\}$ were to equal 1988, say $b_t = 1988$, then the corresponding term in $\{a_k\}$ would be 1979, and conversely; that is to say, the condition that 1988 belongs to $\{a_k\}$ or $\{b_k\}$ implies that some term a_t is equal to 1988 or to 1979. Letting x denote whichever integer it is, we have

$$a_t = a_0 \cdot 8^t + 80 \cdot \frac{8^t - 1}{7} = x,$$

where x is either 1988 or 1979. Solving for a_0, we get

$$a_0 = \left[x - 80 \cdot \frac{8^t - 1}{7}\right] 8^{-t},$$

from which

$$a_k = \left[x - 80 \cdot \frac{8^t - 1}{7}\right] 8^{k-t} + 80 \cdot \frac{8^k - 1}{7}$$

from (1) above. Thus we need to show that this expression is not a perfect square for any $k \geq 0$ when x is either 1988 or 1979.

As t is a nonnegative integer, either $t \geq 1$ or $t = 0$.

Suppose $t \geq 1$. Noting that, for $t \geq 1$, $8^t - 1 = (8-1)(1+8+8^2+\cdots+8^{t-1})$, we obtain, for $k = 0$, that

$$\begin{aligned} a_0 &= \left[x - 80 \cdot \frac{8^t - 1}{7}\right] 8^{-t} \\ &= \frac{x - 80(1 + 8 + 8^2 + \cdots + 8^{t-1})}{8^t}. \end{aligned}$$

But, with $t \geq 1$, this isn't even an integer for $x = 1988$ or 1979, since neither of these values of x is divisible by 8. That is to say, the only way to have 1988 a member of $\{a_k\}$ or $\{b_k\}$ is to have $t = 0$, and this makes 1988 the very first term in the sequence to which it belongs. Since we know this **does** happen one way or the other, then t must indeed be 0, giving

$$a_k = x \cdot 8^k + 80 \cdot \frac{8^k - 1}{7},$$

and it remains to show that this is not a perfect square for any $k \geq 0$ and $x = 1988$ or 1979.

For $k = 0$, we get $a_0 = x$, which is not a square in either case.

Suppose $k \geq 1$. Then

$$a_k = x \cdot 8^k + 80 \cdot \frac{8^k - 1}{7} = x \cdot 8^k + 80(1 + 8 + \cdots + 8^{k-1}),$$

and modulo 10, we have

$$a_k \equiv x \cdot 8^k;$$

thus for $x = 1988$,

$$a_k \equiv 8 \cdot 8^k,$$

and for $x = 1979$,

$$a_k \equiv 9 \cdot 8^k.$$

Now, there are four cases to consider, depending on the value of x and whether k is odd or even.

(i) *Suppose* $k = 2r$ *and* $x = 1988$ (i.e., $a_k = 8 \cdot 8^k$): Then (mod 10) we have

$$8^k = 8^{2r} \equiv 64^r \equiv 4^r \equiv 4 \quad \text{or} \quad 6$$

(since all powers of 4 end in 4 or 6), in which case $a_k \equiv 8 \cdot 4$ or $8 \cdot 6$, i.e., $a_k \equiv 2$ or 8. Since no square ends in 2 or 8, it follows that a_k is never a square in this case.

(ii) Similarly, for $k = 2r + 1$ and $x = 1979$ (i.e., $a_k \equiv 9 \cdot 8^k$), we have

$$a_k \equiv 9 \cdot 8^{2r+1} \equiv 72 \cdot 8^{2r} \equiv 2 \cdot 4 \quad \text{or} \quad 2 \cdot 6,$$

i.e., 8 or 2, leading to the same conclusion.

(iii) *Suppose* $k = 2r$ *and* $x = 1979$: Then

$$a_k = 1979 \cdot 8^{2r} + 80(1 + 8 + \cdots + 8^{2r-1}),$$

and **modulo 9** we have

$$\begin{aligned} a_k &\equiv (-1)(-1)^{2r} + (-1)(1 - 1 + 1 - 1 + \cdots \text{ ending with } -1) \\ &\equiv -1 + (-1)(0) \equiv 8. \end{aligned}$$

Since no square is congruent to 8 (mod 9), a_k is never a square in this case either.

(iv) Finally *suppose* $k = 2r + 1$ *and* $x = 1988$: Then

$$a_k = 1988 \cdot 8^{2r+1} + 80(1 + 8 + \cdots + 8^{2r}).$$

Since $k \geq 1$, that is $2r + 1 \geq 1$, a_k is divisible by 16:

$$a_k = 16\big[994 \cdot 8^{2r} + 5(1 + 8 + \cdots + 8^{2r})\big].$$

Therefore a_k is a square if and only if the complementary factor

$$q_k = 994 \cdot 8^{2r} + 5(1 + 8 + \cdots + 8^{2r})$$

is a square. Since $k \geq 1$, then $r \geq 0$, and for $r = 0$, we have

$$q_k = 994 + 5(1) = 999,$$

which is not a square. Otherwise $2r \geq 1$, and **modulo 8** we have

$$q_k \equiv 0 + 5(1) \equiv 5.$$

But no square is congruent to 5 (mod 8), and the solution is complete.

An Excursion into the Complex Plane

This essay is based on the splendid paper "A Commuting Formula for Polygonal Paths" by Edward Kitchen (Santa Monica, California).

(i) Let us begin with an intriguing result arising out of the following simple construction.

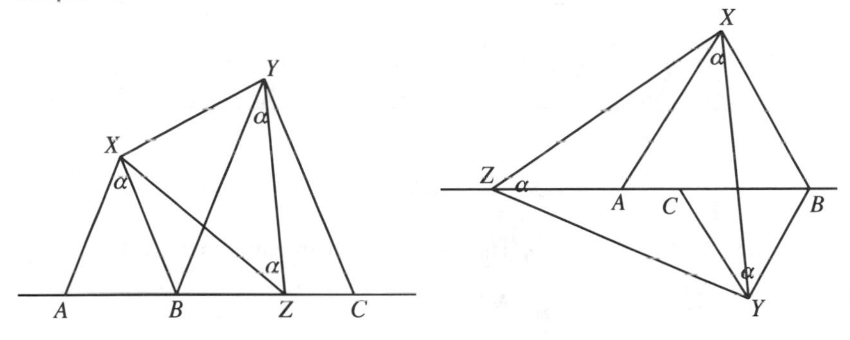

FIGURE 176

FIGURE 177

Suppose A, B, and C are arbitrary points on a straight line, and α is a given angle. Isosceles triangles ABX and BCY, having equal vertical angles α at X and Y, are drawn so that their cyclic orientations A-B-X and B-C-Y are counterclockwise. Then isosceles triangle XYZ, having vertical angle α at Z, is drawn so that its cyclic orientation X-Y-Z is clockwise.

Remarkably, then, the point Z will always be found to be in line with A, B, and C.

To prove this engaging property, suppose the construction is performed in the complex plane. If M is the midpoint of AB, the vector $\mathbf{X}$ is clearly given by the sum of the vectors $\mathbf{A}$, $\mathbf{AM}$, and $\mathbf{MX}$ (figure 178).

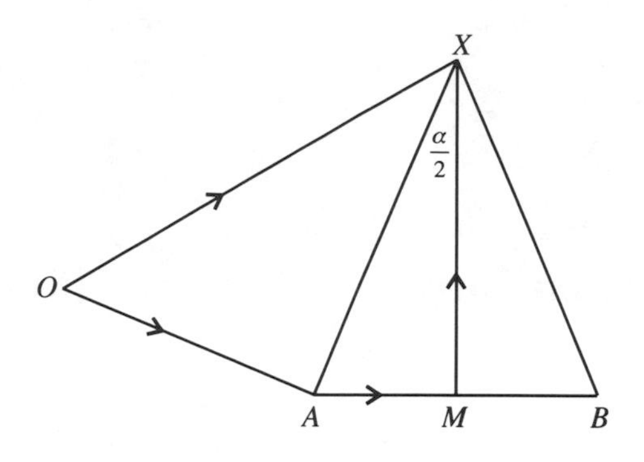

FIGURE 178

Now, since $\triangle ABX$ is isosceles, MX is the perpendicular bisector of AB and bisects angle X. Hence

$$\cot \angle AXM = \cot \frac{\alpha}{2} = \frac{|MX|}{|AM|},$$

giving $|MX| = |AM| \cot \frac{\alpha}{2}$, and, recalling that multiplication by $\mathbf{i}$ $(= \sqrt{-1})$ rotates a vector through a right angle, we have that the vector $\mathbf{MX}$ is given by

$$\mathbf{MX} = \left(\mathbf{AM} \cot \frac{\alpha}{2}\right) \cdot \mathbf{i}.$$

Noting that $\mathbf{AM} = \frac{1}{2}\mathbf{AB} = \frac{1}{2}(\mathbf{B} - \mathbf{A})$, the constructed point X is given by

$$\begin{aligned}
\mathbf{X} &= \mathbf{A} + \mathbf{AM} + \mathbf{MX} \\
&= \mathbf{A} + \frac{1}{2}(\mathbf{B} - \mathbf{A}) + \left[\frac{1}{2}(\mathbf{B} - \mathbf{A}) \cot \frac{\alpha}{2}\right] \cdot \mathbf{i} \\
&= \frac{1}{2}\left[\mathbf{A}\left(1 - \mathbf{i} \cdot \cot \frac{\alpha}{2}\right) + \mathbf{B}\left(1 + \mathbf{i} \cdot \cot \frac{\alpha}{2}\right)\right].
\end{aligned}$$

Letting $\cot \frac{\alpha}{2} = t$, this may be written more simply in the form

$$\mathbf{X} = \frac{1}{2}\big[\mathbf{A}(1 - \mathbf{i}t) + \mathbf{B}(1 + \mathbf{i}t)\big].$$

We need to emphasize that this is a formula for the vertex X of an isosceles triangle constructed on AB so that the orientation A-B-X is **counterclockwise** (for the clockwise orientation, X would need to be reflected in AB, requiring the vector $\mathbf{MX}$ to be subtracted from $\mathbf{A} + \mathbf{AM}$ instead of added).

Similarly, then,

$$\mathbf{Y} = \frac{1}{2}\big[\mathbf{B}(1 - \mathbf{i}t) + \mathbf{C}(1 + \mathbf{i}t)\big].$$

Because X-Y-Z is clockwise, Y-X-Z is counterclockwise, and our formula gives

$$\mathbf{Z} = \frac{1}{2}\left[\mathbf{Y}(1 - \mathrm{i}t) + \mathbf{X}(1 + \mathrm{i}t)\right].$$

Substituting for $\mathbf{Y}$ and $\mathbf{X}$, and observing that $(1 + \mathrm{i}t)(1 - \mathrm{i}t) = 1 - \mathrm{i}^2t^2 = 1 + t^2$, we get

$$\begin{aligned}\mathbf{Z} &= \frac{1}{2}\left\{\frac{1}{2}\left[\mathbf{B}(1 - \mathrm{i}t)^2 + \mathbf{C}(1 + t^2)\right] + \frac{1}{2}\left[\mathbf{A}(1 + t^2) + \mathbf{B}(1 + \mathrm{i}t)^2\right]\right\} \\ &= \frac{1}{4}\left[\mathbf{A}(1 + t^2) + \mathbf{B}(2 - 2t^2) + \mathbf{C}(1 + t^2)\right] \\ &= u\mathbf{A} + v\mathbf{B} + w\mathbf{C},\end{aligned}$$

where the coefficients u, v, and w **add up to 1**:

$$u + v + w = \frac{1}{4}(1 + t^2 + 2 - 2t^2 + 1 + t^2) = 1.$$

But $\mathbf{p} = u\mathbf{A} + v\mathbf{B} + w\mathbf{C}$, with the auxiliary relation $u + v + w = 1$, is the parametric form for points p on the straight line through A, B, and C, and the conclusion follows.

(ii) Neuberg's Theorem

> In figure 179, if X, Y, Z are the centers of squares drawn **outwardly** on the sides of $\triangle ABC$, then the centers P, Q, R of squares drawn **inwardly** on the sides of $\triangle XYZ$ are the midpoints of the sides of the original triangle ABC.

Unfortunately, the descriptions "outwardly and inwardly" do not allow the theorem to be stated in its full generality. We need the more powerful device of a cyclic orientation of the sides of $\triangle ABC$. In this way, as one looks along a side in the direction of its orientation, the sides of the line may be distinguished as **right** and **left**.

Also, the cyclic order of $\triangle ABC$ may be used to induce a cyclic order around $\triangle XYZ$:

> if $\triangle ABC$ is oriented A-B-C and the centers of the squares on AB and BC are X and Y, respectively, then the side XY is to be ordered X-Y.

In both figures 179 and 180, $\triangle ABC$ is to be considered oriented A-B-C, leading in each case to the induced X-Y-Z around $\triangle XYZ$. In figure 179, then, the squares are drawn on the **right** side of $\triangle ABC$, while in figure 180 they are drawn on the **left**.

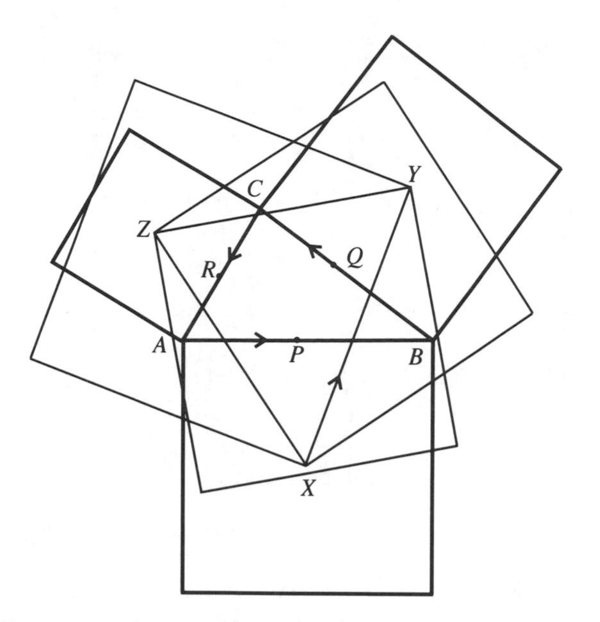

FIGURE 179

A-B-C in triangle ABC induces X-Y-Z in triangle XYZ

Now, the result of the theorem holds in either case, provided the squares on $\triangle XYZ$ are drawn on the **opposite** side to those drawn on $\triangle ABC$: thus, in figure 179, the squares are drawn on the **right** of $\triangle ABC$ and the **left** of $\triangle XYZ$, and vice-versa in figure 180. (In order not to clutter figure 180, only the square on the right of ZX is drawn.)

Let us prove the theorem for figure 179. It suffices to show P is the midpoint of AB, since Q and R provide similar cases.

Observing that the centers X, Y, Z are the vertices of isosceles right-angled triangles drawn on the sides of $\triangle ABC$, our **counterclockwise** formula of section (i) yields

$$\mathbf{Y} = \frac{1}{2}\left[\mathbf{C}(1 - \mathbf{i}t) + \mathbf{B}(1 + \mathbf{i}t)\right],$$

and

$$\mathbf{Z} = \frac{1}{2}\left[\mathbf{A}(1 - \mathbf{i}t) + \mathbf{C}(1 + \mathbf{i}t)\right],$$

where $\alpha = \frac{\pi}{2}$, making $t = \cot \frac{1}{2}\left(\frac{\pi}{2}\right) = 1$. Similarly, for the center P of the

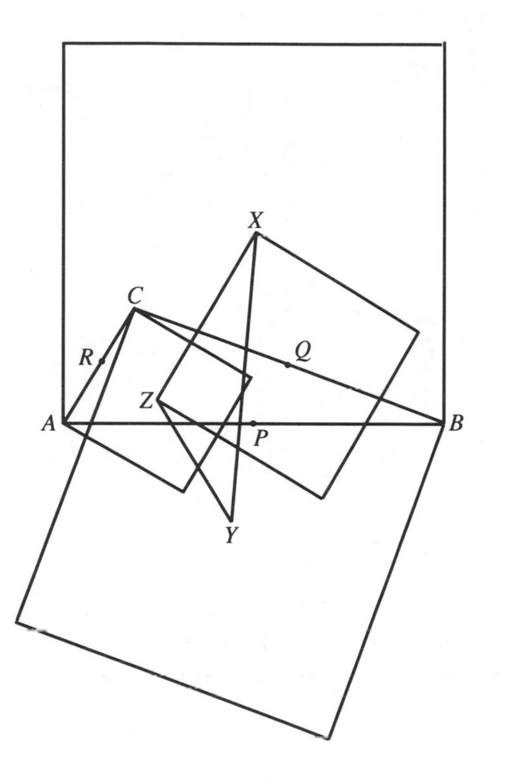

FIGURE 180

square on YZ, we have

$$\begin{aligned}
\mathbf{P} &= \frac{1}{2}\left[\mathbf{Y}(1-\mathbf{i}t)+\mathbf{Z}(1+\mathbf{i}t)\right] \\
&= \frac{1}{4}\left\{\left[\mathbf{C}(1-\mathbf{i})^2+\mathbf{B}(1+1)\right]+\left[\mathbf{A}(1+1)+\mathbf{C}(1+\mathbf{i})^2\right]\right\} \\
&= \frac{1}{4}\left[2\mathbf{A}+2\mathbf{B}+\mathbf{C}(1-2\mathbf{i}-1+1+2\mathbf{i}-1)\right] \\
&= \frac{1}{2}(\mathbf{A}+\mathbf{B}),
\end{aligned}$$

which places P at the midpoint of AB.

(iii) A Delightful Result

Suppose squares $ABCD$ and $BXYZ$ are hinged at B (figure 181). If K and M are their centers and L and N are the midpoints of AX and CZ, prove that $KLMN$ is always a **square**.

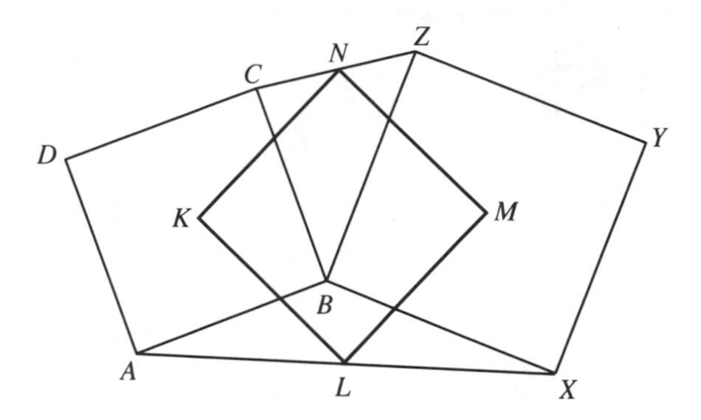

FIGURE 181

Since K and M are the centers of squares drawn on AB and BX, applying Neuberg's theorem to $\triangle ABX$ reveals that the midpoint L is the center of the square drawn on the appropriate side of KM. That is to say, KL and KM, being segments from vertices to the center of a square, are equal and perpendicular.

Similarly, applying Neuberg's theorem to $\triangle CBZ$, we see that N is the center of the square drawn on the other side of KM, making KN and MN equal and perpendicular. But KN and MN are also equal in length to KL and ML since the two squares on opposite sides of KM are the same size. Thus $KLMN$, having four equal sides and a right angle at L, is indeed a square.

(iv) The Anticomplementary Triangle

If A, B, and C are the midpoints of the sides of $\triangle PQR$, then ABC is the **medial** triangle of $\triangle PQR$; conversely, PQR is called the **anticomplementary** triangle of $\triangle ABC$.

Suppose $\triangle ABC$ is oriented A-B-C and equilateral triangles are drawn on its sides to the **right** to yield $\triangle XYZ$ with induced orientation X-Y-Z (figure 182). Now let equilateral triangles be drawn on the sides of $\triangle XYZ$ to the **left** to give $\triangle PQR$; then, remarkably, triangle PQR is the anticomplementary triangle of $\triangle ABC$.

Figure 183 illustrates that the result holds when the equilateral triangles are drawn on the **left** of $\triangle ABC$ and on the **right** of $\triangle XYZ$; in fact the very same triangle PQR is obtained.

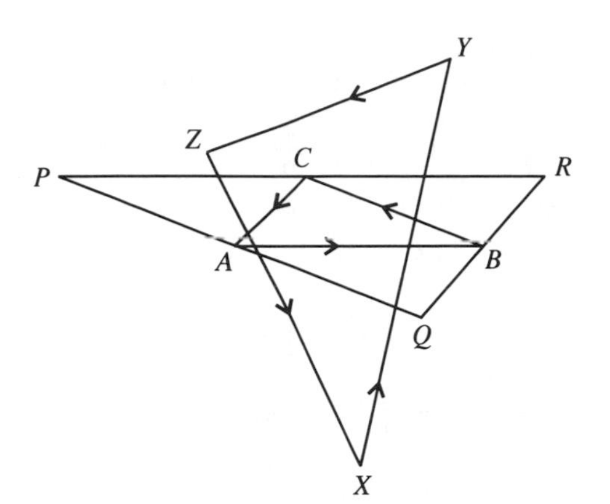

FIGURE 182

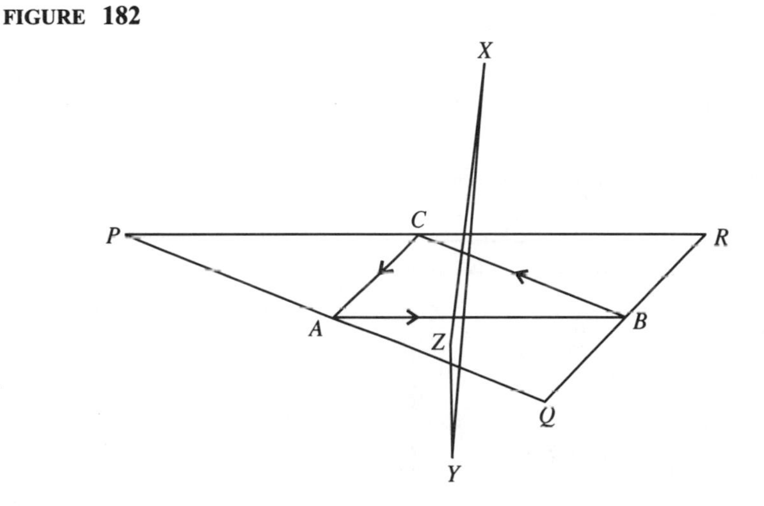

FIGURE 183

The vertical angle α in an equilateral triangle is $60°$ giving

$$t = \cot \frac{\alpha}{2} = \cot 30° = \sqrt{3}.$$

In figure 182, then, our counterclockwise formula yields the six results

$$\mathbf{X} = \frac{1}{2}[\mathbf{B}(1 - \mathbf{i}\sqrt{3}) + \mathbf{A}(1 + \mathbf{i}\sqrt{3})]$$

$$\mathbf{Y} = \frac{1}{2}[\mathbf{C}(1 - \mathbf{i}\sqrt{3}) + \mathbf{B}(1 + \mathbf{i}\sqrt{3})]$$

$$\mathbf{Z} = \frac{1}{2}\left[\mathbf{A}(1 - \mathbf{i}\sqrt{3}) + \mathbf{C}(1 + \mathbf{i}\sqrt{3})\right]$$
$$\mathbf{P} = \frac{1}{2}\left[\mathbf{X}(1 - \mathbf{i}\sqrt{3}) + \mathbf{Y}(1 + \mathbf{i}\sqrt{3})\right]$$
$$\mathbf{Q} = \frac{1}{2}\left[\mathbf{Y}(1 - \mathbf{i}\sqrt{3}) + \mathbf{Z}(1 + \mathbf{i}\sqrt{3})\right]$$
$$\mathbf{R} = \frac{1}{2}\left[\mathbf{Z}(1 - \mathbf{i}\sqrt{3}) + \mathbf{X}(1 + \mathbf{i}\sqrt{3})\right].$$

Thus the midpoint M of PQ is given by

$$\begin{aligned}
\mathbf{M} &= \frac{1}{2}(\mathbf{P} + \mathbf{Q}) \\
&= \frac{1}{4}\left[\mathbf{X}(1 - \mathbf{i}\sqrt{3}) + 2\mathbf{Y} + \mathbf{Z}(1 + \mathbf{i}\sqrt{3})\right] \\
&= \frac{1}{4}\left\{\frac{1}{2}\left[\mathbf{B}(1 - \mathbf{i}\sqrt{3})^2 + \mathbf{A}(1+3)\right] + 2\mathbf{Y} + \frac{1}{2}\left[\mathbf{A}(1+3) + \mathbf{C}(1 + \mathbf{i}\sqrt{3})^2\right]\right\} \\
&= \frac{1}{4}\left[4\mathbf{A} + \frac{1}{2}\mathbf{B}(-2 - 2\mathbf{i}\sqrt{3}) + \frac{1}{2}\mathbf{C}(-2 + 2\mathbf{i}\sqrt{3}) + 2\mathbf{Y}\right] \\
&= \frac{1}{4}\left[4\mathbf{A} - \mathbf{B}(1 + \mathbf{i}\sqrt{3}) - \mathbf{C}(1 - \mathbf{i}\sqrt{3}) + 2\mathbf{Y}\right] \\
&= \frac{1}{4}(4\mathbf{A}) \\
&= \mathbf{A}.
\end{aligned}$$

Similarly for the midpoints of QR and RP.

(v) A Property of Parallelograms

Let parallelogram $ABCD$ be oriented A-B-C-D and let equilateral triangles be drawn on its sides to the **right** to give quadrilateral $XYZW$ with induced orientation X-Y-Z-W (figure 184). Then, if equilateral triangles are drawn on the sides of $XYZW$ to the **left**, the resulting quadrilateral is just $ABCD$ again. Figure 185 illustrates that **right** and **left** can be interchanged.

Referring to figure 184, since AB and CD are equal and parallel in parallelogram $ABCD$, the vectors **AB** and **DC** are the same and we have

$$\mathbf{AB} = \mathbf{DC}$$
$$\mathbf{B} - \mathbf{A} = \mathbf{C} - \mathbf{D}, \quad \text{or} \quad \mathbf{A} - \mathbf{B} = \mathbf{D} - \mathbf{C}.$$

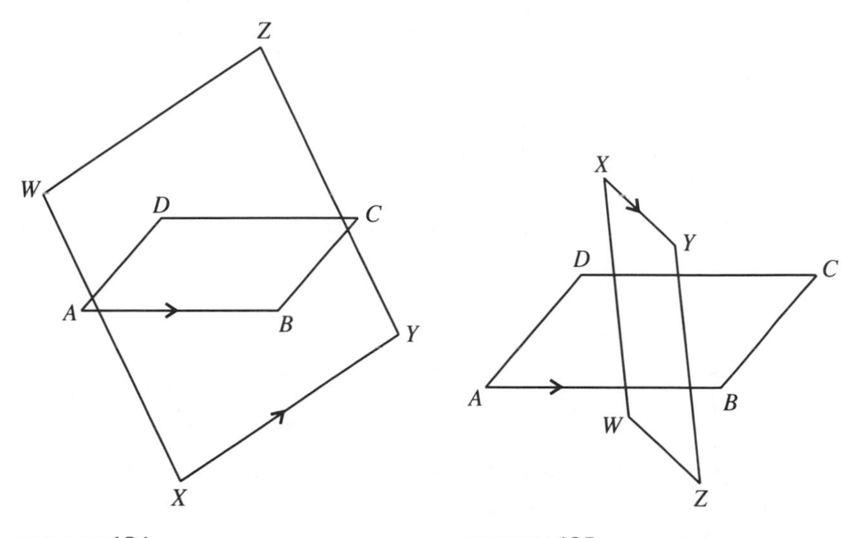

FIGURE 184 FIGURE 185

Now, **X**, **Y**, and **P** are given by exactly the same expressions we just found for them in section (iv):

$$\mathbf{X} = \frac{1}{2}[\mathbf{B}(1 - \mathbf{i}\sqrt{3}) + \mathbf{A}(1 + \mathbf{i}\sqrt{3})]$$
$$\mathbf{Y} = \frac{1}{2}[\mathbf{C}(1 - \mathbf{i}\sqrt{3}) + \mathbf{B}(1 + \mathbf{i}\sqrt{3})]$$
$$\mathbf{P} = \frac{1}{2}[\mathbf{X}(1 - \mathbf{i}\sqrt{3}) + \mathbf{Y}(1 + \mathbf{i}\sqrt{3})]$$

Therefore P is given by

$$\begin{aligned}
\mathbf{P} &= \frac{1}{2}[\mathbf{X}(1 - \mathbf{i}\sqrt{3}) + \mathbf{Y}(1 + \mathbf{i}\sqrt{3})] \\
&= \frac{1}{4}\left\{[\mathbf{B}(1 - \mathbf{i}\sqrt{3})^2 + \mathbf{A}(1+3)] + [\mathbf{C}(1+3) + \mathbf{B}(1 + \mathbf{i}\sqrt{3})^2]\right\} \\
&= \frac{1}{4}[4\mathbf{A} + \mathbf{B}(2-6) + 4\mathbf{C}] \\
&= \mathbf{A} - \mathbf{B} + \mathbf{C} \\
&= \mathbf{D} - \mathbf{C} + \mathbf{C} \text{ (recall } \mathbf{A} - \mathbf{B} = \mathbf{D} - \mathbf{C}) \\
&= \mathbf{D}.
\end{aligned}$$

Similarly for the other vertices of $ABCD$.

Two Problems from the 1990 International Olympiad[†]

1. From a given initial integer $n_0 > 1$, A and B take turns playing a game of constructing a sequence $\{n_0, n_1, n_2, \ldots\}$ as follows:
(i) A plays first, and therefore, as the game proceeds, he decides on the terms $n_1, n_3, n_5, \ldots, n_{2k+1}, \ldots$ by selecting the integer of his choice from the closed interval determined by the previous term and its square:

$$n_{2k} \leq n_{2k+1} \leq n_{2k}^2.$$

(ii) after which B plays n_{2k+2} by selecting a **divisor** of the previous term; not just any divisor, however, but one whose complementary divisor

$$\frac{n_{2k+1}}{n_{2k+2}}$$

is **a power of a prime number.**

For example, a game might proceed from $n_0 = 5$ as follows:

$$5,\ 24,\ 6,\ 28,\ 4,\ 12, \ldots :$$

A's 24 is properly chosen from the closed interval $[5, 25]$, B's 6 divided into 24 correctly gives a power of a prime (4), A chose his 28 from $[6, 36]$, B's 4 gives $28/4 = 7$, a power of a prime, and so forth.

B wins by playing the number 1 *and A wins by playing* 1990. Determine the values of the initial terms n_0 for which there is a winning strategy

(a) for A,
(b) for B,
(c) for neither player.

---

† (*Crux Mathematicorum,* 1990, 194)

(a) Probably the most evident consequence of the rules is that, with 1990 lying between 44^2 and 45^2, A would be able to win with 1990 on his very first move when n_0 is any of the numbers $\{45, 46, \ldots, 1990\}$; and since n_0 is in effect an involuntary move by B, A would be able to win on his **next** move whenever he can force B to play a number from $[45, 1990]$. Since we want to classify all initial values $2, 3, \ldots,$ let's start our investigation at the beginning.

When $n_0 = 2$**,** A must play either 2, 3, or 4, to each of which B can reply with his winning 1 (yielding either the 3-term sequence 2, 2, 1; 2, 3, 1; or 2, 4, 1). Thus B wins whenever he has the chance to play the number 2.

Now, clearly A must avoid playing any integer which is itself a power of a prime, thus eliminating

$$\{2, 3, 4, 5, 7, 8, 9, 11, 13, 16, 17, 19, 23, 25, 27, 29, 31, 37\}.$$

When $n_0 = 3$**,** requiring A to play from the closed interval $[3, 9]$, his only hope is to play 6; however, B can reply with a winning 2 (giving $3, 6, 2, \ldots$), and so again B wins when $n_0 = 3$.

For $n_0 = 4$**,** A's options increase to the interval $[4, 16]$, containing the new hopefuls $\{10, 12, 14, 15\}$ (the others up to 16 being immediate losers); however, B can still win every time as follows (recall B wins with 2 or 3):

$$(4, \underline{10}, 2, \ldots),\ (4, \underline{12}, 3, \ldots), (4, \underline{14}, 2, \ldots),\ (4, \underline{15}, 3, \ldots).$$

Similarly, $n_0 = 5$ is another winner for B (who we have seen wins with 2, 3, and 4): A can now also play 18, 20, 21, 22, and 24, but to no avail:

$$(5, \underline{18}, 2, \ldots),\ (5, \underline{20}, 4, \ldots), (5, \underline{21}, 3, \ldots),$$
$$(5, \underline{22}, 2, \ldots),\ (5, \underline{24}, 3, \ldots).$$

At this point, however, things start looking up for A.

For $n_0 = 6$**,** A's new options are $\{26, 28, 30, 33, 34, 35, 36\}$, none of which is any good except 30:

$$(6, \underline{26}, 2, \ldots),\ (6, \underline{28}, 4, \ldots),\ (6, \underline{33}, 3, \ldots),$$
$$(6, \underline{34}, 2, \ldots),\ (6, \underline{35}, 5, \ldots),\ (6, \underline{36}, 4, \ldots).$$

However, by playing $30 = 2 \cdot 3 \cdot 5$, A restricts B's replies to $\{2 \cdot 3, 2 \cdot 5, 3 \cdot 5\} = \{6, 10, 15\}$ (in order to leave a complementary divisor which is a power of a prime, B must play an integer that is the product of two of the primes 2, 3, 5). Consequently, no matter which reply is chosen by B, A can again play 30 and stave off a loss. As we shall see, it would be a mistake for B to play either 10 or 15, and so, with proper play, the game would proceed indefinitely $6, 30, 6, 30, \ldots,$

For $n_0 = 7$**,** A again has the option of playing 30, thus assuring himself of at least a draw. An easy check shows that none of his other new options $\{38, 39, 40, 42, 44, 45, 46, 48\}$ leads to a win; in fact, they are all losers except 42 which, like 30, gives A a draw. Thus A would again accept the draw provided by 30.

Now things take a decided turn in A's favor. The reason is that, at $n_0 = 8$, A is able to play the **key number** $\underline{60} = 2 \cdot 2 \cdot 3 \cdot 5$. Not being able to leave more than one of the primes to constitute the required power of a prime, B must play either

$$2 \cdot 2 \cdot 3 = 12, \; 3 \cdot 5 = 15, \; 2 \cdot 2 \cdot 5 = 20, \quad \text{or} \; 2 \cdot 3 \cdot 5 = 30.$$

In any case, then, A is able to continue with another key number $\underline{105} = 3 \cdot 5 \cdot 7$, which similarly forces B to play from $\{15, 21, 35\}$, each of which allows A to play $\underline{210} = 2 \cdot 3 \cdot 5 \cdot 7$. From here B must choose from $\{30, 42, 70, 105\}$, from which he would certainly choose either 30 or 42 since anything over 45 is fatal, permitting A to play his next key number $\underline{385} = 5 \cdot 7 \cdot 11$. This restricts B to $\{35, 55, 77\}$, taking A to his final key number $\underline{1155} = 3 \cdot 5 \cdot 7 \cdot 11$. This forces B over 44 for all his options, which are $\{105, 165, 231, 385\}$, after which A wins with 1990. Thus A can get into his winning sequence $\{60, 105, 210, 385, 1155, 1990\}$ as soon as n_0 becomes 8, making A a winner for all n_0 from 8 to 1990.

(b) When n_0 exceeds 1990, A's options take on enormous proportions and it is extremely unlikely that all of them would be outright losers; in fact, even a draw would be a lot for B to hope for. Accordingly, let's direct our efforts towards an inductive proof that A wins for all $n_0 \geq 8$.

To this end, suppose that A wins for each n_0 in $\{8, 9, \ldots, 4t\}$, where $t \geq 497$ (making $4t \geq 1988$). We know this is true for $t = 497$. Thus A can go on to win if he can force B to play any of these integers $\{8, 9, \ldots, 4t\}$ at any time during the game.

Let p and q be the two greatest unequal prime numbers that are less than $4t$. By Bertrand's Postulate there is a prime number between t and $2t$ and another between $2t$ and $4t$, and for $t \geq 497$, both would trivially clear the minimum value of 8 to make it into A's winning set $\{8, 9, \ldots, 4t\}$. Hence the product pq must exceed $t \cdot 2t$ which, for $t \geq 497$, is much bigger than $4(t+1)$, and since each of p and q is less than $4t$, we have

$$4t < 4(t+1) < pq < (4t)^2,$$

making pq eligible for A's first move when n_0 is any integer between $4t$ and pq itself. Now, when A plays pq, B's only reply is one of the primes p, q, each of which leads to a win for A. Thus pq is another key number for A and is available to A for all n_0 in the range $\{8, 9, \ldots, pq\}$. Since $pq > 4(t+1)$, then A's string of winning numbers can be extended to $4(t+1)$ whenever it goes to $4t$, and by induction it follows that A wins for all $n_0 \geq 8$.

In summary, then,

(i) B wins for $n_0 = 2$, 3, 4, and 5,

(ii) they draw for $n_0 = 6$ and 7, and

(iii) A wins for $n_0 \geq 8$.

2. Suppose a set S of $2n-1$ different points, $n \geq 3$, is marked around a circle. Each pair of points of S determines two arcs on the circle whose interiors contain various numbers of points of S, depending on the pair under consideration. In figure 186, (X, Y) cuts off arcs containing 2 and 3 points. What is the *minimum* positive integer k such that *every* subset of k points of S contains two that cut off an arc containing *exactly n points of S in its interior?*

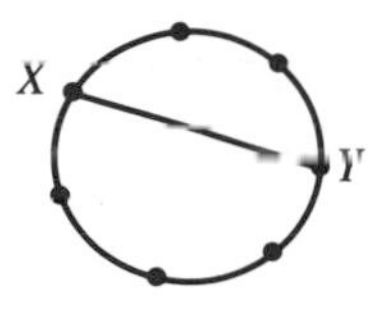

FIGURE 186

(a) If the interior of one of the arcs determined by two points of S has exactly n points of S in its interior, then the interior of the complementary arc must contain the remaining $(2n-1)-n-2 = n-3$ points of S, and conversely. Consequently, if the points of S are numbered $1, 2, \ldots, 2n-1$ in order around the circle, any two points whose numbers differ by $\pm(n-2) \pmod{2n-1}$ provide the desired $(n, n-3)$ split of the points of S; for example, one of the arcs between the points 1 and $n-1$ contains the $n-3$ points $\{2, 3, \ldots, n-2\}$, and the other contains a prized set of n points $\{n, n+1, \ldots, 2n-1\}$. Thus the pairs of points that determine the desired arcs are $\pmod{2n-1}$

$$(1, n-1), (2, n), (3, n+1), \ldots, (2n-1, n-2);$$

we seek the minimum size k of a subset of S which is sure to contain both members of at least one of these $2n-1$ pairs.

Now, the key to the problem is the observation that the chords which join these pairs of points go together into cycles—sometimes there is only one cycle, as in the case of $n = 4$; otherwise, surprisingly, there are always *exactly three,* as for $n = 5$ (figure 187). Clearly, a subset T of S will determine a required arc if and only if it contains *two consecutive* points of one of these cycles. Thus, for $n = 4$, a subset of three points can avoid consecutive points by taking every second one around the cycle, e.g., $T = \{1, 5, 2\}$, but no matter what 4 points might be selected, it is impossible to avoid taking two consecutive points on the cycle. Hence, for $n = 4$, the minimum k is n itself, and since $2n - 1$ is odd, it is pretty obvious that this is always the case when there is only one cycle. For $n = 5$, we could fail to determine a desired arc by taking only one point from each of the 3 cycles, but any 4-subset would have to contain two from some cycle, implying a consecutive pair, since each cycle contains only 3 points altogether. Thus, for $n = 5$, k is again 4, but this time k is just $n - 1$, which we shall see is typical of the case when there are 3 cycles.

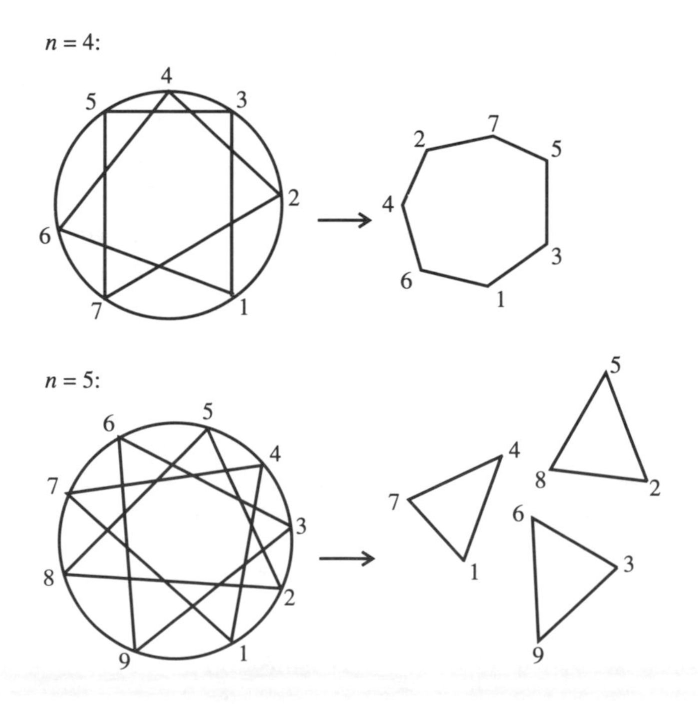

FIGURE 187

(b) Let's go right to the central issue and show that there are always either one or exactly 3 cycles. Since the numbers at the ends of a chord in question always differ by $\pm(n-2) \pmod{2n-1}$, these chords are obtained by starting at any point of S and marching around the circle with a step of size $n-2$. In such a march, one might string out the entire $2n-1$ points of S in a single cycle (the pigeonhole principle implies that you can't go beyond $2n-1$ steps without revisiting some point of S and thus closing a cycle), but it is conceivable that one might return prematurely to a point of S and simply go around the same small cycle again and again without picking up any new points of S; in the latter case, to continue obtaining the desired chords, one would start another march at a new point of S. Therefore, suppose that, as we march around, the *first* cycle that is closed contains r points and is closed after going around the circle m times. Since each point after the first has a chord entering it and also one leaving it, the only way of closing a cycle is by returning to the initial point itself (clearly no point of S can be a distance of $\pm(n-2)$ from more than two other points of S); thus the number m must be a whole number. For example, for $n = 8$, the step is of size $8 - 2 = 6$, and the $2n - 1 = 15$ points of S provide 3 cycles of length 5, each carrying us twice around the cycle when it is traced (figure 188).

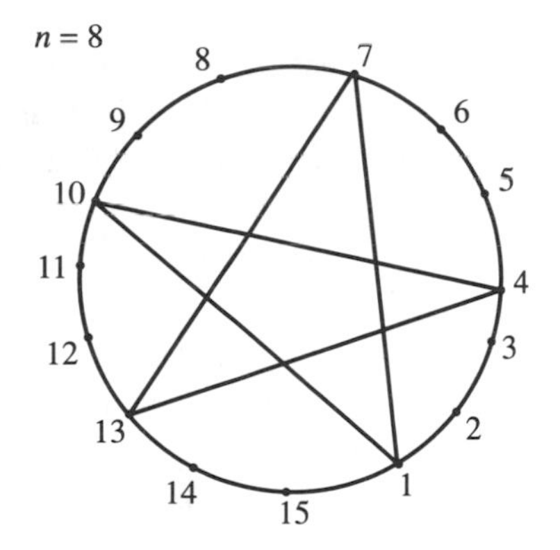

FIGURE 188

We can count the number of points around the circle that are passed over in tracing a cycle by counting the steps that are taken. However, we must be careful because the last point of a step is also the first point in the next step. If we count the $n-2$ points consisting of the initial point and the $n-3$ interior points of a step, leaving the last point of the step to be counted as the initial point of the next step, we will account for each of the points exactly once and obtain a proper total. For $n = 8$, we observe that each of the 5 steps around a cycle covers 6 points for a total of 30, which is just twice the 15 points in the set S. In general, tracing a cycle involves passing over a total of $r(n-2)$ points of S (in taking r steps of size $n-2$) and also over $m(2n-1)$ points (in going around the circle m times), and we have

$$r(n-2) = m(2n-1).$$

Consequently, if we didn't pick up all $2n-1$ points of S in this first cycle, i.e., if $r < 2n-1$, this equation implies that $n-2$ must be bigger than m. But clearly $n-2$ must divide $m(2n-1)$, and being bigger than m, some nontrivial factor of $n-2$ must divide $2n-1$; that is to say, when $r < 2n-1$, $n-2$ and $2n-1$ are *not* relatively prime. However, it is easy to see that if $n-2$ and $2n-1$ have a nontrivial common divisor, it can only be the number 3:

suppose $(n-2, 2n-1) = d$; then $d|n-2$ implies $d|2n-4$, and we have $d|(2n-1)-(2n-4) = 3$, making $d = 1$ or 3.

Therefore, when $(n-2, 2n-1) = 1$, it must be that $r = 2n-1$ (not the contradictory $r < 2n-1$), implying that our march generates an all-inclusive single cycle.

Now, if our first cycle C, of length r, does not pick up all the points of S, then, by shifting C around the circle to have it begin at any new point of S, we obtain another cycle whose vertices would "parallel" those of C at a fixed "distance" around the circle from the corresponding vertices of C, thus picking up r new points of S. So long as an unclaimed point of S remains, this procedure can be continued. Since no two cycles can intersect (recall no point of S is at a distance $n-2$ from more than two other points of S), each cycle must claim a subset of r *new* points of S. But, eventually, every point of S is taken up into some cycle, that is, each point of S belongs to one and only one r-cycle, and therefore it follows that the points of S can be partitioned into subsets of size r, implying that r must be a divisor of $2n-1$. If t is the number of cycles produced, then

$$rt = 2n-1.$$

From the earlier equation, $r(n-2) = m(2n-1)$, then, we get $r(n-2) = mrt$, and $n-2 = mt$. Therefore the greatest common divisor $(n-2, 2n-1)$ is given by

$$(mt, rt) = t \cdot (m, r).$$

When $(n-2, 2n-1) = 1$, then both t and (m, r) must be 1, and as noted above, only a single cycle results.

Suppose that $(n-2, 2n-1) = 3$

In this case, for some relatively prime positive integers p and q, we have $n-2 = 3p$ and $2n-1 = 3q$, i.e.,

$$p = \frac{1}{3}(n-2), \quad \text{and} \quad q = \frac{1}{3}(2n-1).$$

Now consider a march around the circle with our usual step of size $n-2$, starting anywhere, and proceeding until we have gone around the circle exactly p times. The total number of points passed over around the circle would be

$$p(2n-1) = \left[\frac{1}{3}(n-2)\right](2n-1) = (n-2)\left[\frac{1}{3}(2n-1)\right] = (n-2)q,$$

showing that exactly q steps of size $n-2$ would have been taken. Now, going around the circle a whole number of times would bring us back to the starting point and therefore our path would be closed, implying that it is either a single cycle of length q or a shorter cycle of some length j which we have gone around some i times. In any case, taking j steps each of the i times around the cycle (including the possibility of q steps just once), would require a total of ij steps, making

$$q = ij.$$

On the other hand, suppose each tracing of our cycle carries us around the circle v times; in going around the cycle i times, then, we would have gone around the circle a total of vi times, giving

$$p = vi.$$

Thus the geatest common divisor

$$(p, q) = (vi, ij) = i \cdot (v, j) \geq i.$$

But $(p, q) = 1$, and hence $i \leq 1$, making $i = 1$. Thus the cycle in our path is traced only once, implying that the path is a single cycle of length

$$j = ij = q = \frac{1}{3}(2n-1),$$

containing one-third of the points of S. Hence, when $(n-2, 2n-1) = 3$, there are exactly 3 cycles. (It is similarly proved in general that going around a set of n points on a circle with a step of size s generates $d = (n, s)$ cycles of length s/d.)

Finally we come to the easy completion of the solution.

(a) Suppose there is only one cycle. From the $2n - 1$ points in the cycle, we can take every second point $\{1, 3, 5, \ldots, 2n - 3\}$ without selecting two that are consecutive, for a total of $n - 1$ points. However, any n-subset must contain a consecutive pair and provide a desired arc. Hence, when $n - 2$ and $2n - 1$ are relatively prime, there is just one cycle and $k = n$.

(b) Suppose there are 3 cycles. Each cycle has length $\frac{1}{3}(2n - 1) = q$, which is clearly an odd number. Letting $q = 2t - 1$, then, we can take as many as $t - 1$ points from each cycle without taking two that are consecutive on a cycle, for a total of $3t - 3$ points. Now, $2n - 1 = 3q = 6t - 3$, giving $n = 3t - 1$ and $3t - 3 = n - 2$. However, a subset of $3t - 2$ or more points of S, i.e., $n - 1$ or more, would have to contain t points of one of the cycles (since its length is $2t - 1$) and thus a pair of adjacent points on that cycle. Hence, when $(n - 2, 2n - 1) = 3$, we have 3 cycles and $k = n - 1$. In this case, $3 | n - 2$, and we have that

$k = n$, *except when* $n \equiv 2 \pmod{3}$, *in which case* $k = n - 1$.

Exercises

These exercises are generally much easier than most of the problems we have been considering and they are included just for your amusement. They are questions, some slightly revised, that occurred on various Grade 13 Problems Papers of Ontario, a defunct examination which used to be taken by candidates writing for university scholarships involving mathematics.

1. (1960). If the lengths of the sides of a triangle are in the ratios $3:7:8$, show that its angles are in arithmetic progression.

2. (1947). A candidate at an examination writes four papers. If the maximum number of marks obtainable on each paper is m, show that the number of ways of obtaining a total of $2m$ marks is $\frac{1}{3}(m+1)(2m^2+4m+3)$.

3. (1945). Find the coefficient of x^6yzt^5 in the expansion of

$$(x+y+z+t)^{10}(x+z)(x^2+y^2).$$

4. (1950). Prove that on the axis of any parabola there is a certain point K which has the property that, if a chord PQ of the parabola is drawn through it, then

$$\frac{1}{PK^2}+\frac{1}{QK^2}$$

is constant for all positions of the chord.

5. (1949). Find the roots of

$$17x^4+36x^3-14x^2-4x+1=0,$$

given that they are in harmonic progression.

6. (1936). A and B are arbitrary points inside a given circle K. Construct a circle through A and B to touch K.

7. (1938). Reduce

$$\frac{\sqrt{3-\sqrt{5}}}{\sqrt{2}+\sqrt{7-3\sqrt{5}}}$$

to its simplest surd form.

8. (1937). A quadratic expression in x is positive except for the range $-2 \leq x \leq 1$, and a second quadratic expression is negative except for the range $-1 \leq x \leq 4$. Each expression has the value 60 when $x = 3$. For what other value of x are these expressions equal?

9. (1940). Find the equation of the locus of a point P which moves so that the tangents from P to the circle $x^2 + y^2 = a^2$ cut off a segment of length $2a$ on the line $x = a$.

10. (1937). Show that for any positive integer p and any integer $s > 1$ there are p consecutive odd numbers whose sum is p^s, and find the first of these numbers.

11. (1945). Find a polynomial $f(x)$ of degree five such that $f(x) - 1$ is divisible by $(x-1)^3$ and $f(x)$ is itself divisible by x^3.

12. (1958). A uniform cylindrical tub can be filled with water from the cold water tap in 14 minutes, and drained completely through a hole in the bottom in 21 minutes. When both hot and cold water taps are turned on and the plug removed from the hole, the tub fills in 12.6 minutes. How long would it take the hot water tap to fill the tub by itself?

13. (1934). A, B, C, and D are arbitrary fixed points in 3-space and P is a variable point. Show that $PA^2 + PB^2 + PC^2 + PD^2$ is a minimum for P at the midpoint of the segment which joins the midpoints of AC and BD.

14. (1951). Show that the equation

$$\frac{1}{x+2} + \frac{1}{y+2} = \frac{1}{2} + \frac{1}{z+2}$$

is not satisfied by any set of positive integers x, y, z, in which $x \geq 4$. Hence find all solutions (x, y, z) in positive integers.

15. (a) (1956). If an ellipse M rolls without slipping on an identical fixed ellipse E, starting with them touching at a pair of vertices, prove that the locus of a focus of M is a circle with center at a focus of E.

(b) (1956). If a parabola M rolls without slipping on an identical fixed parabola P, starting with them touching at their vertices, prove that the locus of the focus of M is the directrix of P. (More interestingly, the locus of the vertex is a cissoid of Diocles, a curve famous for providing a solution to the ancient problem of doubling the cube.)

16. (1934). Prove that

$$\sin 1^\circ + \sin 3^\circ + \sin 5^\circ + \cdots + \sin 99^\circ = \frac{\sin^2 50^\circ}{\sin 1^\circ}.$$

17. (1942). Of the 9! numbers formed by permuting the digits 1, 2, 3, 4, 5, 6, 7, 8, 9, how many are between 678000000 and 859000000?

18. (a) (1959). Show that the segment of a tangent to a hyperbola which is intercepted between the asymptotes is bisected by the point of contact and that, with the asymptotes, such a segment forms a triangle of constant area.

(b) (1954). A straight line cuts a hyperbola at the points Q and R and its asymptotes at P and S. Prove the midpoint of QR is also the midpoint of PS.

19. (1961). Given that one side of a triangle is twice as long as another side, show that the angle opposite the longer of these sides is greater than twice the angle opposite the lesser side.

20. (1935). If I_1, I_2, I_3 are the centers of the escribed circles of $\triangle ABC$, prove that the area of

$$\triangle I_1 I_2 I_3 = \frac{abc}{2r}.$$

21. (1943). A triangle ABC is such that $3AB = 2AC$. Also, a point D on BC is such that $BD = 2DC$ and $AD = BC$. Show that

$$\tan \frac{1}{2} \angle ADB = \sqrt{\frac{5}{19}}.$$

22. (1950). Let x be a number between -1 and 1. Noting that $n^2 = n(n+1) - n$, find a formula for the sum of the infinite series

$$S = 1 + 2^2 x + 3^2 x^2 + \cdots + n^2 x^{n-1} + \cdots.$$

23. (1942). If $m = \csc\theta - \sin\theta$ and $n = \sec\theta - \cos\theta$, show that

$$m^{2/3} + n^{2/3} = (mn)^{-2/3}.$$

24. (1961). If s_n denotes the sum of the first n positive integers, find the sum of the infinite series

$$S = \frac{s_1}{1} + \frac{s_2}{2} + \frac{s_3}{4} + \cdots + \frac{s_n}{2^{n-1}} + \cdots.$$

25. (1941). Find the smallest positive value of

$$4\tan^{-1}\frac{1}{5} - \tan^{-1}\frac{1}{70} + \tan^{-1}\frac{1}{99}.$$

Solutions to the Exercises

1. (1960). The key to this problem is the recognition that the angles of a triangle are in arithmetic progression if and only if one of them is 60°:

> if an angle is 60°, the amount a second angle is less than 60° is precisely the amount the third angle must exceed 60° in order to make the sum 180° (in general, three numbers are in arithmetic progression if and only if one of them is equal to their average).

In a scalene triangle, then, the 60° angle must be the mid-sized angle and therefore be opposite the side of intermediate length. Thus if the sides are of lengths $3k$, $7k$ and $8k$, we need only show that the angle x opposite the side of length $7k$ is 60°. But this is immediate from the law of cosines:

$$49k^2 = 9k^2 + 64k^2 - 2 \cdot 3k \cdot 8k \cos x,$$

$$\cos x = \frac{73k^2 - 49k^2}{48k^2} = \frac{24}{48} = \frac{1}{2}.$$

2. (1947). The exponents in the generating function

$$f(x) = 1 + x + x^2 + x^3 + \cdots + x^m$$

are the possible scores on a paper. We may consider scores of a, b, c, and d on the four papers to be registered in the product $f(x)^4$ by the term $x^{a+b+c+d}$:

$$\begin{aligned} f(x)^4 &= (1 + x + x^2 + \cdots + x^m)(1 + x + x^2 + \cdots + x^m)\times \\ &\qquad (1 + x + x^2 + \cdots + x^m)(1 + x + x^2 + \cdots + x^m) \\ &= \cdots + (x^a)(x^b)(x^c)(x^d) + \cdots \\ &= \sum x^{a+b+c+d}. \end{aligned}$$

From this viewpoint, the number of ways of accumulating a total score of $2m$ is the number of ways of generating the term x^{2m} in this product, making the required number simply the coefficient of x^{2m} in $f(x)^4$. Letting $[x^n]g(x)$ denote the coefficient of x^n in the function $g(x)$, we want to determine the number $[x^{2m}]f(x)^4$.

Now

$$\begin{aligned} f(x)^4 &= (1+x+x^2+x^3+\cdots+x^m)^4 \\ &= \left(\frac{1-x^{m+1}}{1-x}\right)^4 \\ &= (1-x^{m+1})^4(1-x)^{-4}, \end{aligned}$$

and so the required number is

$$\begin{aligned} &= [x^{2m}](1-x^{m+1})^4(1-x)^{-4}, \\ &= [x^{2m}](1-4x^{m+1}+6x^{2m+2}-\cdots)(1-x)^{-4} \\ &= [x^{2m}](1-4x^{m+1})(1-x)^{-4} \end{aligned}$$

(higher powers of x^{m+1} do not contribute to the coefficient of x^{2m})

$$\begin{aligned} &= [x^{2m}](1-4x^{m+1})\sum_{i\geq 0}\binom{4+i-1}{i}x^i \\ &= [x^{2m}](1-4x^{m+1})\sum_{i\geq 0}\binom{i+3}{3}x^i \\ &= \binom{2m+3}{3}-4\binom{m+2}{3} \\ &= \frac{(2m+3)(2m+2)(2m+1)}{6}-4\cdot\frac{(m+2)(m+1)m}{6} \\ &= \frac{2(m+1)}{6}(4m^2+8m+3-2m^2-4m) \\ &= \frac{m+1}{3}(2m^2+4m+3), \end{aligned}$$

as required.

3. (1945). Since the term x^6yzt^5 contains only a single y, in multiplying out the product $(x+y+z+t)^{10}(x+z)(x^2+y^2)$, from the factor (x^2+y^2) it must be the x^2 that is taken toward the formation of a term in x^6yzt^5. Thus

$$\begin{aligned} &[x^6yzt^5](x+y+z+t)^{10}(x+z)(x^2+y^2) \\ =&[x^4yzt^5](x+y+z+t)^{10}(x+z) \end{aligned}$$

$$
\begin{aligned}
&=[x^4yzt^5](x+y+z+t)^{10}\cdot x+[x^4yzt^5](x+y+z+t)^{10}\cdot z\\
&=[x^3yzt^5](x+y+z+t)^{10}+[x^4yt^5](x+y+z+t)^{10}\\
&=\binom{10}{3}\binom{7}{1}\binom{6}{1}\binom{5}{5}+\binom{10}{4}\binom{6}{1}\binom{5}{5}\\
&=\frac{10\cdot 9\cdot 8}{3\cdot 2}\cdot 7\cdot 6\cdot 1+\frac{10\cdot 9\cdot 8\cdot 7}{4\cdot 3\cdot 2}\cdot 6\cdot 1\\
&=5040+1260\\
&=6300.
\end{aligned}
$$

4. (1950). Let the parabola be $y^2 = x$, let $P(a^2, a)$ be any point on it and $K(k, 0)$ be a point on its axis (figure 189). Then

$$\frac{1}{PK^2} = \frac{1}{(a^2-k)^2+a^2}.$$

The slope m of PK is $m = a/(a^2 - k)$, implying $a^2 - k = a/m$, and therefore

$$\frac{1}{PK^2} = \frac{1}{\frac{a^2}{m^2}+a^2} = \frac{m^2}{a^2(1+m^2)}.$$

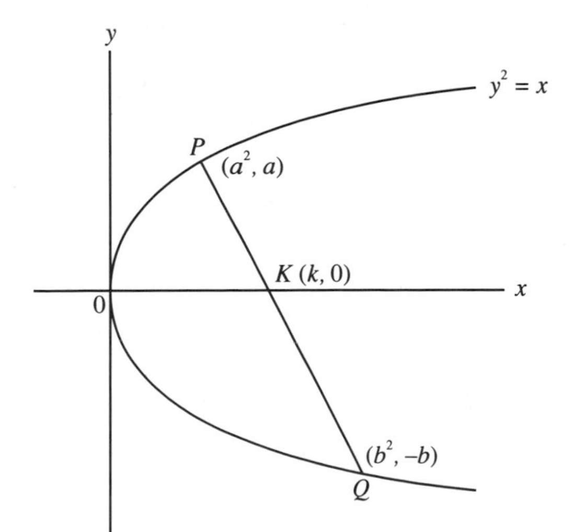

FIGURE 189

Let PK extend into chord PQ and that Q is $(b^2, -b)$. Then

$$\text{slope } QK = m = \frac{b}{k - b^2},$$

implying $k - b^2 = \frac{b}{m}$, giving

$$\frac{1}{QK^2} = \frac{1}{(k-b^2)^2 + b^2} = \frac{1}{\frac{b^2}{m^2} + b^2} = \frac{m^2}{b^2(1+m^2)}.$$

Then

$$N = \frac{1}{PK^2} + \frac{1}{QK^2} = \frac{m^2}{1+m^2}\left(\frac{1}{a^2} + \frac{1}{b^2}\right) = \frac{(a^2+b^2)m^2}{a^2b^2(1+m^2)}.$$

Now

$$\text{slope } PQ = m = \frac{a+b}{a^2-b^2} = \frac{1}{a-b},$$

implying

$$N = \frac{\dfrac{a^2+b^2}{(a-b)^2}}{a^2b^2\left[1 + \dfrac{1}{(a-b)^2}\right]} = \frac{a^2+b^2}{a^2b^2\left[(a-b)^2+1\right]}.$$

Also $m = a/(a^2 - k)$ implies $a^2 - k = a/m$, giving

$$k = a^2 - \frac{a}{m} = a^2 - a(a-b) = ab.$$

Thus we want to show that there is a value of ab which makes N a constant.

One way this could happen is if

$$\frac{a^2+b^2}{(a-b)^2+1} = 1,$$

making

$$N = \frac{1}{a^2b^2} = \frac{1}{k^2},$$

a constant for all chords through K. In this case, we would have

$$a^2 + b^2 = (a-b)^2 + 1 = a^2 + b^2 - 2ab + 1,$$

giving $ab = \frac{1}{2}$ and $N = \frac{1}{k^2} = 4$. Thus N *is* constant for all chords through the point $K(\frac{1}{2}, 0)$, and, after observing that the focus of our parabola is $(\frac{1}{4}, 0)$, the required point for the parabola $y^2 = 4px$ is $K(2p, 0)$.

5. (1949). Since the roots are in harmonic progression, by definition their reciprocals are in arithmetic progression. Thus, let $y = \frac{1}{x}$ and let the four values of y be $a - 3d$, $a - d$, $a + d$, and $a + 3d$. Then $x = \frac{1}{y}$ and these values of y are the roots of the equation

$$\frac{17}{y^4} + \frac{36}{y^3} - \frac{14}{y^2} - \frac{4}{y} + 1 = 0,$$

that is,

$$y^4 - 4y^3 - 14y^2 + 36y + 17 = 0.$$

Thus

$$\text{the sum of the roots } = 4a = 4, \quad \text{and } a = 1,$$

and the roots are $1-3d$, $1-d$, $1+d$, and $1+3d$. Also the sum of the products of the roots taken two at a time is

$$(1 - 4d + 3d^2) + (1 - 2d - 3d^2) + (1 - 9d^2) + (1 - d^2) \\ +(1 + 2d - 3d^2) + (1 + 4d + 3d^2) = -14,$$

giving

$$-10d^2 + 6 = -14,$$
$$d^2 = 2,$$
$$d = \pm\sqrt{2}.$$

Both these values of d give the same set of roots,

$$y = 1 \pm 3\sqrt{2} \quad \text{and} \quad 1 \pm \sqrt{2}.$$

Hence the required roots x are the reciprocals

$$\frac{1}{1 \pm 3\sqrt{2}}, \ \frac{1}{1 \pm \sqrt{2}}, \quad \text{i.e.,} \quad \frac{-1 \pm 3\sqrt{2}}{17} \ \text{and} \ -1 \pm \sqrt{2}.$$

6. (1936). Let the desired circle touch K at T and let the common tangent at T meet the chord XY, through A and B, at the exterior point P (figure 190). Then

$$PT^2 = PX \cdot PY = PA \cdot PB,$$

implying

$$\frac{PX}{PA} = \frac{PB}{PY}.$$

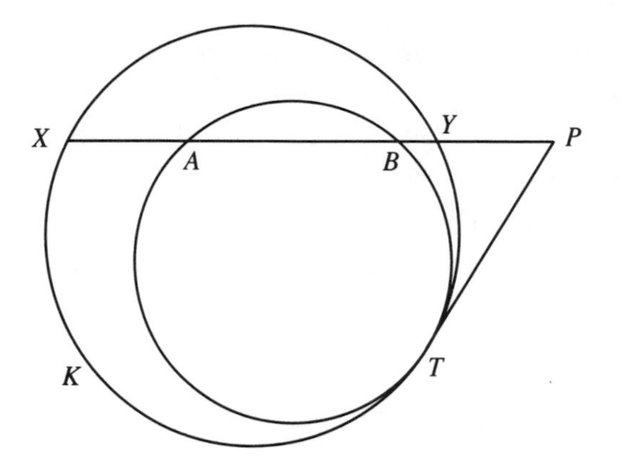

FIGURE 190

Subtracting 1 from each side yields

$$\frac{PX - PA}{PA} = \frac{PB - PY}{PY},$$

that is,

$$\frac{XA}{AP} = \frac{BY}{YP},$$

and we have

$$\frac{XA}{YB} = \frac{AP}{PY}.$$

That is to say, P can be found on XY extended as the point which divides AY externally in the known ratio $XA : YB$.

Thus, after first locating P, draw the tangent to K from P to get T and then draw the circle through A, B, and T.

7. (1938). Observe that $6 - 2\sqrt{5} = (\sqrt{5} - 1)^2$ and $9 - 4\sqrt{5} = (\sqrt{5} - 2)^2$.

We have

$$\begin{aligned}\frac{\sqrt{3-\sqrt{5}}}{\sqrt{2}+\sqrt{7-3\sqrt{5}}} &= \frac{\sqrt{3-\sqrt{5}}}{\sqrt{2}+\sqrt{7-3\sqrt{5}}} \cdot \frac{\sqrt{2}-\sqrt{7-3\sqrt{5}}}{\sqrt{2}-\sqrt{7-3\sqrt{5}}} \\ &= \frac{\sqrt{6-2\sqrt{5}} - \sqrt{36-16\sqrt{5}}}{3\sqrt{5}-5} \\ &= \frac{\sqrt{5}-1-2(\sqrt{5}-2)}{3\sqrt{5}-5}\end{aligned}$$

$$= \frac{3-\sqrt{5}}{3\sqrt{5}-5}$$
$$= \frac{3-\sqrt{5}}{3\sqrt{5}-5} \cdot \frac{3\sqrt{5}+5}{3\sqrt{5}+5}$$
$$= \frac{4\sqrt{5}}{20} = \frac{\sqrt{5}}{5}.$$

8. (1937). For some value a, the first expression must be $a(x-1)(x+2)$, and for some value b, the second expression must be $b(x-4)(x+1)$. Since, for $x=3$, we get

$$a(2)(5) = 60 = b(-1)(4),$$

it follows that $a=6$ and $b=-15$. Hence the expressions are equal when

$$6(x-1)(x+2) = -15(x-4)(x+1),$$
$$2(x-1)(x+2) = -5(x-4)(x+1),$$
$$7x^2 - 13x - 24 = 0$$
$$(x-3)(7x+8) = 0,$$

giving $x=3$ and $-\frac{8}{7}$.

9. (1940). Let $Q(a,k)$ and $R(a,k+2a)$ be any two points on $x=a$ which are a distance $2a$ apart and let the tangents from Q and R to the given circle meet at $P(x,y)$ (figure 191). We want the locus of P.

A non-vertical straight line through Q has equation

$$y-k = m(x-a), \text{ i.e., } mx - y + k - am = 0, \; m \text{ finite.}$$

To be a tangent to the given circle, we must have

$$\frac{k-am}{\sqrt{m^2+1}} = \pm a,$$
$$k - am = \pm a\sqrt{m^2+1},$$
$$k^2 - 2amk + a^2m^2 = a^2m^2 + a^2,$$
$$\frac{k^2-a^2}{2ak} = m,$$

implying the equation of the tangent is

$$\frac{k^2-a^2}{2ak}x - y + k - \frac{k^2-a^2}{2k} = 0$$

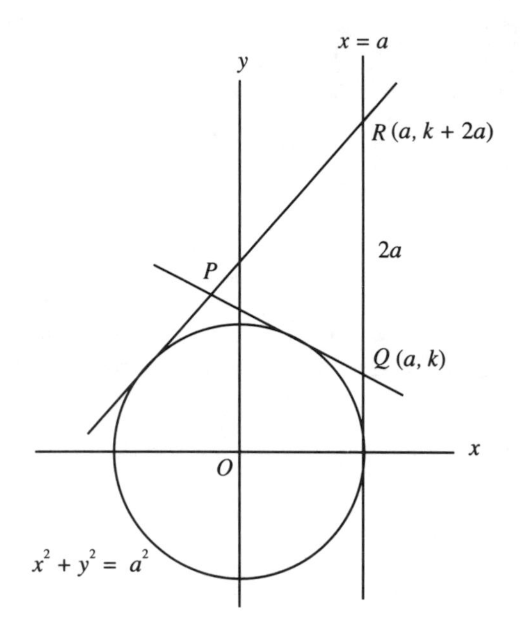

FIGURE 191

$$(k^2 - a^2)x - 2aky + 2ak^2 - ak^2 + a^3 = 0$$

$$(k^2 - a^2)x - 2aky + ak^2 + a^3 = 0. \qquad (1)$$

Similarly, the tangent through R has equation

$$\left[(k+2a)^2 - a^2\right]x - 2a(k+2a)y + a(k+2a)^2 + a^3 = 0. \qquad (2)$$

Thus the equation of the locus in question is obtained by eliminating k from equations (1) and (2).

Subtracting (1) from (2) gives

$$(4ak + 4a^2)x - 4a^2y + 4a^2k + 4a^3 = 0,$$

which, divided by $4a$, yields

$$(k + a)x - ay + ak + a^2 = 0,$$

giving

$$k(x + a) = a(y - x - a) \quad \text{and} \quad k = \frac{a(y - x - a)}{x + a}.$$

Substituting in (1), we get

$$\left[\frac{a^2(y-x-a)^2}{(x+a)^2}-a^2\right]x-2ay\frac{a(y-x-a)}{x+a}+a\frac{a^2(y-x-a)^2}{(x+a)^2}+a^3=0,$$

which, after considerable but straightforward simplification, reduces to

$$y^2=2ax+2a^2.$$

Thus the locus consists of all points on the parabola having vertex at $(a,0)$ and the x-axis for its axis *except* for the points given by $x=\pm a$, which require special interpretation to be admitted and are better rejected altogether.

10. (1937). (a) If p is odd, say $2n+1$, then we wish to solve the following equation for a:

$$(a-2n)+\cdots+(a-4)+(a-2)+a+(a+2)+(a+4)+\cdots+(a+2n)=p^s$$

that is, $(2n+1)a=p^s$, and since $(2n+1)=p$, we may take $a=p^{s-1}$, which is odd, and begin at the number

$$a-2n=p^{s-1}-(p-1)=p^{s-1}-p+1.$$

(b) If p is even, say $2n$, we similarly wish to solve

$$\begin{aligned}(a-2n)+\cdots+(a-4)+(a-2)+a+(a+2)+&(a+4)\\ +\cdots+(a+2n-2)&=p^s\\ 2na-2n&=p^s\\ 2n(a-1)=p(a-1)&=p^s\\ a&=p^{s-1}+1,\end{aligned}$$

which is odd, and again having the series start at the same number $p^{s-1}-p+1$:

$$a-2n=a-p=p^{s-1}-p+1.$$

11. (1945). Let

$$\begin{aligned}f(x)&=(x-1)^3(ax^2+bx+c)+1\\ &=(x^3-3x^2+3x-1)(ax^2+bx+c)+1.\end{aligned}$$

Since $f(x)$ is divisible by x^3, the terms in x^2 and x, as well as the absolute term, must vanish. Thus

$$\begin{aligned} &\text{the absolute term} = -c + 1 = 0, && \text{giving} \quad c = 1, \\ &\text{the coefficient of } x = 3c - b = 3 - b = 0, && \text{giving} \quad b = 3. \\ &\text{the coefficient of } x^2 = -3 + 3b - a = 0, && \text{giving} \quad a = 6, \end{aligned}$$

and making

$$\begin{aligned} f(x) &= (x^3 - 3x^2 + 3x - 1)(6x^2 + 3x + 1) + 1 \\ &= 6x^5 - 15x^4 + 10x^3. \end{aligned}$$

12. (1958). Let V denote the volume of the tub and let the hot water tap by itself take x minutes to fill the tub. Thus the hot water tap supplies water at the rate of $\frac{V}{x}$ units per minute, the cold water tap supplies water at the rate of $\frac{V}{14}$ units per minute, while the hole in the bottom drains water at the rate of $\frac{V}{21}$ units per minute. Thus

$$\frac{V}{x} + \frac{V}{14} - \frac{V}{21} = \frac{V}{21.6},$$

$$\frac{1}{x} = \frac{1}{21.6} + \frac{1}{21} - \frac{1}{14} = \frac{1}{126}(10 + 6 - 9) = \frac{7}{126},$$

giving

$$x = \frac{126}{7} = 18 \text{ minutes.}$$

13. (1934). Let the midpoints of AC and BD be X and Y and the midpoint of XY be Z (figure 192). Our solution is based on the theorem that the sum of the squares of two sides of a triangle is equal to twice the square of the

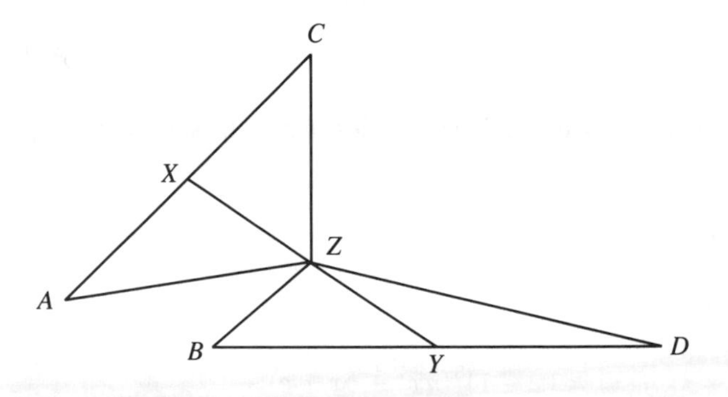

FIGURE 192

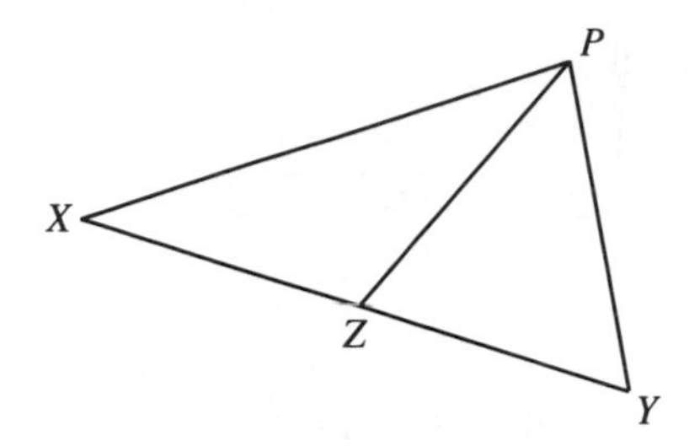

FIGURE 193

median to the third side increased by one-half the square of the third side. When the point P is at Z, we have both

$$ZA^2 + ZC^2 = 2ZX^2 + \tfrac{1}{2}AC^2, \quad \text{and} \quad ZB^2 + ZD^2 = 2ZY^2 + \tfrac{1}{2}BD^2,$$

and therefore

$$\begin{aligned} ZA^2 + ZB^2 + ZC^2 + ZD^2 & \\ &= 2ZX^2 + 2ZY^2 + \frac{1}{2}(AC^2 + BD^2) \\ &= 4ZX^2 + \frac{1}{2}(AC^2 + BD^2) \quad \text{(recall } Z \text{ bisects } XY) \\ &= XY^2 + \frac{1}{2}(AC^2 + BD^2). \end{aligned}$$

For P in general position, we have similarly that

$$\begin{aligned} PA^2 + PB^2 + PC^2 + PD^2 &= (PA^2 + PC^2) + (PB^2 + PD^2) \\ &= 2PX^2 + \frac{1}{2}AC^2 + 2PY^2 + \frac{1}{2}BD^2 \\ &= 2(PX^2 + PY^2) + \frac{1}{2}(AC^2 + BD^2). \end{aligned}$$

In order to show a minimum occurs when P is at Z, then, we need to show that

$$2(PX^2 + PY^2) > XY^2. \tag{1}$$

But, in $\triangle PXY$ (figure 193), PZ is a median and we have

$$PX^2 + PY^2 = 2PZ^2 + \frac{1}{2}XY^2 > \frac{1}{2}XY^2,$$

from which (1) follows immediately, completing the solution.

14. (1951). If $x \geq 4$, then

$$\frac{1}{2}+\frac{1}{z+2}=\frac{1}{x+2}+\frac{1}{y+2} \leq \frac{1}{6}+\frac{1}{y+2},$$

giving

$$\frac{1}{2}-\frac{1}{6}=\frac{1}{3} \leq \frac{1}{y+2}-\frac{1}{z+2}.$$

But $y \geq 1$, implying $y+2 \geq 3$, and $\frac{1}{y+2} \leq \frac{1}{3}$. Hence

$$\frac{1}{y+2}-\frac{1}{z+2}<\frac{1}{3},$$

a contradiction, and it follows that $x < 4$.

Now, if both x and y are greater than 1, then each of $\frac{1}{x+2}$ and $\frac{1}{y+2}$ is less than or equal to $\frac{1}{4}$, making it impossible for $\frac{1}{x+2}+\frac{1}{y+2}$ to exceed $\frac{1}{2}$, which it must do in order to be equal to $\frac{1}{2}+\frac{1}{z+2}$. Hence either one or both of x and y must equal 1. Also, the equation is symmetrical in x and y, implying a limit of 3 on the value of y as well as on x, and that the values of x and y can be interchanged for the same value of z.

For $x = 1$ and $y = 1$, we get $z = 4$, and the solution $(x, y, z) = (1, 1, 4)$; also

$$x = 1, \; y = 2, \quad \text{gives} \quad (1, 2, 10);$$

$$x = 1, \; y = 3, \quad \text{gives} \quad (1, 3, 28);$$

and interchanging x and y we also have the additional two solutions

$$(2, 1, 4) \quad \text{and} \quad (3, 1, 28).$$

15. (1956). (a) In all positions, M is the reflection of E in the common tangent at their point of contact P. Therefore the focal radii to P in M are respectively the same lengths as those to P in E. However, in figure 194, the reflector property of an ellipse gives $\angle x = \angle y$, and the reflection of E in the tangent gives $\angle x = \angle z$. Thus $y = z$ and APA' is straight. Hence at every position $AA' = r_1 + r_2 = 2a$, the length of the major axis, implying the locus of A' is the circle $A(2a)$. Similarly B' traces the circle $B(2a)$.

(b) This is similar to part (a). The solution is based on the following property:

> the perpendicular from the focus of a parabola to a tangent meets the tangent at the point where it crosses the tangent at the vertex.

The details are left as an additional exercise.

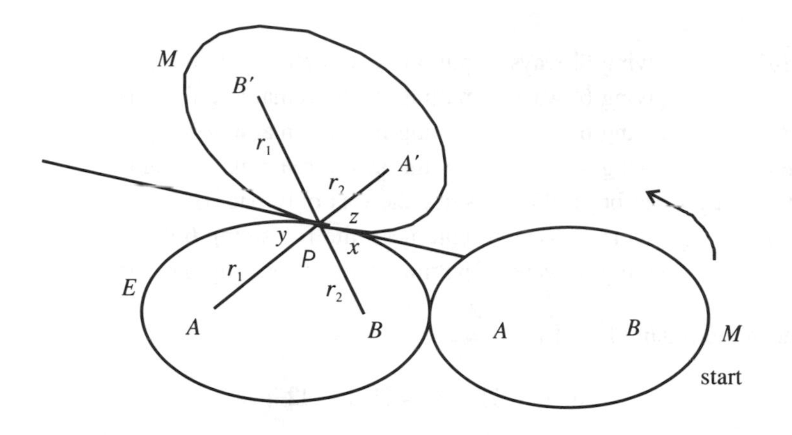

FIGURE 194

16. (1934). Clearing of fractions, we need to show that
$\sin 1^\circ \sin 1^\circ + \sin 1^\circ \sin 3^\circ + \sin 1^\circ \sin 5^\circ + \cdots + \sin 1^\circ \sin 99^\circ = \sin^2 50^\circ$.
But this is immediate from $\sin A \sin B = -\frac{1}{2}\left[\cos(A+B) - \cos(A-B)\right]$:

$$\begin{aligned}
\text{the left side } &= -\frac{1}{2}\left[(\cos 2^\circ - \cos 0^\circ) + (\cos 4^\circ - \cos 2^\circ)\right. \\
&\qquad \left. + (\cos 6^\circ - \cos 4^\circ) + \cdots + (\cos 100^\circ - \cos 98^\circ)\right] \\
&= -\frac{1}{2}(\cos 100^\circ - \cos 0^\circ) \\
&= -\frac{1}{2}\left[(1 - 2\sin^2 50^\circ) - 1\right] \\
&= \sin^2 50^\circ.
\end{aligned}$$

17. (1942). The initial digits of the integers in question can be classified nicely into 16 types in each of which the remaining digits may be appended in any order:

$678\ldots,$ giving 6! ways of putting in the remaining 6 digits,
$679\ldots,$ giving 6! ways of putting in the remaining 6 digits,
$68\ldots,$ giving 7! ways of putting in the remaining 7 digits,
$69\ldots,$ giving 7! ways of putting in the remaining 7 digits,
$7\ldots,$ giving 8! ways of putting in the remaining 8 digits,
$81\ldots,$ giving 7! ways of putting in the remaining 7 digits,
$82\ldots,$ giving 7! ways of putting in the remaining 7 digits,
$83\ldots,$ giving 7! ways of putting in the remaining 7 digits,
$84\ldots,$ giving 7! ways of putting in the remaining 7 digits,

851 . . . , giving 6! ways of putting in the remaining 6 digits,
852 . . . , giving 6! ways of putting in the remaining 6 digits,
853 . . . , giving 6! ways of putting in the remaining 6 digits,
854 . . . , giving 6! ways of putting in the remaining 6 digits,
(no integer can begin 855 . . . since there is only one 5)
856 . . . , giving 6! ways of putting in the remaining 6 digits,
857 . . . , giving 6! ways of putting in the remaining 6 digits.

Hence the required total is

$$\begin{aligned} 8 \cdot 6! + 6 \cdot 7! + 8! &= 6!(8 + 42 + 56) \\ &= 720 \cdot 106 = 76320. \end{aligned}$$

18. (a) (1959). If $P(s,t)$ is a point on the hyperbola $b^2x^2 - a^2y^2 = a^2b^2$, the equation of the tangent QPR is

$$b^2 sx - a^2 ty = a^2 b^2 \quad \text{(figure 195).}$$

Solving with the equation of the asymptotes, $b^2x^2 - a^2y^2 = 0$, we have

$$x = \frac{a^2b^2 + a^2ty}{b^2 s},$$

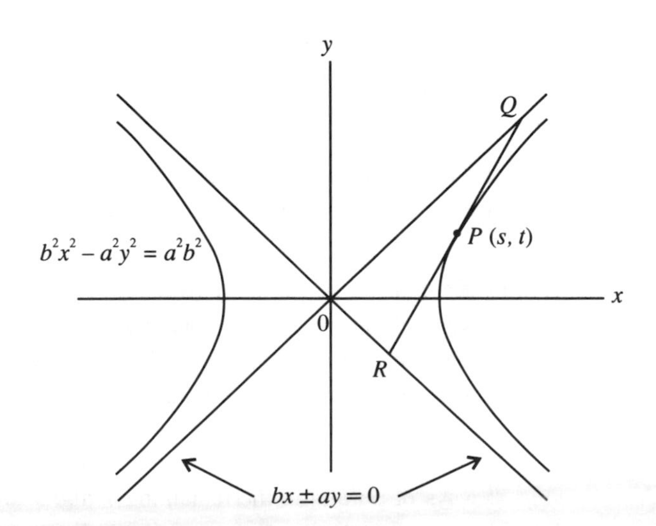

FIGURE **195**

therefore

$$b^2\left(\frac{a^2b^2+a^2ty}{b^2s}\right)^2 - a^2y^2 = 0,$$

$$b\left(\frac{a^2b^2+a^2ty}{b^2s}\right) = \pm ay,$$

$$ab^2 + aty = \pm bsy,$$

$$ab^2 = y(\pm bs - at),$$

giving

$$y = \frac{ab^2}{bs-at} \quad \text{or} \quad \frac{ab^2}{-bs-at}.$$

Hence the ordinate of the midpoint of QR is

$$\begin{aligned} Y &= \frac{1}{2}\left(\frac{ab^2}{bs-at} - \frac{ab^2}{bs+at}\right) \\ &= \frac{ab^2}{2}\left(\frac{bs+at-bs+at}{b^2s^2-a^2t^2}\right) \\ &= \frac{a^2b^2t}{b^2s^2-a^2t^2}. \end{aligned}$$

But (s,t) satisfies the equation of the hyperbola, giving $b^2s^2 - a^2t^2 = a^2b^2$, and implying $Y = t$.

Substituting the values of y in the equation of the asymptotes, similar calculations give Q and R to have the coordinates

$$Q\left(\frac{a^2b}{bs-at}, \frac{ab^2}{bs-at}\right) \quad \text{and} \quad R\left(\frac{a^2b}{bs+at}, \frac{-ab^2}{bs+at}\right).$$

Accordingly, it is easily checked that the abscissa of the midpoint of QR is t, making the midpoint P, as required. Also, the area of $\triangle QOR$ is easily found to be the constant ab for all values of s and t.

(b) (1954). Solving the equations $b^2x^2 - a^2y^2 = a^2b^2$ and $y = mx + k$ for the coordinates of Q and R (figure 196), we have

$$b^2x^2 - a^2(m^2x^2 + 2mkx + k^2) = a^2b^2,$$

$$(b^2 - a^2m^2)x^2 - 2a^2mkx - a^2(k^2 + b^2) = 0, \tag{1}$$

giving the x-coordinate of the midpoint M of QR to be

$$\frac{1}{2}\left(\frac{2a^2mk}{b^2-a^2m^2}\right) = \frac{a^2mk}{b^2-a^2m^2}.$$

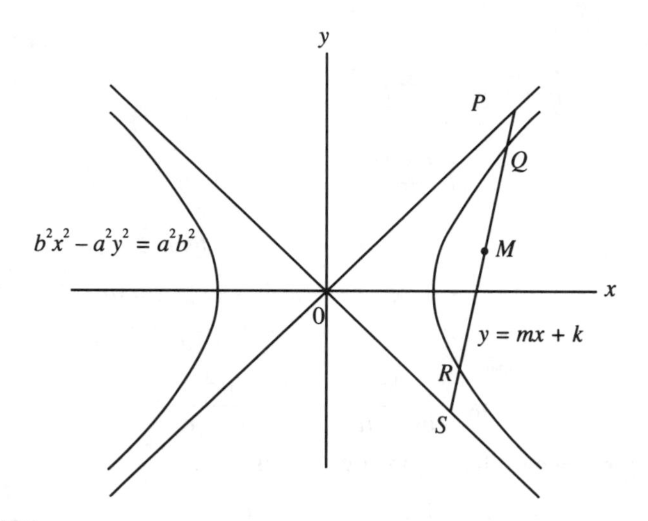

FIGURE 196

Solving the equations $b^2x^2 - a^2y^2 = 0$ and $y = mx + k$ for the midpoint of PS clearly does not alter the coefficients of x^2 and x in the equation corresponding to (1) (only the absolute term is changed), and so the same x-coordinate is obtained for the midpoint. Similarly for the y-coordinates.

19. (1961). In figure 197, we would like to show that $\phi > \theta$. By the law of sines, we have

$$\frac{x}{\sin\theta} = \frac{2x}{\sin 2\phi} \quad \left(= \frac{x}{\sin\phi\cos\phi} \right),$$

implying

$$\sin\theta = \sin\phi\cos\phi.$$

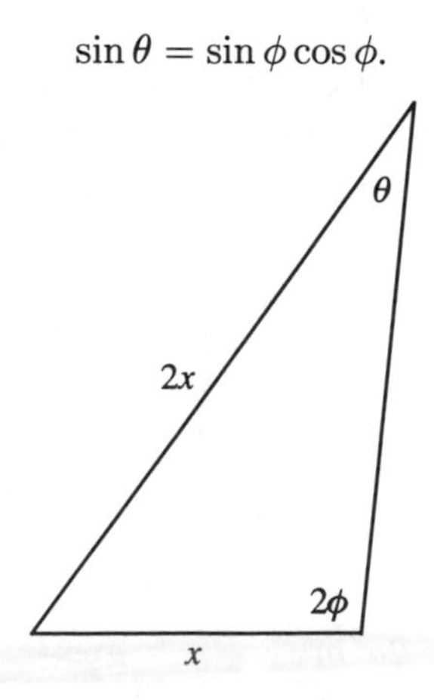

FIGURE 197

Now $\phi > 0$, implying $\cos\phi < 1$, and therefore

$$\sin\theta < \sin\phi.$$

But because $2\phi < 180°$ (in the triangle), then $\phi < 90°$, and since the sine function is increasing in the range $(0°, 90°)$, $\sin\theta < \sin\phi$ implies $\theta < \phi$.

20. (1935). Recall the standard relations

1. $AL = s$, the semiperimeter of triangle ABC,
2. the area Δ of triangle ABC is given by rs and by $\sqrt{s(s-a)(s-b)(s-c)}$,
3. (from similar triangles, see the figure)

$$\frac{r_1}{AL} = \frac{r_1}{s} = \frac{r}{s-a},$$

giving

$$r_1 = \frac{rs}{s-a} = \frac{\Delta}{s-a},$$

and similarly for r_2 and r_3.

From figure 198, we have that

$$\Delta ABI_1 = \frac{1}{2}cr_1 \quad \text{and} \quad \Delta ACI_1 = \frac{1}{2}br_1,$$

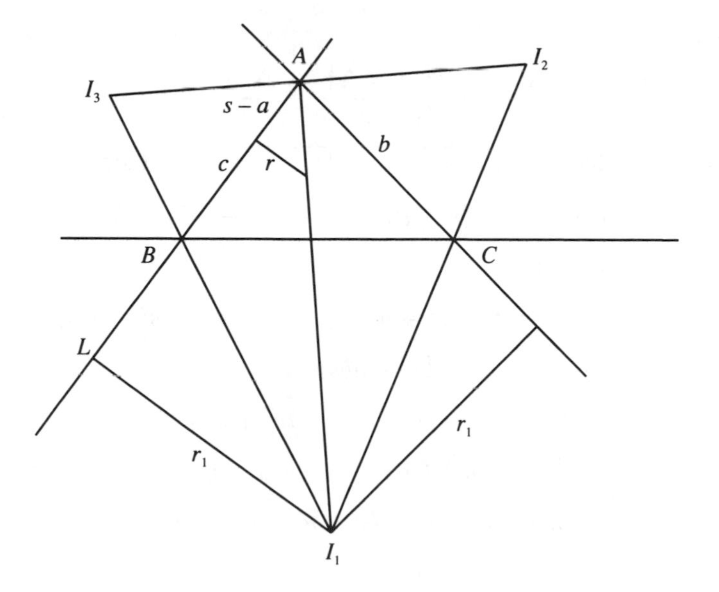

FIGURE 198

and hence, adding, we get $ABI_1C = \frac{1}{2}r_1(b+c)$; similarly for $ABCI_2$ and $ACBI_3$, and we have altogether that

$$\begin{aligned}
\Delta I_1I_2I_3 &= ABI_1C + ABCI_2 + ACBI_3 - 2\Delta ABC \\
&= \left[\frac{1}{2}r_1(b+c)\right] + \left[\frac{1}{2}r_2(a+c)\right] + \left[\frac{1}{2}r_3(a+b)\right] - 2\Delta \\
&= \frac{1}{2}\big[r_1(2s-a) + r_2(2s-b) + r_3(2s-c)\big] - 2\Delta \\
&= \frac{1}{2}\Big\{r_1\big[s+(s-a)\big] + r_2\big[s+(s-b)\big] + r_3\big[s+(s-c)\big]\Big\} - 2\Delta \\
&= \frac{1}{2}\big[s(r_1+r_2+r_3) + 3rs\big] - 2rs \text{ (recall } r_1(s-a) = rs = \Delta\text{, etc.)} \\
&= \frac{1}{2}s(r_1+r_2+r_3-r).
\end{aligned}$$

Thus we want to show that

$$\frac{1}{2}s(r_1+r_2+r_3-r) = \frac{abc}{2r},$$

that is,

$$r_1+r_2+r_3-r = \frac{abc}{\Delta},$$

or equivalently,

$$\frac{\Delta}{s-a} + \frac{\Delta}{s-b} + \frac{\Delta}{s-c} - \frac{\Delta}{s} = \frac{abc}{\Delta},$$

or finally,

$$\frac{\Delta^2}{s-a} + \frac{\Delta^2}{s-b} + \frac{\Delta^2}{s-c} - \frac{\Delta^2}{s} = abc.$$

Recalling that $\Delta = \sqrt{s(s-a)(s-b)(s-c)}$, the left side is

$$\begin{aligned}
&s(s-b)(s-c) + s(s-a)(s-c) + s(s-a)(s-b) \\
&\quad - (s-a)(s-b)(s-c) \\
&\qquad = (s^3 - bs^2 - cs^2 + bcs) + (s^3 - as^2 - cs^2 + acs) \\
&\qquad\quad + (s^3 - as^2 - bs^2 + abs) \\
&\qquad\quad - (s^3 - bs^2 - cs^2 + bcs - as^2 + abs + acs - abc) \\
&\qquad = 2s^3 - s^2(2a+2b+2c-a-b-c) \\
&\qquad\quad + s(bc+ac+ab-bc-ac-ab) + abc \\
&\qquad = 2s^3 - s^2(2s) + abc \\
&\qquad = abc.
\end{aligned}$$

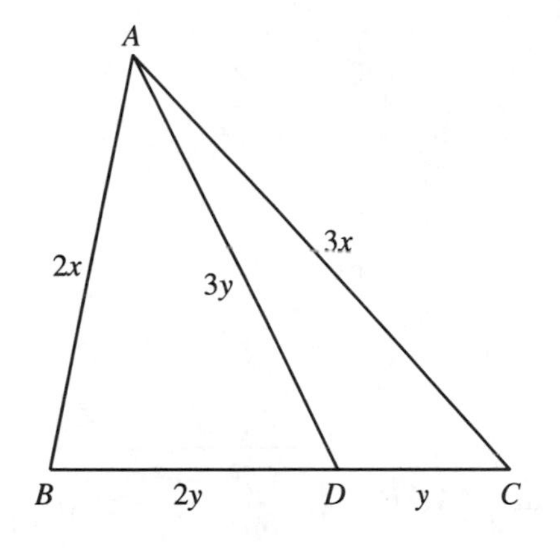

FIGURE 199

21. (1943). In $\triangle ABD$ (figure 199), the semiperimeter $s = x + \frac{5}{2}y$, and therefore

$$\tan\frac{1}{2}\angle ADB = \sqrt{\frac{(s-a)(s-b)}{s(s-d)}} = \sqrt{\frac{(x+\frac{y}{2})(x-\frac{y}{2})}{(x+\frac{5}{2}y)(\frac{5}{2}y-x)}}$$

$$= \sqrt{\frac{x^2-\frac{y^2}{4}}{\frac{25}{4}y^2-x^2}} = \sqrt{\frac{4x^2-y^2}{25y^2-4x^2}}.$$

Now in $\triangle ABD$, the law of cosines gives

$$4x^2 = 4y^2 + 9y^2 - 12y^2\cos\angle ADB,$$

implying

$$\cos\angle ADB = \frac{13y^2-4x^2}{12y^2};$$

similarly, from $\triangle ADC$ we get

$$9x^2 = 9y^2 + y^2 - 6y^2\cos\angle ADC,$$

and

$$\cos\angle ADC = \frac{9x^2-10y^2}{6y^2}.$$

But $\cos\angle ADB = -\cos\angle ADC$ (supplementary angles), and so

$$\frac{13y^2-4x^2}{12y^2} = \frac{9x^2-10y^2}{6y^2},$$

$$13y^2 - 4x^2 = 18x^2 - 20y^2,$$

$$33y^2 = 22x^2,$$
$$3y^2 = 2x^2,$$

giving

$$6y^2 = 4x^2.$$

Hence

$$\tan\frac{1}{2}\angle ADB = \sqrt{\frac{6y^2 - y^2}{25y^2 - 6y^2}} = \sqrt{\frac{5}{19}}.$$

22. (1950).

$$\begin{aligned} S &= 1 + 2^2x + 3^2x^2 + \cdots + n^2x^{n-1} + \cdots \\ &= (1\cdot 2 - 1) + (2\cdot 3 - 2)x + (3\cdot 4 - 3)x^2 \\ &\quad + \cdots + [n(n+1) - n]x^{n-1} + \cdots. \end{aligned}$$

Now, the order of these terms may be rearranged provided the series is absolutely convergent for $-1 < x < 1$. This *is* the case, but at this point let us simply assume it. Then

$$\begin{aligned} S = {} & [1\cdot 2 + 2\cdot 3x + 3\cdot 4x^2 + \cdots + n(n+1)x^{n-1} + \cdots] \\ & - (1 + 2x + 3x^2 + \cdots + nx^{n-1} + \cdots), \end{aligned}$$

which is the derivative of $(1+2x+3x^2+\cdots+nx^{n-1}+\cdots)$ minus the series itself. Since $1 + 2x + 3x^2 + \cdots + nx^{n-1} + \cdots = (1-x)^{-2}$, we have

$$\begin{aligned} S &= \frac{d(1-x)^{-2}}{dx} - (1-x)^{-2} \\ &= \frac{2}{(1-x)^3} - \frac{1}{(1-x)^2} \\ &= \frac{2-(1-x)}{(1-x)^3} \\ &= \frac{1+x}{(1-x)^3}. \end{aligned}$$

To justify our assumption concerning convergence, we need to check this result. We have

$$\begin{aligned}[x^{n-1}](1+x)(1-x)^{-3} &= [x^{n-1}](1+x)\sum_{i\geq 0}\binom{i+2}{2}x^i \\ &= \binom{n+1}{2}+\binom{n}{2} = \frac{(n+1)n+n(n-1)}{2} \\ &= \frac{2n^2}{2} = n^2,\end{aligned}$$

as desired.

23. (1942). We have

$$\begin{aligned}m = \csc\theta - \sin\theta &= \frac{1}{\sin\theta} - \sin\theta \\ &= \frac{1-\sin^2\theta}{\sin\theta} = \frac{\cos^2\theta}{\sin\theta},\end{aligned}$$

and

$$\begin{aligned}n = \sec\theta - \cos\theta &= \frac{1}{\cos\theta} - \cos\theta \\ &= \frac{1-\cos^2\theta}{\cos\theta} = \frac{\sin^2\theta}{\cos\theta}.\end{aligned}$$

Therefore $mn = \sin\theta\cos\theta$.

We want to show that

$$\left(\frac{\cos^2\theta}{\sin\theta}\right)^{2/3} + \left(\frac{\sin^2\theta}{\cos\theta}\right)^{2/3} = \left(\frac{1}{\sin\theta\cos\theta}\right)^{2/3},$$

that is,

$$\frac{\cos^{4/3}\theta}{\sin^{2/3}\theta} + \frac{\sin^{4/3}\theta}{\cos^{2/3}\theta} = \frac{1}{\sin^{2/3}\theta\cos^{2/3}\theta},$$

or

$$\frac{\cos^{4/3}\theta\cos^{2/3}\theta + \sin^{4/3}\theta\sin^{2/3}\theta}{\sin^{2/3}\theta\cos^{2/3}\theta} = \frac{1}{\sin^{2/3}\theta\cos^{2/3}\theta}.$$

But, since

$$\cos^{4/3}\theta\cos^{2/3}\theta + \sin^{4/3}\theta\sin^{2/3}\theta = \cos^2\theta + \sin^2\theta = 1,$$

this is clearly so, and because all the steps in our derivation are reversible, the desired conclusion follows.

24. (1961). $S_n = \dfrac{n(n+1)}{2}$, and therefore

$$
\begin{aligned}
S &= \sum_{n\geq 1} \frac{n(n+1)}{2^n} \\
&= \frac{d}{dx}\left(\sum_{n\geq 1} nx^{n+1}\right) \qquad \text{evaluated at } x = \tfrac{1}{2}, \\
&= \frac{d}{dx}(x^2 + 2x^3 + 3x^4 + \cdots) \qquad \text{at } x = \tfrac{1}{2}, \\
&= \frac{d}{dx}\left[x^2(1 + 2x + 3x^2 + \cdots)\right] \qquad \text{at } x = \tfrac{1}{2}, \\
&= \frac{d\left[x^2(1-x)^{-2}\right]}{dx} \qquad \text{at } x = \tfrac{1}{2}, \\
&= \left[2x(1-x)^{-2} + x^2(-2)(1-x)^{-3}(-1)\right] \qquad \text{at } x = \tfrac{1}{2}, \\
&= 1 \cdot \left(\frac{1}{2}\right)^{-2} + \frac{1}{4} \cdot 2\left(\frac{1}{2}\right)^{-3} \\
&= 4 + 4 \\
&= 8.
\end{aligned}
$$

25. (1941). Let $\tan^{-1}\frac{1}{5} = x$, $\tan^{-1}\frac{1}{70} = y$, and $\tan^{-1}\frac{1}{99} = z$. Then we want the smallest value of $4x - y + z = 4x - (y - z)$.

Now,

$$
\begin{aligned}
\tan 4x = \frac{2\tan 2x}{1 - \tan^2 2x} &= \frac{2\left(\dfrac{2\tan x}{1 - \tan^2 x}\right)}{1 - \left(\dfrac{2\tan x}{1 - \tan^2 x}\right)^2} \\
&= \frac{4\tan x(1 - \tan^2 x)}{(1 - \tan^2 x)^2 - 4\tan^2 x} \\
&= \frac{4(\frac{1}{5})(1 - \frac{1}{25})}{(1 - \frac{1}{25})^2 - 4(25)} \\
&= \frac{4(5)(25-1)}{(25-1)^2 - 4(25)} \\
&= \frac{480}{476} = \frac{120}{119}.
\end{aligned}
$$

Also,

$$\tan(y-z) = \frac{\tan y - \tan z}{1 + \tan y \tan z} = \frac{\frac{1}{70} - \frac{1}{99}}{1 + \frac{1}{70} \cdot \frac{1}{99}} = \frac{99 - 70}{70 \cdot 99 + 1}$$
$$= \frac{29}{6931} = \frac{1}{239}.$$

Therefore

$$\tan(4x - y + z) = \tan[4x - (y - z)] = \frac{\tan 4x - \tan(y-z)}{1 + \tan 4x \tan(y-z)}$$
$$= \frac{\frac{120}{119} - \frac{1}{239}}{1 + \frac{120}{119} \cdot \frac{1}{239}} = \frac{120 \cdot 239 - 119}{119 \cdot 239 + 120}$$
$$= \frac{28561}{28561} = 1.$$

Hence the desired

$$4x - y + z = \frac{\pi}{4}.$$